Complying with TSCA
Inventory Requirements

Complying with TSCA Inventory Requirements

A Guide with Step-by-Step Processes for Chemical Manufacturers, Processors, and Importers

Chan B. Thanawalla
Consultant, TSCA/Organic Chemicals

WILEY-
INTERSCIENCE
A John Wiley & Sons, Inc., Publication

For ordering and customer service, call 1-800-CALL-WILEY.

Library of Congress Cataloging-in-Publication Data:

ISBN: 0-471-21481-7

Printed in the United States of America.

10 9 8 7 6 5 4 3 2 1

To
My mother, Taraben,
who sacrificed her life pleasures so that her children may have healthy lives and a
bright future,
and
my wife, Indira, and my daughter, Sandhya,
for their help and encouragement throughout this work.

Contents

Chapter 1

Introduction

A large task facing any industry today is coping with the increasingly shorter time window for bringing its new products to market, and the chemical industry is no exception. For chemical industry, whether it serves the manufacturing industry or consumers directly, coping with the problem of the shortening market window has, in recent years, become compounded because, unlike other industries, it must also comply with a number of government regulations relating to chemicals. For instance, consider the following. When market evaluation of a new laboratory-made chemical product deems it promising for commercialization, sometimes it is the regulatory constraints of the new products' manufacture or import that slow its commercialization process or worse yet bring it to a halt. If only one could identify and deal with regulatory aspects of a new product while it is still in the developmental stage, problems with the pace of its commercialization could be avoided. Typically, however, regulatory issues of a new product are not examined until after its composition has been decided and its process development evaluated. While, in some cases and for some regulations, regulatory issues of a new product may be best examined following the new product development, if chemists, chemical engineers, and other professionals involved in the new product developmental efforts were made aware of the Toxic Substances Control Act (TSCA) regulatory issues, they could make prudent choices in selecting raw materials, processes and possibly even the new product composition, such that commercialization process of the new product might face fewer regulatory compliance hurdles. Second, it is again the same professionals who typically assign both a precise composition to a chemical product and a corresponding Chemical Abstract Service Registry Number (CASRN) to it. Since the chemical composition and CASRN together form the core elements of the chemical's compliance with TSCA, apprising the professionals of TSCA regulatory issues could help improve the overall pace of commercialization. Finally, if the process of assessing compliance of chemicals against TSCA by the regulatory compliance staff were streamlined and made user-friendly, that too could help speed up the commercialization process. These thoughts prompted me to write this book as a guide for my fellow chemists and other professionals that are engaged in the new product development and in assessing TSCA compliance of chemicals that their organizations process, manufacture, or import.

Besides inventory requirements, the TSCA, an all-encompassing law for the chemical industry, has many provisions that a chemical company must comply with. For example, manufacturers and importers of certain chemicals already listed on the TSCA inventory must periodically report current data regarding production volume, plant site information, and site-limited status to the USEPA in order to comply with the TSCA Inventory Update Rule. Also any person who manufactures, processes, or distributes in commerce any chemical substance or mixture must keep records of significant adverse reactions to health or the environment that allegedly were caused by the chemical to comply with yet another provision of the TSCA In writing this book, however, I have focused *only* on the TSCA inventory requirements. This is because, for most chemicals and especially new chemicals, compliance of a chemical with TSCA inventory requirements is crucial, and this determines if a chemical company can manufacture, import, process, or use a chemical in its operations in the first place. A chemical's compliance with TSCA inventory requirements therefore forms a basis upon which additional compliance with most of the other provisions of the TSCA becomes relevant or is assessed.

This guide is an up-to-date and one-stop source for all the information one might need to both learn TSCA regulations as they relate to TSCA inventory requirements and conduct assessment of chemicals vis-à-vis TSCA inventory requirements. In keeping with these objectives, I have provided a brief but pertinent discussion of the TSCA that avoids, where possible, legal jargon for ready understanding of its inventory requirements, obviating any need to look up Code of Federal Regulations or Federal Register. Should there be a need, however, to delve into greater details of some TSCA regulations, I have reproduced the regulations in appendixes or have provided key references to them. I have also included the necessary EPA guidance documents and prescribed forms in appendixes that would make it unnecessary to go on-line to retrieve them. Accordingly, Chapters 2 through 4 provide a brief overview of the TSCA with a particular emphasis on the terms and provisions of the TSCA as they relate to TSCA inventory requirements. Following the overview of the TSCA, Chapter 5 gives step-by-step user-friendly processes, both in textual and schematic forms, that a chemical processor, manufacturer, or an importer can use to check the compliance of chemicals against TSCA inventory requirements. Finally, the appendixes comprise details of the Act and some of its sections, prescribed forms and instruction manual of the USEPA that enable one to comply with the TSCA inventory requirements, and some of the guidance documents that help understand some of the TSCA provisions vis-à-vis the underlying chemistry of the chemical substances.

I trust that this guide would serve as a training tool in assessing TSCA inventory compliance of chemicals for the uninitiated as long as they have a basic knowledge of chemistry and an understanding of the use of the chemicals they are processing, manufacturing, or importing. And, for those who are familiar with the TSCA, I hope that the guide would serve as a handy reference manual and the processes given in Chapter 5 of this guide as standard operating procedures.

I have made every effort to make this guide an up-to-date and one-stop source for those engaged in assessing TSCA inventory compliance of chemicals. Nevertheless, I welcome comments or suggestions from my readers at *cbtict@aol.com.*

Chapter 2

Toxic Substance Control Act (TSCA)

Mounting public concerns over some specific hazardous chemicals such as kepone, vinyl chloride, and PCBs led the U.S. Congress to enact the Toxic Substances Control Act (TSCA)[1] in 1976. The Congress gave the Environmental Protection Agency (EPA) authority to regulate not only those chemical substances and mixtures that "*present* an unreasonable risk of injury to health and environment," but also the manufacture, processing, distribution in commerce, use, or disposal of any chemical substance and mixture that *may* present such unreasonable risk. In providing such an authority under the TSCA, the Congress also asked the Administrator (of the EPA) to consider the environmental, economic, and social impact of any action that the Administrator might take to administer the TSCA.

A key aspect of the TSCA is its Premanufacture Notification (PMN). Under this provision, any substance, not on the list of commercial or imported substances of August 30, 1980, and its subsequent supplements, is considered a new substance, and anyone intending to manufacture or import a new substance must submit a PMN 90 days prior to its manufacture or import. The PMN facilitates the EPA to determine if the manufacture (which includes import under the TSCA definition of "manufacture"), processing, distribution in commerce, use, or disposal of a new substance *may* present an unreasonable risk of injury to health and environment.

If a substance poses an unreasonable risk of injury to health and environment, the EPA can prohibit the manufacture, processing, or distribution of the substance; limit the amount of the substance that can be manufactured, processed, or distributed; prohibit the use of the substance in a particular application; limit the concentration of the substance during the manufacture, processing, or distribution; regulate the methods of disposal of the substance; require manufacturers to maintain records of the substance's manufacture; and conduct tests on the substance to assure compliance with the TSCA.

The TSCA also addresses issues involving the cleanup, handling, and disposal of polychlorinated biphenyls (PCBs) and the manufacture and use of chlorinated fluorocarbons (CFCs).

2.1 DEFINITION OF CERTAIN TERMS UNDER TSCA[2]

Article means any manufactured item:
- That is formed in a specific shape or design during manufacture,
- That has end use function(s) dependent in whole or in part upon its shape or design during end use, and
- That has either no change of chemical composition during its end use, or only those changes of composition having no commercial purpose, separate from that of the article, except that fluids and particles are not considered articles regardless of shape or design.

Biopolymer means a polymer directly produced by living or once-living cells or cellular components

By-product means a chemical substance produced without a separate commercial intent during manufacturing, processing, use, or disposal of another chemical substance(s) or mixture(s)

Chemical name means the scientific designation of a chemical substance in accordance with the nomenclature system developed by the International Union of Pure and Applied Chemistry or the Chemical Abstracts Service's rules of nomenclature, or a name, which will clearly identify a chemical substance for the purpose of conducting a hazard evaluation.

Chemical substance means "any organic or inorganic substance of a particular molecular identity, including any combination of such substances occurring in whole or in part as a result of a chemical reaction or occurring in nature, and any element or uncombined radical." The definition excludes:
- Any mixture,
- Any pesticide (as defined by the Federal Insecticide, Fungicide, and Rodenticide Act) when manufactured, processed, or distributed in commerce for use as a pesticide,
- Tobacco or any tobacco product, but not including any derivative product,
- Any source material, special nuclear material or byproduct material (as such terms are defined in the Atomic Energy Act of 1954 and regulations issued under such Act),

- Any pistol, firearm, revolver, shells, and cartridges, and
- Any food, food additive, drug, cosmetic or device (as such terms are defined in section 201 of the Federal Food, Drug, and Cosmetic Act) when manufactured, processed, or distributed in commerce for use as a food, food additive, drug, cosmetic, or device.

Commerce means trade, traffic, transportation, or other commerce:

- Between a place in a state and any place outside of such state or
- That affects trade, traffic, transportation, or commerce.

"Distribute in commerce" and "distribution in commerce" when used with regard to a chemical substance or mixture or article containing a substance or mixture, mean:

- To sell, or the sale of, the substance, mixture, or article in commerce,
- To introduce or deliver for introduction into commerce,
- The introduction or delivery for introduction into commerce of the substance, mixture, or article, or
- To hold, or the holding of, the substance, mixture, or article after its introduction into commerce.

Environment includes water, air, and land, and the interrelationship that exists among and between water, air, and land and all living things.

Environmental transformation product means any chemical substance resulting from the action of environmental processes on a parent compound that changes the molecular identity of a parent compound.

Importer means any person who imports a chemical substance or a chemical substance as part of a mixture or article, into the customs territory of the United States. "Importer" includes the person primarily liable for the payment of any duties on the merchandise or an authorized agent acting on his or her behalf. The term also includes, as appropriate, the consignee, the importer of record, the actual owner if an actual owner's declaration and superseding bond have been filed in accordance with 19 CFR 141.20 (Appendix 7.2) or the transferee if the right to withdraw merchandise in a bonded warehouse has been transferred in accordance with subpart C of 19 CFR 144 (Appendix 7.3)

Impurity means a chemical substance, which is unintentionally *present* with another chemical substance

Instant photographic film article means a self-developing photographic film article designed so that all the chemical substances contained in the article, including the chemical substances required to process the film, remain sealed during distribution and use.

Known to or reasonably ascertainable by means all information in a person's possession or control, plus all information that a reasonable person similarly situated might be expected to possess, control, or know.

Manufacture means to import into the customs territory of the United States, produce, or manufacture.

Manufacture for commercial purposes means:

- To import, produce, or manufacture with the purpose of obtaining an immediate or eventual commercial advantage for the manufacturer, and includes among other things such "manufacture" of any amount of a chemical substance or mixture:

  For commercial distribution, including for test marketing, or

  For use by the manufacturer, including use for product research and development, or as an intermediate.

- Manufacture for commercial purpose also applies to substances that are produced coincidentally during the manufacture, processing, use, or distribution of another substance or the mixture, including both by-products that are separated from that other substance or mixture and impurities that remain in that substance or mixture. Such by-products and impurities may or may not have commercial value. They are none-theless produced for the purpose of obtaining a commercial advantage, since they are part of the manufacture of a chemical product for a commercial purpose.

Master inventory file means the EPA's comprehensive list of chemical substances, which constitute the Chemical Substance Inventory compiled under section 8(b) of the TSCA.

Metabolite means a chemical entity produced by one or more enzymatic or nonenzymatic reactions as a result of exposure of an organism to a chemical substance.

Mixture means any combination of two or more chemical substances if the combination does not occur in nature and is not, in whole or in part, the result of a chemical reaction; except that such a term does include any combination that occurs, in whole or in part, as a result of a chemical reaction if none of the chemical substances comprising the combination is a new chemical substance and if the combination could have been manufactured for commercial purposes without a chemical reaction at the time the chemical substances comprising the combination were combined.

New chemical substance means any chemical substance that is not included in the chemical substance list, called the Master Inventory File (MIF), compiled, maintained, and kept up to date by the EPA. The MIF includes both the confidential and nonconfidential chemical identities (the latter is the public version of the TSCA Inventory, called simply "TSCA Inventory").

Non-isolated intermediate means any chemical substance that is not intentionally removed from the equipment in which it is manufactured, including the reaction vessel in which the chemical substance is manufactured, equipment that is ancillary to the reaction vessel, and any equipment through which the substance passes during a continuous flow process, but not including tanks or other vessels in which the substance is stored after its manufacture. Mechanical or gravity transfer through a closed system is not considered to be intentional removal, but storage or transfer to shipping containers "isolates" the substance by removing it from process equipment in which it manufactured.

Number-average molecular weight means the arithmetic average (mean) of the molecular weight of all the molecules in a polymer.

Oligomer means a polymer molecule consisting of only a few monomer units (dimer, trimer, tetramer).

Peel-apart film article means a self-developing film article consisting of a positive image receiving sheet, a light sensitive negative sheet, and a sealed reagent pod containing a developer reagent and designed so that all the chemical substances required to develop or process the film will not remain sealed within the article during the development of the film.

Person includes any individual firm, company, corporation, joint venture, partnership, sole proprietorship, association, or any other business entity; any state or political subdivision thereof; any municipality; any interstate body; and any department, agency, or instrumentality of the federal government.

Polymer means a chemical substance consisting of molecules characterized by the sequence of one or more types of monomer units and comprising a simple weight majority of molecules containing at least three monomer units, which are covalently bound to at least one other monomer unit or other reactant and which consist of less than a simple weight majority of molecules of the same molecular weight. Such molecules must be distributed over a range of molecular weights wherein differences in the molecular weight are primarily attributable to differences in the number of monomer units.

Process for commercial purpose means the preparation of a chemical substance or mixture, after its manufacture, for distribution in commerce:

- In the same form or physical state as, or in different form or physical state from, that in which it was received by the person preparing such substance or mixture, or
- As part of an article containing the chemical substance or mixture.

Reasonably anticipated means that a knowledgeable person would expect a given physical or chemical composition or characteristic to occur based on such factors as the nature of the precursors used to manufacture the polymer, the type of reaction, the type of manufacturing process, the products produced in polymerization, the intended uses of the substance, or associated use conditions.

Site means a contiguous property unit. Property divided only by a public right-of-way is one site. There may be more than one manufacturing plant on a single site.

Small manufacturer or importer, for the purposes of manufacturing and processing notices (TSCA, section 5), means any manufacturer or importer whose total annual sales, when combined with those of its parent company, if any, are less than $40 million. For the purposes of reporting and recordkeeping requirements (TSCA, section 8(a)), however, small manufacturer or importer is one that meets either of the following standards:

- First Standard: A manufacturer or importer of a substance is small if its total annual sales (subject to the EPA's inflation index adjustment described below), when combined with those of its parent company, if any, are less than $40 million. However, if the annual production or importation volume of a particular substance at any individual site owned or controlled by the manufacturer or importer is greater than 45,400 kilograms (100,000 pounds), the manufacturer or importer will not qualify as small for purposes of reporting on the production or importation of that substance at that site, unless the manufacturer or importer qualifies as small under Second Standard of this definition

- Second Standard: A manufacturer or importer of a substance is small if its total annual sales, when combined with those of its parent company, if any, are less that $4 million, regardless of the quantity of substance produced or imported by that manufacturer or importer.

- Inflation Index: The EPA will make use of the Producer Price Index for Chemical and Allied Products, as compiled by the U.S. Bureau of Labor Statistics, for purposes of determining the need to adjust the total annual sales values and determining new sales values when adjustments are made. The EPA may adjust the total annual sales values whenever the Agency deems it necessary to do so, provided that the Producer Price Index for Chemical and Allied Products has changed more than 20 percent since either the most recent previous change in sales values or the date of promulgation of this rule, whichever is later. The EPA will provide Federal Register notification when changing the total annual sales values.

Small quantities for research and development, as defined in §704.3 specifying reporting and recordkeeping under section 8(a), means quantities of a chemical substance manufactured, imported, or processed—or proposed to be manufactured, imported, or processed:

- That are no greater than reasonably necessary for such purpose and additionally, per §710.2, specifying inventory reporting regulations under section 8(a),

- That are used by or directly under the supervision of a technically qualified individual(s).

Also, according to §710.2, specifying inventory-reporting regulations under section 8(a), any chemical substance manufactured, imported, or processed in quantities less than 1000 pounds annually will be presumed to be manufactured, imported, or processed for research and development purposes. No person may report for the inventory any chemical substance in such quantities unless that person can certify that the substance was not manufactured, imported, or processed for research and development as defined above.

State means any state of the United States, the District of Columbia, the Commonwealth of Puerto Rico, the Virgin Islands, Guam, the Canal Zone, American Samoa, the Northern Mariana Islands, or any other territory or possession of the United States.

Test marketing means the distribution in commerce of no more than a predetermined amount of a chemical substance, mixture, article containing that chemical substance or mixture, or a mixture containing that substance by a manufacturer or processor, to no more than a defined number of potential customers to explore market capability in a competitive situation during a predetermined testing period prior to broader distribution of that chemical substance, mixture, or article in commerce.

Total annual sales means the total annual revenue (in dollars) generated by the sale of all products of a company. Total annual sales must include the total annual sales revenue of all sites owned or controlled by that company and the total annual sales revenue of that company's subsidiaries and foreign or domestic parent company, if any.

2.2 PROVISIONS OF THE TSCA

There are several sections of the TSCA that give directions to the EPA in exercising its authority to regulate chemical substances or mixtures. Given below are descriptions of some of these sections that the author believes are relevant to complying with TSCA inventory requirements[1]:

2.2.1 Testing Requirements

Section 4 of the TSCA provides the EPA to require testing (or additional testing if the existing test data are insufficient) of a chemical substance or a mixture if the Administrator finds that it may present an unreasonable risk of injury to health or the environment. Further the provision states that the cost of testing must be borne by the manufacturers and/or the processors of the chemical substance or a mixture and that such a cost may be shared among them.

2.2.2 Listing New Chemical Substances or Substances with "Significant New Use" in the Inventory

Section 5 of the TSCA requires that any person intending to manufacture or process a new chemical substance must give the Administrator 90 days' prior notice. This Premanufacture Notice (PMN) enables the EPA to determine if the new chemical presents or will present unreasonable risk of injury to health or the environment *before its commercial manufacture or processing.* The PMN also provides the EPA with an opportunity to determine if the Agency has sufficient test data to reasonably evaluate health and environmental effects of the new chemical substance, and in most cases the EPA obviously does and therefore reviews the notices without seeking additional test data. If the Agency doesn't have sufficient test data on the new chemical substance, section 5 of the TSCA further provides the EPA with means to prohibit or limit manufacture, processing, and distribution in commerce, use or disposal of the new chemical substance. Under one of these provisions, namely section 5(e), if the Agency finds that there is insufficient information available to reasonably evaluate health and environmental effects of the PMN substance, and in view of the latter the substance *may* present an unreasonable risk to health and environment, the EPA may issue a Consent Order to control the manufacture, processing, distribution, use or disposal of the new chemical substance. Consent Orders, when signed by PMN submitters, bind the manufacturers to manufacture, process, distribute, use, or dispose of the PMN substance under the terms and conditions of the Order. Consent Orders remain in effect until such time when the EPA receives sufficient test data to reasonably evaluate health and environmental effects of the PMN substance and find that it does not present an unreasonable risk to injury to health or the environment. The PMN substance, which had been subject of a Consent Order, does get listed in the TSCA inventory and that allows others to manufacture the substance without being required to notify the EPA.

Also, under the provision of this section, if the Agency finds that certain use of a chemical listed in the TSCA inventory with a prior Consent Order may significantly increase human or environmental exposure that the Agency did not evaluate heretofore, the Agency can issue a Significant New Use Rule (SNUR). In cases where the EPA does not issue a Consent Order, the Agency may propose an expedited SNUR when potential new use, not identified in the PMN, could result in increased exposure to or releases of the chemical, and in turn an unreasonable risk to health or the environment. The SNUR requires that anyone intending to manufacture or process a chemical for a use identified by the Agency as a "significant new use," must file a Significant New Use Notice (SNUN). The SNUN is comparable to a PMN in that it enables the EPA to determine if the new use of an inventory-listed chemical presents or will present unreasonable risk of injury to health or the environment *before the commercial manufacture or processing* of the SNUN chemical. Unlike the PMN, the SNUN applies to processors as well as to manufacturers and importers.

2.2.2.1 Chemical Substances Excluded from the Inventory

The following chemical substances are not only not subject to PMN requirements but also are excluded from the inventory. These substances are considered manufactured or processed for a commercial purpose; they are not manufactured or processed for distribution in commerce as chemical substances per se and have no commercial purpose separate from the substance, mixture, or an article of which they may be a part:

2.2.2.1.1   Any impurity.

2.2.2.1.2   Any by-product. A by-product that has commercial value only to municipal or private organizations that:

- Burn it as a fuel,
- Dispose of it as a waste, including in a landfill or for enriching soil, or
- Extract from it component chemical substances that have commercial value and may be reported for the inventory but are not be subject to PMN requirements if not included.

2.2.2.1.3   Any chemical substance resulting from a chemical reaction that occurs incidental to exposure of another chemical substance, mixture, or article to environmental factors such as air, moisture, microbial organisms, or sunlight.

2.2.2.1.4   Non-isolated intermediates.

2.2.2.1.5   Any chemical substance resulting from a chemical reaction that occurs incidental to storage of another chemical substance, mixture, or article.

2.2.2.1.6   Any chemical substance resulting from a chemical reaction that occurs when

- A stabilizer, colorant, odorant, antioxidant, filler, solvent, carrier, surfactant, plasticizer, corrosion inhibitor, antifoamer or de-foamer, dispersant, precipitation inhibitor, binder, emulsifier, de-emulsifier, dewatering agent, agglomerating agent, adhesion promoter, flow modifier, pH neutralizer, sequesterant, coagulant, flocculant, fire retardant, lubricant, chelating agent, or quality control reagent functions as intended, or
- A chemical substance, solely intended to impart specific physicochemical characteristic, functions as intended.

2.2.2.1.7   Any chemical substance resulting from a chemical reaction that occurs upon end use of other chemical substances, mixtures, or articles such as adhesives, paints, miscellaneous cleansers, or other housekeeping products, fuels, and fuel additives, water-softening and treatment agents, photographic films, batteries, matches, and safety flares, and that is not itself manufactured for distribution in commerce or for use as an intermediate.

2.2.2.2 New Chemical Exemptions

Section 5 of the TSCA further permits the Agency, *without application prior to manufacture or import*, to exempt the following categories of new chemicals from some or all of the PMN requirements: These are:

2.2.2.2.1 Research and development[3] (R&D). R&D includes synthesis of new chemical substances for analysis, experimentation, or research on new or existing chemical substances, including product development activities. R&D may include tests of physical, chemical, production, and performance characteristics of a substance.

Submission of a PMN for a new chemical (NC) is not required if it is manufactured or imported in small quantities solely for R&D. Small quantities are those not greater than reasonably necessary for R&D. To qualify for the exemption, R&D activities must be conducted under the supervision of technically qualified individuals who are notified of any health risks associated with the use of the NC. Manufacturers and importers who distribute an R&D substance to other persons must provide those persons with written notification of known hazards and of the requirement that the substance be used solely for R&D.

The following R&D records must be retained until five years after they are developed:

- Information reviewed to determine the need to make any notification of risk,
- Documentation of the nature and method of risk notification,
- Documentation of prudent lab practices, if used instead of risk notification and evaluation,
- Records regarding the chemical identity of the substance to the extent known if an R&D substance is manufactured at greater than 100 kilograms per year,
- Production volume, and
- Disposition of the R&D chemical.

Quantities of the chemical substance, or of mixture or articles containing the chemical substance, remaining after completion of research and development activities may be disposed of as a waste in accordance with applicable federal, state, or local regulations, or used for the following commercial purposes:

- Burning it as a fuel, and
- Reacting or otherwise processing it to form other chemical substances for commercial purposes, including extracting component chemical substance.

2.2.2.2.2 Polymer. The EPA grants an exemption from certain requirements of the PMN for the manufacture of certain polymers[4,5,6] that:

- Meet the "polymer" definition,
- Are not one of the ineligible polymers,
- Meet the general criteria of polymers, and if the manufacturer of such substances,
- Submit a one-time annual report, and
- Keep records.

2.2.2.2.2.1 Polymer definition. The EPA defines a polymer as follows:

Polymer means a chemical substance consisting of molecules characterized by the sequence of one or more types of monomer units and comprising a simple weight majority of molecules containing at least 3 monomer units which are covalently bound to at least one other monomer unit or other reactant and which consists of less than a simple weight majority of molecules of the same molecular weight. Such molecules must be distributed over a range of molecular weights wherein differences in the molecular weight are primarily attributable to differences in the number of monomer units.

2.2.2.2.2.2 Ineligible polymers.

2.2.2.2.2.2.1 Any cationic polymer or a polymer that is reasonably anticipated to become cationic in natural aquatic environment, unless:

- The polymer is a solid that is not soluble or dispersible in water and will be used only in the solid form, and
- The combined (total) equivalent weight of cationic functionality of the polymer is equal to or greater than 5000 daltons,

2.2.2.2.2.2.2 Polymers that do not contain, as an integral part of their composition, at least two of the following elements: carbon, hydrogen, nitrogen, oxygen, silicon, and sulfur,

2.2.2.2.2.2.3 Polymers that contain, as an integral part of their composition, except as impurities:

- Any elements other than carbon, hydrogen, nitrogen, oxygen, silicon, and sulfur,
- Any ions other than Na^+, Mg^{2+}, Al^{3+}, K^+, Ca^{2+}, Cl^-, Br^-, or I^-,
- Fluorine, chlorine, bromine, and iodine covalently bound to carbon, and
- Less than 0.2 weight percent of any combination of the following elements: boron, copper, iron, lithium, manganese, nickel, phosphorous, tin, titanium, zinc, and zirconium,

2.2.2.2.2.2.4 Polymers that are designed or are reasonably anticipated to sub-stantially degrade, decompose, or depolymerize, including those polymers that could substantially decompose after manufacture and use, even though they are not actually intended to do so. The EPA defines degradation, decomposition, or depolymerization as chemical changes including but not limited to oxida-tion, hydrolysis, attack by solvent, heat, light, or microbial action,

2.2.2.2.2.2.5 Polymers that are manufactured from chemical substances in-cluding monomers, initiators, and chain transfer agents, for example, which are not listed in the MIF,

Exception: Polymers that have a reactant incorporated at a level less than two weight percent of the dry weight of the polymer when the identity of the polymer, for the purposes of PMN, does not include the identity of the reactant (this is the so-called Two Percent Rule[6]), and

2.2.2.2.2.2.6 Polymers that are water-absorbing with number average mo-lecular weight[6] equal to or greater than 10,000.

2.2.2.2.2.3 General criteria of polymer. Provided that a chemical substance can be considered a polymer under the polymer definition and is not excludable as one of the ineligible polymers, the substance must also meet one of the fol-lowing three general criteria to be eligible for polymer exemption.

2.2.2.2.2.3.1 The polymer:

- Must have number average molecular weight of between 1000 and 10,000 daltons.
- Must have less than 10% oligomers with a number average molecular weight of 500 daltons and less than 25% oligomers with a number average molecular weight of 1000 daltons.
- Must have no reactive functional groups unless it meets one of the follow-ing three criteria:
- The polymer contains only the following reactive functional groups: carboxylic acid groups, aliphatic hydroxyl groups, unconjugated olefinic groups that are considered "ordinary," meaning that they are not specifi-cally activated either by being part of a larger functional group (e.g., a vinyl ether) or by other activating influences (e.g., strongly electron-withdrawing sulfone group with which the olefinic groups interact); butenedioic acid groups; those conjugated olefinic groups contained in naturally occurring fats, oils, and carboxylic acids; blocked isocyanates (including ketoxime-blocked isocyanates); thiols; unconjugated nitrile groups; and halogens (except that reactive halogen-containing groups such as benzylic or allylic halides cannot be included).

- The polymer has a combined (total) reactive group equivalent weight greater than or equal to 1000 for the following reactive functional groups: acid halides, acid anhydrides, aldehydes, hemiacetals; methylolamides, -amines, or -ureas; alkoxysilanes with alkoxy greater than C_2-alkoxysilanes; allyl ethers; conjugated olefins, cyanates, epoxides, imines, or unsubstituted positions ortho or para to phenolic hydroxyl.
- If any reactive functional groups are not included above, the combined (total) reactive group equivalent weight is greater than or equal to 5000.

2.2.2.2.2.3.2 The polymer:

- Must have number average molecular weight equal to or in excess of 10,000 daltons, and
- Must have less than 2% oligomers with a number average molecular weight of 500 daltons and less than 5% oligomers with a number average molecular weight of 1000 daltons

2.2.2.2.2.3.3 The polymer is a polyester manufactured exclusively from a list[13] of approved monomers and reactants (some of the listed substances may not be listed in the MIF).

2.2.2.2.2.4 One-time report. Manufacturers and importers must submit to the EPA an annual report giving their name, name of their technical contact, site at which the polymer(s) was (were) manufactured or imported, and the number of manufactured or imported polymers, which they believed were exemptible under criteria cited above. If a manufacturer submits information, which the manufacturer claims to be confidential business information, the manufacturer must clearly identify the information at the time of submission by bracketing, circling, or underlining it and stamping it "Confidential" or some other appropriate designation. Any information not claimed confidential at the time of submission might be made available to the public without further notice. Such one-time reports are due by 31 January of the year for the first time manufacture or import of polymer(s) of the prior year.

2.2.2.2.2.5 Recordkeeping. Manufacturers and importers of the polymer(s), which they believed were exemptible under criteria cited above, must keep records for a period of five years from the date of commencement of manufacture or import. The records must include the following:

2.2.2.2.2.5.1 The monomers and reactants used in the manufacture and their CAS, PMN, or EPA accession numbers.

2.2.2.2.2.5.2 Weight percents of monomers used at both above and below two percent levels of the dry weight of the polymer.

2.2.2.2.2.5.3 Method used to determine the weight percent levels of the monomers and reactants and supporting analytical data, if any.

2.2.2.2.2.5.4 Theoretical or analytical data supporting manufacturer's or importer's determination that the polymer is eligible for polymer exemption.

2.2.2.2.2.5.5 Production or importation records showing when the polymer was first manufactured or imported and the production or importation volumes in the first three years.

2.2.2.2.3 Instant or "peel-apart" film article (also called "Polaroid" exemption). The EPA grants exemption from the PMN requirements for the manufacture and processing of new chemical substances used in or for the manufacture or processing of instant photographic and peel-apart film articles.

2.2.2.2.3.1 To manufacture a new chemical substance under the terms of this exemption,[7] a manufacturer of instant photographic film or peel-apart film article must submit, when manufacture begins, a fee of $2500 and an exemption notice, that:

- Identifies the manufacturer, sites, and locations where the new chemical substance and instant photographic or peel-apart film articles will be manufactured and processed,
- Identifies the new substance (as class 1 or 2 substance) by its CAS or IUPAC chemical name and its CAS number (if known), its by-products, and its impurities, and the photographic article containing the new substance,
- Describes physicochemical properties of the new substance,
- Provides an estimate of the anticipated maximum annual production volume,
- Provides any test data of the new substance's health and environmental effects that are known to or reasonably ascertainable by the manufacturer,
- Describes methods used to control and treat wastewater or discharge to a publicly owned treatment works (POTW) or other receiving body of water and identify the POTW or receiving body of water, and
- Certifies that the manufacturer is familiar with the terms of the exemption and that the manufacture, processing, distribution, use, and disposal of the new chemical will comply with those terms.

2.2.2.2.3.2 Other requirements under the terms of the exemption are as follows:

2.2.2.2.3.2.1 The manufacturer must comply with:

- Conditions of manufacture and processing in the special production area,
- Conditions of processing outside the special production area,
- Incorporation of photographic articles into instant photographic and peel-apart film articles,
- Environmental release and waste treatment, and
- Recordkeeping.

2.2.2.2.3.2.2 The manufacturer must limit the manufacture and processing of a new chemical to the site(s) listed in the exemption.

2.2.2.2.3.2.3 The manufacturer must not distribute in commerce or use a peel-apart film article containing a new chemical substance until submission of a PMN and until the review period for the notice has ended without EPA action to prevent distribution or use.

2.2.2.2.4 New chemical imported as part of an article. If a new chemical is a part of an imported article, it is exempt from PMN requirements.

2.2.2.3 New Chemical Exemptions Requiring Prior Application

Finally, section 5 of the TSCA permits the Agency, *upon application prior to manufacture or import*, to exempt certain groups of new chemicals and manufacture or processing of new chemicals under certain conditions from some or all of the PMN requirements. These exemptions are:

- Chemical substances manufactured in quantities of 10,000 kilograms or less per year (Low Volume Exemption, LVE),
- Chemical substances with low environmental releases and human exposure (LoREX), and
- Test Marketing Exemption (TME)

2.2.2.3.1 Low Volume Exemption (LVE). The EPA grants LVE for the manufacture of a chemical substance in quantities of 10,000 kilograms or less per year under the following terms[8]:

2.2.2.3.1.1 To seek an LVE, a manufacturer or importer must submit, using the PMN form, a notice of intent to manufacture 30 days before manufacture begins. The notice must include:

- Manufacturer's identity,
- Chemical identity,
- Impurities,
- Known synonyms or trade names, and
- By-products

Unlike PMN, no fee needs be remitted for LVE exemption notice.

2.2.2.3.1.2 In assessing risk of production volume, the EPA will assume that the manufacturer intends to manufacture or import the chemical at an annual volume of 10,000 kilograms per year. Manufacturers or importers, who intend to produce or import less than 10,000 kilograms, may specify lesser annual production or importation volume in the appropriate box on the PMN form and mark the adjacent binding option box. Manufacturers or importers who opt to specify annual production or importation volume below 10,000 kilograms and who mark the binding option box cannot manufacture or import more than the specific annual amount of the exempted substance unless a new exemption notice for a higher annual production or importation volume (up to 10,000 kilograms) is approved by the EPA.

2.2.2.3.1.3 A manufacturer or importer needs to provide, for manufacturer-controlled sites, identity of the manufacturing sites, process description and worker exposure, and environmental release information, and for manufacturing sites not controlled by the manufacturer, processing and use operation description, estimated number of processing and use sites, and worker exposure/ environmental release information. The manufacturer or importer need not provide information on worker exposure and environmental release if such information is not known or not readily available to the manufacturer. However, in the absence of such information, the EPA will generally apply "bounding estimates," that is, exposure estimates higher than those incurred by persons in the population with the highest exposure to account for the uncertainties in actual exposure and release scenarios.

2.2.2.3.1.4 The manufacturer must clearly indicate on the first page of the PMN that the submission is for LVE.

2.2.2.3.1.5 The manufacturer's or importer's certifications must state that:

- The manufacturer or importer intends to manufacture or import the new chemical substance for commercial purposes,
- The manufacturer or importer is familiar with the terms of this exemption and will comply with those terms,
- The new chemical substance for which the LVE is submitted meets all applicable exemption conditions, and
- The manufacturer or importer intends to commence manufacture or import for commercial purposes within one year of the date of the expiration of the 30-day review period.

2.2.2.3.1.6 If any information is claimed confidential, the manufacturer or importer must submit a second copy of the notice, with all information claimed as confidential deleted.

2.2.2.3.1.7 Upon expiration of the 30-day review of the LVE application, if EPA has taken no action, the manufacturer may consider the exemption approved and begin to manufacture or import the new chemical substance. Unlike standard PMN applications, the LVE application does not require the manufacturer or importer to file an NOC. *Because of this difference the LVE chemicals are not placed in the Inventory and therefore remain "new" and subject to the standard PMN, should they no longer qualify for an LVE.*

2.2.2.3.1.8 The EPA will not approve any subsequent LVE exemption application, by another manufacturer or importer, unless it can determine that the potential human exposure to, and environmental release of, the new chemical substance at the higher aggregate production or importation volume will not present an unreasonable risk of injury to human health or the environment.

2.2.2.3.1.9 Any person who manufactures an LVE chemical must submit a new LVE notice before manufacturing the LVE chemical not approved for the following in previous LVE notice:

- At a different site,
- For a new use,
- For employing new human exposure and environmental controls,
- For manufacturing a different physical form of the LVE substance, and
- At an increased annual volume. However, the new LVE notice need only include identity and location of the new site, name and telephone number of the technical contact, new worker protection and environmental release controls, new use information, and new certifications, besides the EPA-designated exemption number to the previous LVE application.

2.2.2.3.1.10 A manufacturer may, without submitting a new LVE notice, manufacture the LVE chemical at a site not listed in its exemption application under the following conditions:

2.2.2.3.1.10.1 The magnitude, frequency, and duration of exposure of individual workers to the LVE chemical at the new manufacturing site are equal to or less than the magnitude, frequency, and duration of exposure of the individual workers to the LVE chemical at the manufacturing site for which the EPA performed its risk assessment in the original LVE application,

2.2.2.3.1.10.2 Either:

- At the new manufacturing site, the manufacturer does not release to surface waters any of the LVE substance, or any waste streams containing the LVE substance, or
- At the new manufacturing site, the manufacturer maintains surface water concentrations of the LVE chemical, resulting from direct or indirect dis-

charges from the manufacturing site, at or below 1 part per billion, or at or below an alternative concentration level approved by the Agency in writing, or surface water concentrations of at or below 1 part per billion as calculated using methods prescribed in §§721.90 and 721.91(Appendix 7.4).

2.2.2.3.1.10.3 The manufacturer shall notify the EPA of any new manufacturing site no later than 30 days after the commencement of the manufacture of the LVE chemical at the new manufacturing site giving the following information:

- The EPA-designated exemption number of the LVE chemical, manufacturer's identity, the street address of the new manufacturing site, the date on which manufacture commenced at the new site, the name and telephone number of a technical contact at the new site, any claim of confidentiality, and a statement that the notice is an amendment to the original LVE application,
- The notice may be submitted on the EPA form 7710-56 "Notice of Commencement" (Appendix 7.5). However, the manufacturer must add the statement that the notice is an amendment to the original LVE application, and
- The notice must contain an original signature of an authorized official of the manufacturer.

2.2.2.3.1.11 The manufacturer of the LVE chemicals must notify processors and industrial users that the LVE substance can be used only for the use specified in the LVE application and also inform them of any controls specified in the LVE application.

2.2.2.3.1.12 The manufacturer or importer of LVE chemical must maintain records at the manufacturing site or site of importation for a period of five years. These records must include:

- Annual production and import volume, and
- Documentation showing compliance with applicable requirements and restrictions.

2.2.2.3.2 Chemical substances with Low Environmental Releases and Human Exposure (LoREX). Requirements to seek LoREX exemption are the same as those listed above for LVE, with exceptions listed in 2.2.2.3.2.6 below. Also, no fee needs be remitted for LoREX exemption application. EPA grants exemption from PMN requirements for the manufacture of chemical substances with low environmental releases and human exposure under certain terms[8]. LoREX exemption applies to "any manufacturer of a new chemical substance satisfying all of the following low environmental release and low human exposure eligibility criteria:

- Consumers and general population,
- Workers,
- Ambient groundwater,
- Incineration, and
- Land or groundwater.

2.2.2.3.2.1 For exposure of consumers and the general population to the new chemical substance during all manufacturing, processing, distribution in commerce, use, and disposal of the substance:

- No dermal exposure,
- No inhalation exposure {except as described in 2.2.2.3.2.4 below}, and
- Exposure in drinking water no greater than a 1 milligram per year (estimated average dosage resulting from drinking water exposure in streams from the maximum allowable concentration level from ambient surface water releases established under 2.2.2.3.2.3 below or a higher concentration authorized by the EPA under 2.2.2.3.2.3 below)

2.2.2.3.2.2 For exposure of workers to the new chemical substance during all manufacturing, processing, distribution in commerce, use, and disposal of the substance:

- No dermal exposure (this criterion is met if adequate dermal exposure controls are used in accordance with applicable EPA guidance) and
- No inhalation exposure (this criterion is met if adequate inhalation exposure controls are used in accordance with applicable EPA guidance).

2.2.2.3.2.3 For ambient surface water releases, no releases resulting in surface water concentration above 1 part per billion, calculated using methods prescribed in §§721.90 and 721.91 (Appendix 7.4), unless the EPA has approved a higher surface water concentration. To obtain this exemption, relevant and scientifically valid data must be submitted to the EPA in a LoREX exemption notice on the substance or a close structural analogue of the substance that demonstrates that the new substance will not present an unreasonable risk of injury to aquatic species or human health at the higher concentration.

2.2.2.3.2.4 For ambient air releases from incineration, no releases of the new chemical substance above 1 microgram per cubic meter maximum annual average concentration, calculated using the formula:

(kg/day of release after treatment) × (number of release days per year)
× (9.68×10^{-6}) micrograms per cubic meter.

2.2.2.3.2.5 No releases to groundwater, to land, or to a landfill, unless the manufacturer has demonstrated to the EPA's satisfaction in a LoREX exemption notice that the new substance has negligible groundwater migration potential.

2.2.2.3.2.6 In addition to the foregoing terms, there are other terms of LoREX exemption, all of which are identical to those given for LVE chemicals above, that also apply to LoREX exemption with the following exceptions:

- Unlike that stated under 2.2.2.3.1.2 on production volume for the LVE application above, the manufacturers submitting a notice for LoREX exemption must list the maximum amount to be manufactured during the first year of production and the estimated maximum amount to be manufactured during any 12-month period during the first three years of production.
- Unlike that stated under 2.2.2.3.1.4 on the clear indication for an LVE application above, the manufacturers submitting a notice for LoREX exemption, must clearly state on the first page of the PMN that the submission is for LoREX
- Item 4 under 2.2.2.3.1.5 on the manufacturer's certification for an LVE application above is not required for manufacturers submitting a notice for LoREX exemption.

2.2.2.3.3 Test marketing exemption (TME). The TME is sought by filing a PMN form. Unlike PMN, however, no fee needs to be remitted with the notice for TME. Any person may apply for an exemption to manufacture or import a new chemical substance for test marketing. The EPA may grant the exemption if the person demonstrates that the chemical substance will not present an unreasonable risk of injury to health or to the environment as a result of the test marketing. Application for test marketing exemption should contain the following information:

- All existing data regarding health and environmental effects of the chemical substance, including physicochemical properties, or in the absence of such data, a discussion of toxicity based on structure-activity relationships and relevant data on chemical analogues.
- The maximum quantity of the chemical substance that the applicant will manufacture or import for test marketing.
- The maximum number of persons who may be provided the chemical substance during the test marketing.
- The maximum number of persons who may be exposed to the chemical substance as a result of test marketing, including information regarding duration and route of such exposure.
- A description of test marketing activity, including its length and how it can

be distinguished from full-scale commercial production and research and development.

Following the receipt of the application for exemption, the EPA will publish in the Federal Register a notice containing a summary of the information provided in the application, to the extent it has not been claimed confidential

No later than 45 days after the EPA receives an application for exemption, the Agency will either approve or deny the application and publish a notice in Federal Register giving reasons for its approval or denial

In approving an application for exemption, the EPA may impose restrictions necessary to ensure that the substance will not present an unreasonable risk of injury to health and the environment as a result of test marketing.

2.2.3   Regulation of Hazardous Chemical Substances and Mixture

Section 6 of the TSCA states that if the Administrator finds that the manufacture, processing, use, or disposal of a chemical substance or mixture presents or will present an unreasonable risk of injury to health or environment, the Administrator will apply one or more of the following requirements to such substance or mixture to the extent necessary to protect against such risk:

- Prohibit the manufacturing, processing, or distribution in commerce of such substance or mixture, and
- Limit the amount of such substance or mixture, which may be manufactured, processed, or distributed in commerce.

In addition to the above, there are other requirements that deal with the recordkeeping of the process of manufacture, label, quality control, method(s) of use of the substance and mixture, and so on, that the Administrator can apply.

2.2.4   Recordkeeping

Unless a chemical manufacturer or processor qualifies as a small manufacturer or importer, as defined under 2.1 above, section 8(a) of the TSCA requires that anyone that manufactures or processes or proposes to manufacture or process chemical substance, will maintain records and submit reports, as the Administrator may reasonably require, with respect to the following:

- Name, identity, and molecular structure of each chemical substance or mixture manufactured or processed. Include impurities, by-products, and substances to be disposed of that result from such activities.
- Amounts of each chemical substance or mixture manufactured or processed. If the chemical substance or mixture has more than one use, the amounts manufactured or processed must be classified for each use.

- All data concerning environmental and health effects of chemical substance or mixture manufactured or processed.
- Number of workers exposed or likely to be exposed to chemical substance or mixture in their work place and the duration of such exposure

2.2.5 TSCA Inventory

Section 8(b) of the TSCA requires the EPA to compile, keep current, and publish a list ("inventory") of each chemical substance that is manufactured or processed in the United States for commercial purposes.

2.2.5.1 *Scope of TSCA Inventory*

2.2.5.1.1 Chemical substances subject to inclusion in the inventory. Only chemical substances that are manufactured, imported, or processed for a commercial purpose, as defined in Section 2.1, are subject to inclusion in the Inventory

2.2.5.1.2 Naturally occurring chemical substances automatically included. Any chemical substance is naturally occurring:

- That is unprocessed or processed only by manual, mechanical, or gravitational means; by dissolution in water; by floatation; or by heating solely to remove water; or
- That is extracted from air by any means.

These chemical substances are automatically included in the inventory under the category "Naturally Occurring Chemical Substances." Examples of such substances are raw agricultural commodities; water, air, natural gas, and crude oil; and rocks and minerals

2.2.5.1.3 Substances excluded by definition. The following substances are excluded from the inventory:

- Any substance that is not considered a "chemical substance" as defined under 2.1 above,
- Any mixture,
- Any chemical substance that is manufactured, imported, or processed solely in small quantities for research and development (see Section 2.1 for definition of "small quantities for research and development"),
- Any chemical substance not manufactured, imported, or processed for a commercial process since January 1, 1975

In addition to the above, there are other chemicals listed in 2.2.2.1 above that are also excluded from the inventory.

2.2.6   Inspections and Subpoenas

Section 11 of the TSCA provides that the Administrator and any duly designated representative of the Administrator may inspect any establishment, facility, or other premises in which chemical substances, mixtures, or products are manufactured, processed, stored, or held before or after their distribution in commerce and any conveyance being used to transport chemical substances, mixtures, such products or such articles in connection with the distribution in commerce.

2.2.7   Exports

Section 12 of the TSCA provides that:

- Other than all provisions of section 8, the TSCA shall not apply (but see exception below) to any chemical substance, mixture, or to an article containing a chemical substance or mixture, if (1) it can be shown that such substance, mixture, or article is being manufactured, processed, or distributed in commerce for export from the United States, unless such substance, mixture, or article was, in fact, manufactured, processed, or distributed in commerce *for use* in the United States, and (2) such substance, mixture, or article (when distributed in commerce), or any container in which it is enclosed (when so distributed), bears a stamp or label stating that such substance, mixture, or article is intended for export (Exception: if the Administrator finds that the substance, mixture, or article will present an unreasonable risk of injury to health within the United States or to the environment in the United States, then the substance, mixture, or article is not exempt from the TSCA and the Administrator may require testing of the substance, mixture, or article to determine the risk).
- Exporters must notify the EPA before shipment abroad of any substance, identified as being subject to export notification in the EPA's "Chemicals on Reporting Rules" (CORR) list. The CORR list is available from the TSCA Assistance Information Service (telephone: 202-554-1404) or on-line at *www.epa.gov/docs/CORR*. Such a notice should include (1) the name of the regulated chemical as it appears in the CORR list, (2) the name and address of the exporter, (3) the country (countries) of import, (4) the date(s) of export or intended export, and (5) the section (4, 5, 6, or 7) of the TSCA under which the EPA has taken action.

2.2.8   Entry into Customs Territory of the United States

Section 13 of the TSCA requires the Secretary of Treasury to refuse entry into the customs territory of the United States of a chemical substance, mixture, or article if it does not comply with rules in effect under the TSCA, or if it is offered for entry in violation of the TSCA or rules or orders. To implement this provision, the Secretary of Treasury, under the regulations issued by the U.S. Customs Service[9], requires an importer to sign the following statement for each import of chemical substance subject to the TSCA: "I certify that all chemical substances in this shipment comply with all applicable rules or orders under TSCA and that I am not offering a chemical substance for entry in violation of TSCA or any applicable rule or order under TSCA" as a "Positive Certification," or the following statement for each import of chemical substance not subject to TSCA: "I certify that all chemicals in this shipment are not subject to TSCA" as a "Negative Certification."

2.2.9 Prohibited Acts

Section 15 of TSCA states that it is unlawful for any person to:

- Fail or refuse to comply with provisions of the TSCA's sections 4, 5, or the Asbestos Hazard Emergency Response,
- Use for commercial purposes a chemical substance or mixture that such person knew or had reason to know was manufactured, processed, or distributed in commerce in violation of the provisions of the TSCA's sections 5 and 6, or section 7, which deals with imminently hazardous chemical substance or mixture or any article containing such a substance or mixture,
- Fail or refuse to (1) establish or maintain records, (2) submit reports, notices, or other information, or (3) permit access to or copying records, or
- Fail or refuse to permit entry or inspection as required under the TSCA's section 11.

2.2.10 Penalties

Section 16 of the TSCA provides the following penalties:

2.2.10.1 Civil

- Any person who violates provision(s) of the TSCA's section 15 will be liable to the United States for a civil penalty in an amount not to exceed $27,500 for each violation. Each day such a violation continues will constitute a separate violation.
- In determining the amount of a civil penalty, the Administrator will take into account the nature, circumstances, extent, and the gravity of the violation(s) and, with respect to the violator, ability to pay, effect on ability to continue to do business, any history of prior such violations, the degree of culpability, and other such matters as justice may require.

2.2.10.2 *Criminal*

Any person who knowingly or willfully violates the provisions of the TSCA's section 15 will, in addition to or in lieu of any civil penalty which may be imposed as per above, be subject, upon conviction, to a fine of not more than $27,500 for each day of violation, or to imprisonment for not more than one year or both.

2.3   WHO IS SUBJECT TO TSCA COMPLIANCE?

Any person or company that manufactures, processes, distributes in commerce, uses, or disposes of TSCA-regulated chemical substances and mixtures is subject to TSCA compliance. Since "manufacture," as defined under the TSCA, also includes import, importers of chemical substances and mixtures are also subject to TSCA compliance.

The term "use" has not been defined under the TSCA. But, to avoid possible violation of certain provision of the TSCA's section 15 described under 2.2.9 above, any person that processes or distributes in commerce TSCA-regulated chemical substances and mixtures, is considered as "using" them and therefore should consider himself subject to TSCA compliance.

Similarly the term "dispose" or "disposal" has not been defined under the TSCA. But, to avoid possible violation of provisions of section 11 described under 2.2.6 and section 12 described under 2.2.7 above, any person involved in manufacture, processing, or distribution in commerce of any substance, mixture, or article *for export only* should consider himself subject to TSCA compliance.

Chapter 3

CAS Registry Numbers and TCSA Inventory Chemicals

3.1 WHAT IS A CAS REGISTRY NUMBER (CASRN)?

In automating its indexing back in early 1960s, the Chemical Abstract Service (CAS) began using a computer-assigned registry number to uniquely identify each chemical substance as an aid in processing chemical information using computers. As of January 1, 2001, the database of CASRN (Chemical Abstract Service Registry Number) identifies over 28 million chemical substances. All types of organic and inorganic substances are covered, including biochemical compounds, alloys, protein and nucleic acid sequences, coordination compounds, minerals, mixtures, polymers, and salts.

3.2 CASRN AND TSCA INVENTORY CHEMICALS

As indicated above, CAS identifies each chemical substance with a computer-assigned CASRN in order to process information associated with that chemical. Typically CAS generates such CASRN for a chemical when they first encounter information regarding it in global chemical and allied literature. Most chemicals, newly identified by a CASRN, are likely not commercial; that is, they are not produced on a large scale for sale.

Manufacturers, importers, and processors of chemicals for commercial purpose in the United States are subject to the TSCA. The TSCA Inventory is a list of chemicals that the U.S. EPA has generated and kept current according to information they have received from the manufacturers, importers, and processors of chemicals for commercial purposes under the provisions of the TSCA.

Given the foregoing background of the CASRN and of the TSCA Inventory, it should be noted that whereas all TSCA inventory–listed chemicals carry a CASRN (even though some of them have been assigned a CASRN but held confidential, others may not have one assigned to them, or still others may have been assigned one but the CASRN may not be known to public), not all chemicals carrying a CASRN are on the TSCA Inventory (even though some of the chemicals may be commercial in other parts of the world).

Chemicals new to the global chemical and allied literature are assigned a CASRN by CAS, and if they are to be commercialized in the United States, they must be listed in the TSCA inventory through the Premanufacture Notification (PMN) process of the U.S. EPA. The assignment of CASRNs to chemicals that are new to the global chemical and allied literature and will be commercialized in the United States can be best handled via the PMN process, since assignment of CASRN by the CAS is part of the process.

Chapter 4

TSCA Inventory Requirements

Since the enactment of the TSCA in 1976, the EPA has promulgated rules, issued orders, and provided guidelines clarifying the rules and orders to implement the provisions of the TSCA. Primary among these rules, orders, and guidelines are those that relate to compiling a list (TSCA Inventory) of the chemical substances already commercially used in the United States, providing means both to list chemical substances yet to be commercialized and to make changes in the list. These rules and orders, for the purposes of this book, are the TSCA inventory requirements.

4.1 TSCA INVENTORY

As mentioned above, section 8(b) of the TSCA requires the EPA to compile, keep current, and publish a list ("inventory") of each chemical substance that is manufactured or processed for commercial purposes in the United States. In 1977 the EPA issued Inventory Reporting Regulations[10] and an instruction manual for manufacturers to report their existing commercial chemical substances for the TSCA Inventory. In 1979 the EPA provided another opportunity to manufacturers to report additional chemical substances that they manufacture or process not already listed in the inventory and published Premanufacturing Notification (PMN) Requirements and Review Procedures.[11]

These efforts led the EPA to publish the so-called Initial Inventory in 1979. In years following 1979, the EPA has issued a number of rules, orders, and guidelines that further delineate or clarify its position on reporting of chemicals not already on the Inventory or correcting of those on the Inventory. In 1986 the EPA issued an Inventory Update Rule (IUR)[12] that required a manufacturer or an importer of certain chemicals listed in the Inventory to report their current production volumes, plant site, and site-limited status.

As of January 1997 there were about 75,000 commercially manufactured, processed, or imported chemicals listed in the TSCA Inventory, and hundreds more of such chemicals have been added to it each year since. The public version of the TSCA Inventory serves as a starting point for identifying which chemicals can be considered "existing" and which cannot.

The public version of the current TSCA Inventory was published in five volumes in 1985 and supplemented in 1990. Volume 1 lists the chemicals in ascending order of their CAS registry numbers. Some of the listed chemicals carry a "flag" such as a letter(s), P, S, or XU, an asterisk (*), or a dagger (†). Chemicals carrying "P" indicate that they were added to the inventory via a PMN process, those carrying "S" indicate that they are subject to a proposed or final significant new use rule (SNUR), those carrying "XU" indicate they are exempt from reporting under the IUR, those with an asterisk (*) are substances with unknown or variable composition, and finally those with a dagger (†) are substances of unknown or variable composition, complex reaction products, and biological materials (UVCB). Volume 1 also has two appendixes: appendix A titled "Chemical Substance Definitions" and appendix B titled "Generic Names for Confidential Identities."

Volumes 2 and 3 list the chemicals by their CAS index, Preferred or Generic Names in alphabetical order. Some these chemicals carry a "flag" as described above for volume 1.

Volume 4 classifies chemicals by their molecular formula. Each molecular formula, listed in alphabetical order of the elements, identifies all substances having the same molecular formula along with their CAS registry number and any applicable flag. Volume 5 lists UVCB chemicals and chemicals subject to a proposed or final SNUR.

A copy of the TSCA inventory can be purchased from the Government Printing Office (GPO) or the National Technical Information Service (NTIS)

> GPO (202) 512-1800
>   1985 TSCA Inventory (report form)
>   Order No. 055-000-00254-1
>   1990 Supplement (report form)
>   Order No. 055-000-00361-1
> NTIS (703) 487-4650
>   TSCA Inventory
>   9-Track magnetic computer tape
>   Order No. PB95503108
>   CD ROM
>   Order No. PB94502168

4.2 BONA FIDE INTENT TO MANUFACTURE (BIFM) REQUESTS

The EPA maintains and continually updates the Master Inventory File (MIF), which comprises both the chemical substances listed in the public version of the TSCA Inventory and the chemical substances whose chemical identity are claimed as confidential in the confidential version of the TSCA Inventory. Any person who intends to manufacture or import a chemical not listed in the public version of the TSCA Inventory and who can demonstrate a "bona fide intent" for doing so may ask the EPA to determine whether the substance in question is listed in the confidential version of the TSCA Inventory as a means of determining if the substance is listed in the MIF. The EPA will answer such an inquiry only if it determines that the person has a bona fide intent to manufacture or import the chemical substance for commercial purposes.

As mentioned in Section 2.1 above, chemicals not listed in the MIF are considered "new" and therefore are required to be listed in the MIF through a PMN process before they can be commercialized.

4.3 THE PREMANUFACTURE NOTICE (PMN) PROCESS

As mentioned before, section 5 of the TSCA requires that any person intending to manufacture or process a new chemical substance must give the Administrator 90 days' prior notice. This notice is the Premanufacture Notice (PMN), and it comprises a 13-page EPA 7710-25 form (Appendix 7.6). The PMN enables the EPA to determine if the new chemical presents or will present unreasonable risk of injury to health or the environment *before its commercial manufacture or processing*. If the Agency takes no regulatory action by the end of the 90-day review period, the manufacturer or importer may begin commercial manufacture or importation of the new chemical immediately without the Agency's approval. However, within 30 days of completion of the manufacture or importation of the new chemical, the manufacturer or importer must submit a Notice of Commencement (NOC) with the EPA to certify that the manufacture of the new chemical has indeed begun or that the new chemical has indeed been imported. For the purposes of this book, therefore, filing of the PMN followed by filing of the NOC constitutes "the PMN process." A manufacturer or a processor that does not complete the PMN process for manufacturing or processing a new chemical (i.e., files the PMN but does not file the NOC within 30 days after beginning the manufacture or importation of the new chemical) is in violation of TSCA's section 5. Filing the SNUN requires filling out the same PMN form with certain additional information, but unlike PMN, this does not need to be followed up with filing of a NOC.

4.4 INVENTORY CORRECTION

Back in 1979 the EPA sought and received from chemical manufacturers and processors information regarding chemicals then in commerce. This information was used in formulating the Initial Inventory. Some of such information was erroneous, and the errors were not realized until years later. Thus subsequently the EPA has provided an inventory correction process,[13] and its main points are as follows:

Typically now the company that provided information for the Initial Inventory can begin the process of correcting the inventory. If the company that provided information for the Initial Inventory does not or does not intend to manufacture or import the chemical but is willing to provide the needed documentation for correcting the inventory to a third party, the latter can initiate the inventory correction process. The EPA entertains such third-party inventory correction requests only in rare cases.

The EPA will consider correcting the Inventory for the following reasons:

4.4.1 Error in Chemical Identity

The error may be due to:

- Typographical errors and/or
- Incorrect compositional assignment because of unavailable characterization equipment or technique.

4.4.2 Identification of New Components

These components could include isolated intermediates in a reaction product. Requests for inventory correction to EPA must include:

- Copies of documents previously submitted for inclusion of the chemical substance in the Initial Inventory,
- Corrected inventory reporting documents,
- Documentation characterizing new components including those of previously isolated intermediates, if any, and
- Documentation showing that the current commercial manufacture of the chemical substance, including of the isolated intermediate, if any, is identical to that practiced at the time when the final product was reported for inclusion in the Initial Inventory.

Inventory corrections can be made even for chemical substances reported for inclusion in the inventory through a PMN process and the process for requesting such correction is identical to the one described above for correcting the Initial Inventory.

The EPA's response to the requests for inventory correction can be positive or negative. If the response is positive, the correction to the inventory is retroactive to the time of the Initial Inventory reporting. However, such corrections are not reflected in the inventory until after they are published in the Federal Register.

Chapter 5

Processes for Assessing TSCA Inventory Requirements

5.1 PROCESS FOR ASSESSING TSCA INVENTORY REQUIREMENTS OF A CHEMICAL FOR CHEMICAL PROCESSORS (SCHEMATIC DIAGRAM I)

As defined previously, a chemical processor is a person who prepares a chemical substance or mixture, after its manufacture, for distribution in commerce:

- In the same form or physical state as, or in different form or physical state from, that in which it was received by the person preparing such substance or mixture, or
- As part of an article containing the chemical substance or mixture.

Examples of chemical processors include a metal fabricator that uses chemicals to degrease or clean metal and an ink manufacturer that uses chemicals as constituents of inks it produces. In the manufacturing operation of a chemical processor, unlike that of a chemical manufacturer, chemicals may become a part of the product but not enter into any chemical reactions resulting in a chemical or chemicals different from the ones originally used.

The process for assessing TSCA inventory requirements for a chemical processor comprises following steps:

- Obtaining suppliers' information on all chemical raw materials used in processor's operations,
- Checking TSCA inventory requirements compliance of chemical raw materials, and

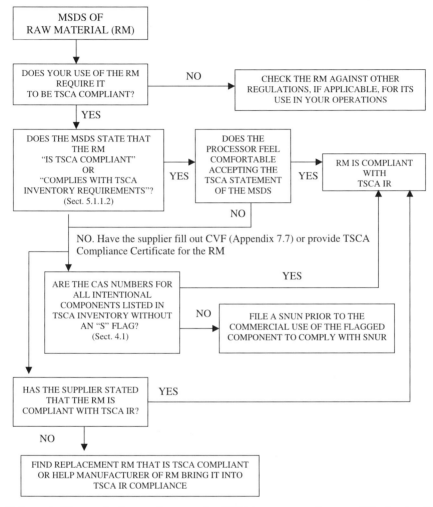

Schematic Diagram I *Process for assessing TSCA inventory requirements (IR) of a chemical raw material (RM) for chemical processors.*

· Periodically rechecking TSCA inventory requirements compliance of chemical raw materials.

5.1.1 Obtaining Suppliers' Information on All Chemical Raw Materials

As a first step toward determining the TSCA inventory requirements compliance of the chemicals, chemical processor should prepare a list or a database of all the chemical raw materials used at the facility. Such a database may contain the following information about these materials:

5.1.1.1   Supplier's Name, Address, and Telephone Number

Chemical supplier may be its manufacturer, distributor, or importer. It is important to recognize which category the supplier belongs to, as that would determine who the processor might have to contact to get the information needed to check the TSCA inventory compliance of the chemical.

5.1.1.2   Supplier's Material Safety Data Sheet (MSDS)

Under the Occupational Safety and Health Act (OSHA) rules, suppliers of chemicals are required to provide a copy of chemical's MSDS to users of the chemical to apprise the recipients of the chemical's safe handling and disposal. While the suppliers are not required to provide information pertaining to composition of the chemical product in the chemical's MSDS, they generally do. When a MSDS contains such information, it is found listed in section 2, and in some cases, also in section 15 of the MSDS. Information pertaining to composition of the chemical product comprises the chemical components, their CAS numbers, and their weight percent in the product. If such compositional information is not found listed in MSDS or if processor is not comfortable with the accuracy of the compositional information in MSDS, the processor will specifically seek the compositional information in writing from the supplier. A convenient form for seeking such information is provided in Compliance Verification Form (CVF) given in Appendix 7.7.

In checking the compositional information of a product, be it in the MSDS or in CVF, it is important to verify the following:

5.1.1.2.1   Does the compositional information of a product account for 100% of its content?

The compositional information must account for 100% of the product. If it does not, the processor should seek the missing information, in writing, from the supplier.

5.1.1.2.2   Is the compositional information of the product accurate?

If the product is a commodity chemical such as acetone and titanium dioxide, or a mixture of commodity chemicals, there is a good chance that the compositional information of the product in terms of its chemical components and their CAS numbers, as provided by the supplier, may be correct. On the other hand, for sake of accuracy, a professional chemist should check compositional information of a specialty chemical or a mixture containing specialty chemical(s). The accuracy (or a lack thereof) of compositional information may relate to (1) appropriate chemical identification of the product or a part thereof and (2) assignment of correct CAS number(s) to the otherwise appropriately identified chemical component(s).

Appropriate chemical identification of the product or a part thereof becomes an important consideration for chemicals that can exist in various isomeric forms. Appropriate chemical identification of the product or a part thereof, and assignment of correct CAS number to chemical components to the otherwise appropriately identified chemical component(s), therefore, require adequate knowledge of the product chemistry.

Since the assignment of a CAS number to a chemical component depends on the chemistry of its manufacture, should the processor determine that CAS number assigned to that chemical component is incorrect, he should bring it to the attention of the manufacturer of the product and have the manufacturer review the assignment of the correct CAS number. The manufacturer must correct the CAS number, if necessary, and issue a revised MSDS or CVF reflecting correct chemical identification, CAS number, or both.

5.1.1.2.3 Which chemical components of a product are intentional and which are impurities?

Intentional chemical components of a product are chemical substances that a manufacturer of the product intends to make, and therefore it is only the manufacturer who can certify whether or not a certain chemical component of its product is intentional. Typically, however, a chemical component present in the largest amount in a product is the intentional component. This is especially true of commodity products such as organic solvents, monomers such as styrene and polymers. Organic intermediates and specialty chemicals that can exist in various isomeric forms should also be examined carefully to determine if one or more of their chemical components may be intentional.

Impurities in a product are obviously those chemical components that are not intentional and typically include small amounts of residual solvent(s) and raw material(s) used to synthesize a product and by-product(s). When a single component of a product with an impurity status constitutes a rather large amount (the actual amount is subject to judgment depending on the product and its components) of the product, it is important to obtain a written certification regarding its impurity status from the manufacturer.

5.1.2 Checking Compliance of Chemical Raw Materials with TSCA Inventory Requirements

As described above, the public version of the TSCA Inventory lists chemical substances by their name and CAS number. It also lists some chemical substances, placed in the TSCA Inventory prior to March 29, 1995, by their generic name and their EPA accession number. When all the supplier's information of their product, as listed in Section 5.1.1 above, is at hand, a chemical processor can check the supplier's product for TSCA inventory compliance by verifying that:

- Section 15 of the supplier's MSDS states that its product is "TSCA compliant," or "complies with TSCA inventory requirements," or
- CAS or EPA accession numbers of all intentional chemical components of the product are listed in the public version of the TSCA inventory without them carrying an "S" flag. If they are, the supplier's product is compliant with TSCA inventory requirements. If a CAS number of intentional com-

ponent carries a "S" flag, then the chemical processor can either look for another raw material that does not have any of its components carry such a flag or file a SNUN (see 5.2.2.8.2 below) prior to the commercial use of the flagged component to comply with the SNUR.

If the processor cannot verify the TSCA Inventory compliance by either of the two means identified above and the product is a chemical used in industrial applications, it is possible that one or more of its intentional chemical components is listed in the confidential version of the TSCA Inventory, or that it is a naturally occurring material supplied without any chemical modification or is a company trade secret for which its CAS number is held as company confidential. Since the processor cannot verify the TSCA compliance status of the supplier's raw material in any of these cases, the processor's recourse is to seek a written TSCA compliance certification for the raw material from its supplier. An acceptable TSCA compliance certificate would be one that is written on supplier's letterhead and signed and dated by a responsible officer of the supplier's organization, and that may read as follows:

All components of (product's name or trade name) are in compliance with the current inventory requirements of the U.S. EPA's Toxic Substances Control Act (TSCA) and the use of (product's name or trade name) is not subject to the Significant New Use Rule (SNUR).

It is important to note that not all commercially sold chemicals have to be compliant with TSCA inventory requirements. This is because the TSCA inventory requirements only apply to commercially manufactured or imported chemicals used in *industrial applications*. See the definition of "chemical substance" in Section 2.1 for other commercially manufactured or imported chemicals that are exempt from the TSCA inventory requirements.

If, at the end of the process, a chemical processor determines that the raw material does not meet the TSCA inventory requirements and no other suitable TSCA compliant replacement raw material can be found, then the processor may choose to work with the manufacturer of the chemical to have the raw material supplier file a PMN to bring the chemical in compliance with the TSCA inventory requirements. Since the EPA's review of a PMN takes at least 90 days, chemical processors should allow sufficient time before initiating their use of chemical raw materials that require working with suppliers for TSCA compliance.

5.1.3 Periodical Rechecking of Compliance of Chemical Raw Materials with TSCA Inventory Requirements

A supplier's product composition may change on account of many factors. Dominant among these are changes in the raw materials used in the product manufacture and in the product's manufacturing process. When compositional changes occur in a product, suppliers issue a new or updated MSDS and send a copy of such MSDS to their customers receiving the product. However, such a practice on part of all suppliers is neither guaranteed nor typical. To guard against possible noncompliance of the TSCA inventory requirements of supplier's products with changed composition, processors should request and review product's current MSDS, seek an updated CVF and a TSCA compliance certificate, if needed, periodically and preferably every year.

5.2 PROCESS FOR ASSESSING TSCA INVENTORY REQUIREMENTS OF A CHEMICAL FOR CHEMICAL MANUFACTURERS (SCHEMATIC DIAGRAM II)

For the process described under this section, "manufacturers" do not include importers.

As mentioned before, chemical manufacturers are organizations that convert chemical raw materials into chemicals different from the raw materials (hereinafter called new chemical, NC) through the use of a chemical reaction or modification. Before embarking on determining if the NC complies with TSCA inventory requirements, manufacturer must make sure that a professional chemist having adequate knowledge of the underlying chemistry has characterized the NC accurately and, where possible, has assigned a correct CAS number to it. For a chemical manufacturer, complying with the TSCA's Inventory requirements means determining if intentional components of the chemical raw materials as well as of the new chemical(s) are listed in either the public or confidential versions of the TSCA inventories.

5.2.1 TSCA Inventory Compliance Check for Chemical Raw Materials

To determine the TSCA compliance for chemical raw materials used in the manufacture of an NC, follow the same process described above for the chemical processors (Schematic Diagram I). If, at the end of the process, a manufacturer determines that the raw material does not meet the TSCA inventory requirements and no other suitable TSCA-compliant replacement raw material can be found, then the chemical manufacturer may choose to find a replacement RM that is compliant with the TSCA IR or work with the manufacturer of the chemical to have the raw material supplier file a PMN to bring the chemical in compliance with the TSCA inventory requirements. Since the EPA's review of a PMN takes at least 90 days, chemical manufacturers should allow sufficient time before initiating their use of chemical raw materials that require working with suppliers for TSCA compliance.

5.2.2 TSCA Inventory Compliance Check for the NC

Before initiating the TSCA inventory compliance of the NC, make sure that the substance has been assigned appropriate chemical structure, chemical

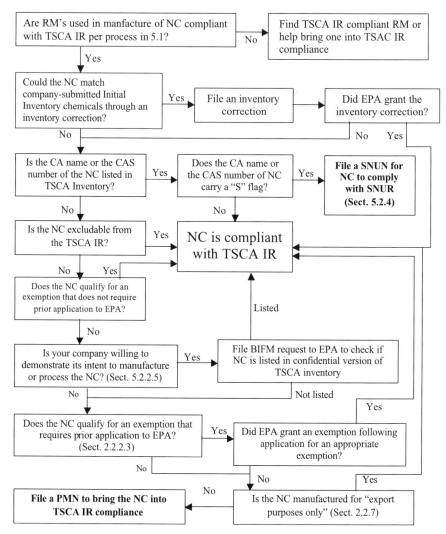

Schematic Diagram II   *Process for assessing TSCA inventory requirements (IR) of a chemical for chemical manufacturers.*

name, and CAS number, if available. Begin the process of assessing the TSCA inventory compliance of the NC as described below:

5.2.2.1   Review Initial Inventory for Correction

Check the chemical structure, chemical name, or the CAS number of the NC against those of each chemical your company might have submitted to the EPA for inclusion in the TSCA Initial Inventory, and determine if inventory correction process might help compliance of the NC with TSCA inventory requirements.

5.2.2.2 Check TSCA Inventory

If inventory correction process cannot help compliance of the NC with the TSCA inventory requirements, check the TSCA inventory to see if the NC is listed either by its chemical name or its CAS number. In doing so, consider as many names as possible to identify the NC.

If the NC is found listed in the TSCA inventory, make sure that it does not carry an "S" flag (see Section 4.1) If the NC is listed without an "S" flag, it is compliant with TSCA inventory requirements and can be immediately manufactured or processed for commercial purposes.

If it carries an "S" flag, then the chemical is subject to an SNUR and requires filing of an SNUN before it can be manufactured or processed (proceed to Section 5.2.4 below).

5.2.2.3 Check TSCA Inventory Exclusions

If the NC is not listed in the TSCA inventory, ensure that the NC is not excludable from the TSCA inventory requirements listed above under the definition of "chemical substance" in Section 2.1. Recognize that an article, as defined by the TSCA, is not a "chemical substance" and is therefore excluded.

5.2.2.4 Check TSCA Inventory Exemptions

If the NC is not excludable, check if the NC qualifies for one of the following exemptions that do not require filing an application with the EPA prior to manufacturing (See 2.2.2-.1 and -.2 for details):

- Research and development (R&D) exemption,
- Polymer exemption,
- Instant or "peel-apart" film article (the so-called Polaroid exemption),
- New chemical imported as part of an article,
- Impurities and by-products,
- Non-isolated intermediates,
- Chemicals incidental to end use or storage, and
- Incidentally formed chemicals upon use of certain additives (the so-called additives exemption).

5.2.2.5 File Bonafide Intent to Manufacture (BIFM) Request

If the NC does not qualify for any of the exemptions, determine whether your company intends to manufacture the NC recognizing that it would require filing certain information (identified below) to the EPA to demonstrate your company's intent to manufacturing it. If your company can demonstrate its intent, then file a BIFM request (Sect. 4.2) to the EPA so that they can check if the NC is listed in the confidential version of the TSCA Inventory. The BIFM request to the EPA should include:

- Specific chemical identity of the NC with its correct CAS name.
- Signed statement that the person intends to manufacture the NC for commercial purpose.
- A brief description of the research and development activities (synthesis, isolation/purification, formulating, product development, process development, end-use application, toxicity testing, etc.) relating to the NC, including the year in which such activities were initiated.
- Description of the major intended use or application of the NC.
- Infrared spectrum or alternative spectrum of the NC that yields a reasonable amount of its structural information.
- Estimated date by which the person intends to submit a PMN for the NC, if the EPA informs the BIFM request submitter that the NC is not in the MIF.
- Address of the company facility where the NC would be manufactured or processed for commercial purpose.
- Brief description of process of manufacture of the NC or of its use.
- If a manufacturer cannot provide required information because a supplier of a reactant used in manufacture of the NC claims the chemical identity of the reactant confidential, the manufacturer of the NC should provide as much information about the manufacture as possible and the supplier of the reactant should provide the identity of the proprietary reactant directly to the EPA along with reference to the manufacturer's BIFM request. If the appropriate supporting document for the BIFM request is not received within 30 days after the EPA receives the request, the latter will be considered incomplete.

The EPA will search the confidential inventory and inform the proposed manufacturer whether the NC is on the confidential inventory within 30 days after receipt of a complete submission of the BIFM request.

5.2.2.6 Check TSCA Inventory Exemptions Requiring Prior Application

If the NC cannot be found listed in the confidential inventory, then check if the NC qualifies for one of the following exemptions that do require filing an application with the EPA prior to manufacturing (see 2.2.2.3 for details):

- Low volume,
- Low release and exposure, and
- Test marketing exemption.

5.2.2.7 Evaluate NC for Export Only Exemption

If the NC does not qualify for any of the exemptions above, check that the NC is being manufactured for the purposes of export only and as such meets the criteria for the "Export Only" exemption (see 2.2.7. above). If the NC can qualify for Export Only exemption, it complies with the TSCA inventory requirements.

5.2.2.8 Allow for a Review Period by EPA

If the NC does not qualify for the Export Only exemption, the NC must be subjected to the PMN process to bring it into compliance with the TSCA inventory requirements. As mentioned before, the PMN process comprises filing a PMN for the NC or an SNUN for a chemical with a significant new use *followed by* an NOC for the PMN chemical to the EPA, assuming that the latter takes no regulatory action within its 90-day PMN or SNUN review period. Again, since the EPA's review of a PMN or an SNUN takes at least 90 days, chemical manufacturer should allow sufficient time before initiating the manufacture of the NC for commercial purposes.

5.2.3 Filing a PMN for the NC

To file a PMN for the NC, make a copy of the PMN form provided in Appendix 7.6 and follow the instructions in the manual (Appendix 7.8) to fill out this form. It is useful to familiarize oneself with the "PMN Review Process" (Appendix 7.10) and with the "Chemical Information Needed for Risk Assessment" (Appendix 7.11). Both documents are issued by the officials at the USEPA to provide guidance in filing PMN submissions. The instruction manual includes a list of contacts at the EPA that a submitter may consult on completing a PMN form.

Following is some additional detailed information about filing a PMN or an SNUN:

5.2.3.1 Fees

Persons submitting a PMN to the EPA must remit for each notice the appropriate fee identified as follows:

5.2.3.1.1 Small business concerns are charged a fee of $100 for each PMN submitted.

5.2.3.1.2 Other businesses remit fees according to the type of PMN:

- A fee of $2,500 for each PMN, consolidated PMN, or significant new use notice (SNUN) submitted, or
- A fee of $1000 for each intermediate PMN.

5.2.3.1.3 Joint submitters of PMN are required to remit appropriate fees identified above regardless of the number of joint submitters for that notice.

5.2.3.1.4 The EPA will refund any fee paid for a PMN whenever the Agency determines:

- That the NC that is the subject of the PMN, intermediate, or exemption notice is not a new chemical as of the date the submission of the notice,
- That, in case of an SNUN, the notice was not required, or
- That the PMN is incomplete.

5.2.3.2 Submission of Health and Environmental Studies[14]

5.2.3.2.1 The PMN must account for all test data, except those described under 5.2.3.3 below, on the new chemical substance in the possession or control of the submitter. This includes test data concerning the new chemical substance in a pure, technical grade, or formulated form.

A full report or standard literature citation must be submitted for the following types of test data:

- Health effects data,
- Ecological effects data,
- Physical and chemical properties data,
- Environmental fate characteristics, and
- Monitoring data and other test data related to human exposure to or environmental release of the chemical substance.

The full report should include the experimental methods and materials, results, discussion and data analysis, conclusions, references, and the name and address of the laboratory that developed the data. A standard literature citation includes author, title, periodical name, date of publication, volume, and page numbers.

If a study, report, or test is incomplete, the submitter must identify the nature and purpose of the study, name and address of the laboratory developing the data, progress to date, types of data collected, significant preliminary results, and anticipated completion date. If a test or experiment is completed before the notice review period ends, the submitter must forward the study, report, or test to the address listed on the notice form within ten days of receiving it, but no later than five days before the end of the review period. If the test or experiment is completed during the last five days of the review period, the submitter must immediately inform its EPA contact for that notice by telephone.

Test data in the submitter's possession or control that are not listed above, do not have to be submitted in a detailed report. A summary of the data is required. If the EPA so requests, a full report must be submitted within ten days of the request, but no later than five days before the end of the review period.

All test data listed above are subject to these requirements regardless of their age, quality, or results

5.2.3.2.2 Other data, except those described under 5.2.3.3 below, concern the health and environmental effects of the new chemical substance that are known to or reasonably ascertainable by the submitter as described below.

The submitter must describe any data from a health and safety study if the data are related to the effects on health or the environment of any manufacture, processing, distribution in commerce, use, or disposal of the new chemical substance, of any mixture or article containing the new chemical substance, or of any combination of such activities. These data include:

- Any data, other than test data, in the submitter's possession or control,
- Any data, as well as test data, that are not in the submitter's possession or control but are known to or are reasonably ascertainable by the submitter. For the purposes of this item, the data are known to or reasonably ascertainable by the submitter if the data are known to any of its employees or other agents who are associated with the research and development, test marketing, or commercial marketing of the substance.

Data that must be described include data concerning the new chemical substance in pure or technical grade, or formulated form. Description of the data must include:

- If the data appeared in the open scientific literature, a standard literature citation that gives the author, title, periodical name, date of publication, volume, and pages,
- If the data are not contained in the open scientific literature, a description of the type of data and summary of the results, if available, and the names and addresses of persons the submitter believes may have the possession or control of the data.

All data under the second item are subject to this requirement, regardless of their age, quality, or results, and regardless of whether they are complete at the time the notice is submitted.

5.2.3.3 *Data Exempt from Submission*

5.2.3.3.1 Certain data previously submitted to EPA are exempt:

- Data previously submitted to EPA with no claims of confidentiality if the notice includes the office or person to whom the data were submitted, the date of submission, and if appropriate, a standard literature citation as specified in 5.2.3.2.1 above.

• Data previously submitted to EPA with a claim of confidentiality must be resubmitted with the notice of any claim of confidentiality.

5.2.3.3.2 Any data relating to the product efficacy do not need to be submitted. This does not exempt the submitter from providing the data specified under 5.2.3.2.1.

5.2.3.3.3 Any data that relates only to exposure of humans or the environment outside the United States are exempt. This does not exclude nonexposure data such as data on health effects (including epidemiological studies), ecological effects, physicochemical properties, or environmental fate characteristics.

5.2.3.4 Common Errors in PMNs

The EPA conducts an initial screening of the PMNs as soon as the notices are received and a comprehensive screening during the 90-day review period. Initial screening involves checking for errors, omissions, inconsistencies, and/or ambiguities. It also focuses on the chemical identity of the reported substance. To date, the EPA has declared 27% of the PMNs incomplete as a result of the initial screening and 5% of the PMNs incomplete as a result of comprehensive screening. The PMNs declared incomplete in the initial screening are returned to the submitters with a checklist of problems that need corrected (but would not explain how to "fix" them), whereas for the PMNs declared incomplete during the 90-day review period, the submitters would receive a letter along with a check list of items that need corrected. The corrected PMNs must be resubmitted in their entirety and will be assigned a new case number. Therefore, before the completed PMN is submitted to the EPA, the submitter should check and ensure that the common errors that would result in the EPA's returning the PMN as an "incomplete notice," are avoided. Some of the common PMN errors are as follows:

5.2.3.4.1 The chemical name is wrong.

5.2.3.4.2 There are inconsistencies:

• The EPA is unable to determine which part of the submitted information is correct.
• That involve two or more chemical identity-related information such as chemical name, CAS number, molecular formula, structural diagram, identity of immediate precursor(s) or monomer(s), or manufacturing process information.
• The CAS, consultant or anyone else is provided with information different from that submitted to the EPA.
• Inappropriate or incorrect use of the Two Percent Rule for polymers[6].

5.2.3.4.3 The structural diagram(s) may be incorrect or insufficient, meaning that the diagrams do not contain all reasonably ascertainable details for class 1 or 2 substances or use abbreviations other than the following allowed ones: Me for methyl, Et for ethyl, Pr for propyl, i-Pr for iso-propyl, Bu for butyl, i-butyl for iso-butyl, s-butyl for sec-butyl, t-bu for tert-butyl, and Ph for phenyl.

5.2.3.4.4 The submitter failed to fill out page 5 for polymers and oligomers.

5.2.3.4.5 The "typical composition" values or ranges for reactants are not > 0%.

5.2.3.4.6 The chemical name and/or available CAS number for precursor or monomer for class 2 substance is missing.

5.2.3.4.7 The free-radical initiators or chain-transfer agents are not listed as reactants.

5.2.3.4.8 Manufacturing information is incomplete, for example, specific name or CAS number and weight for each starting material, including catalysts, solvents, and other nonreactants.

5.2.3.4.9 The consolidated PMN submitters did not use method 1 to obtain correct chemical names for all substances.

5.2.3.4.10 The suppliers of the PMN submitters withhold supporting documents providing confidential identification of reactants or PMN substances.

5.2.3.4.11 Molecular formula are incorrect.

5.2.3.4.12 The generic name is insufficient or deceptive.

5.2.3.5 PMN Review Process and Outcome

The PMN review process is described in detail in Appendix 7.10. Briefly during the first 20 days of the review period, the EPA will declare some PMNs or LVE notices invalid, will drop some PMNs from further review with a letter to submitter stating its concerns for the PMN substance, or will notify the submitter by letter that the PMN substance will have certain restrictions on releases and/or exposures but will not be subject to Consent Order. Also, during the first 20 days, the EPA may determine to deny, grant, or grant conditionally any requests for LVE, LoREX, or test market exemptions. During the remainder of the review period, less than 5% of the notices begin their "standard review" by the EPA notifying the submitter of its possible regulatory action before the review period expires. As a result of the "standard review," the Agency may decide:

- To drop the PMN from further review with or without a letter of its specific concerns.
- To drop the PMN from further review with a letter to submitter stating that the PMN substance will have certain restrictions on releases and/or exposures but will not be subject to Consent Order.
- To issue a Consent Order to prohibit or limit the manufacture, processing, distribution in commerce, use, or disposal of the NC or to prohibit or limit any combination of such activities. Depending on the basis upon which the Agency issues the order, the Agency and the submitter may use one of several "boilerplate" agreements to expeditiously arrive at a final Consent Order. When signed, the final Order binds the submitter of the PMN to manufacture, process, distribute in commerce, use, or dispose of the NC under specified conditions. Along with an issuance of the Consent Order, the Agency would also invariably issue an SNUR.
- To issue a Consent Order that would require the PMN submitter to conduct up-front testing.

5.2.4 Significant New Use Rule (SNUR) and Significant New Use Notice (SNUN)

Once the Consent Order is obtained, the NC is listed on the TSCA Inventory. Any other company may manufacture the chemical without being required to notify the EPA. If, however, EPA determines that a particular use of a chemical already on the TSCA Inventory constitutes or will constitute a significant new use, it can issue a Significant New Use Rule (SNUR). An SNUR requires that anyone intending to manufacture or process a chemical substance for a use that EPA has determined as a significant new use must submit a Significant New Use Notice (SNUN) to EPA at least 90 days before doing so. The SNUN must be submitted on the standard PMN form, and so it requires the same information as a full PMN. The EPA processes an SNUN essentially the same way it does a PMN, but if the Agency does not take any action to regulate the SNUN chemical before 90-day review period has elapsed, the submitter of the SNUN may begin to manufacture, or process the chemical commercially for significant new use.

5.2.5 Filing an NOC

Any person who commences the manufacture of an NC for commercial purpose for which that person previously filed a PMN must notify EPA by submitting a Notice of Commencement (NOC) form (EPA Form 7710-56; Appendix 7.5) within 30 days from the day such manufacture was completed. In addition the NOC filing must accompany the substantiation of any confidential claims that the submitter might have made in filing the PMN. In absence of the required substantiation of the confidential claims, EPA identifies the PMN substance in the public version of TSCA inventory by its chemical name.

Note that before sending an NOC to the USEPA by mail or courier, one should call the OPPT Document Control Office to get the correct physical location, for it may be different from the one given on the NOC form.

5.3 PROCESS FOR ASSESSING TSCA INVENTORY REQUIREMENTS OF A CHEMICAL FOR CHEMICAL IMPORTERS (SCHEMATIC DIAGRAM III)

Chemical importer means any person who imports a chemical substance as part of a mixture or article, into the customs territory of the United States. "Importer" includes the person primarily liable for the payment of any duties on the merchandise or an authorized agent acting on his or her behalf. The term also includes, as appropriate, the consignee, the importer of record, the actual owner if an actual owner's declaration and superseding bond have been filed in accordance with 19 CFR 141.20 (Appendix 7.2) or the transferee, if the right to withdraw merchandise in a bonded warehouse has been transferred in accordance with subpart C of 19 CFR 144 (Appendix 7.3). Process for complying with TSCA inventory requirements for the chemical importers comprises following steps:

5.3.1 Negative Certification

If the chemical does not meet the definition of a chemical substance (see definitions in Section 2.1) and hence not subject to TSCA compliance, the importer must sign a negative certification (see Section 2.2.8).

5.3.2 Positive Certification

If the chemical does meet the definition of a chemical substance and hence is subject to TSCA, check the MSDS of the chemical and look for TSCA compliance statements such as "chemical is TSCA compliant" or "chemical meets TSCA inventory requirements." If such a statement can be found in the MSDS and if the importer feels comfortable accepting the supplier's statement, then the importer can provide a positive certification (see Section 2.2.8) to the Customs and import the chemical substance.

5.3.3 Filing an SNUN

If the TSCA compliance statement is not found in the MSDS or if the importer does not feel comfortable accepting the substance, the importer should ask the supplier to fill out the CVF (Appendix 7.7) and check if the CAS numbers of all the intentional components of the chemical can be found listed in the inventory without an "S" flag. If the chemical can be found so listed, then the importer can provide a positive certification to the Customs to import it. If one or more

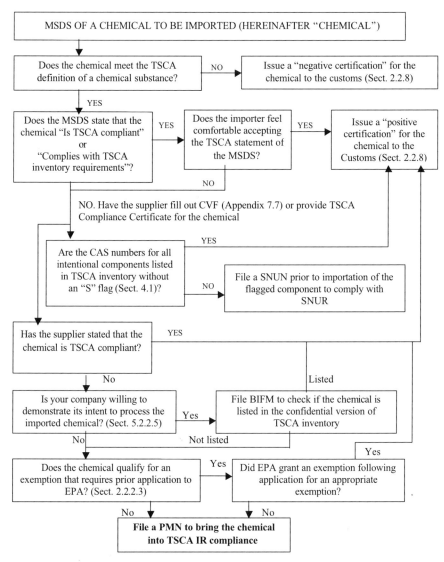

Schematic Diagram III *Process for assessing TSCA inventory requirements of a chemical to be imported for chemical importers.*

components of the chemical carries an "S" flag, then the importer can either file an SNUN for the flagged component to comply with the SNUR or find another chemical that does have all of its components listed in the TSCA Inventory without any of them carrying an "S" flag.

On the CVF, supplier may state that the chemical meets TSCA inventory requirements or that it is TSCA compliant instead of divulging its confidential components or their CAS numbers. If the supplier provided such a statement, the importer may consider it adequate to issue a positive certification.

5.3.4 Filing a BIFM

If CVF does not show that the chemical met TSCA inventory requirements and that CAS numbers of its components are not provided because they are company confidential, the importer may consider filing a BIFM request to the EPA to check if the chemical is listed in the confidential version of the TSCA Inventory. In doing so, the importer must submit the required information to demonstrate an intent to process the imported chemical (see Section 5.2.2.5). Required information for BIFM request by importers is identical to that for manufacturers, except for the following:

5.3.4.1 Description of R&D

In lieu of 5.2.2.5 step 3, an importer that is unable to provide description of R&D activities conducted to date on the chemical from the foreign manufacturer or supplier should submit:

- A statement indicating how long the substance has been in commercial use outside the United States,
- Name of the country in which the chemical has been commercially used, and
- Whether the importer believes that the chemical has already been used commercially, in any country, for the same purpose or application that the importer is intending.

5.3.4.2 Chemical Processing Facility

In lieu of 5.2.2.5 step 7, the address of the facility under the control of the importer at which processing of the chemical would likely occur, if any.

5.3.4.3 Commercial Use of Chemical

In lieu of 5.2.2.5 step 8, a brief description of how the importer is most likely to process or use the substance for a commercial purpose. If the chemical is not expected to be processed or used at any facility under the importer's control, a statement to this effect must be included along with the a description of how the substance will be processed or used at sites controlled by others, if this information is known or reasonably ascertained.

5.3.4.4 Supporting Documents of Foreign Manufacturer

In lieu of 5.2.2.5 step 9, if an importer cannot provide the chemical identity information because its foreign manufacturer or supplier claims is confidential, the foreign manufacturer or supplier must supply the information directly to EPA and reference the importer's notice. If the appropriate supporting document from the foreign party is not received within 30 days after EPA receives importer's BIFM request, the request will be considered incomplete.

If the EPA notifies the importer that the chemical is listed in the confidential inventory, the importer can file a positive certification for the chemical to the Customs. Chemicals not listed in the confidential inventory are discussed next.

5.3.5   Filing for an Exemption

If the chemical to be imported is in small enough volume to qualify for an LVE, could meet the requirements of a LoREX, or could qualify for test market exemption, the importer may consider filing a notice for an appropriate exemption. Importers can a file a positive certification for the chemical to the Customs if EPA has granted the notice for an appropriate exemption.

If, however, the chemical cannot qualify for one of the exemptions or if EPA does not grant the exemption, the importer must file a PMN to bring the chemical into compliance with TSCA inventory requirements before the chemical is imported into the United States.

Chapter 6

References

1. 40 Code of Federal Register (CFR) §700–789 (2000); 15 USC §§2601–2629 as amended through June 1996 (Appendix 7.1).
2. Reproduced as such or rephrased from various sections in 40 CFR §700–789 (2000).
3. 40 CFR §720.36 and §720.38 (2000).
4. 40 CFR §723.250 (2000).
5. 60 Federal Register 16316.
6. See relevant sections of *Polymer Exemption Guidance Manual* (Appendix 7.9) for definitions, details and examples used by the EPA in providing this exemption.
7. 40 CFR §723.175 (2000).
8. 40 CFR §723.50 (2000).
9. 40 CFR §707.20(b)(2)(i) and (ii).
10. 42 Federal Register 64572.
11. 44 Federal Register 28558.
12. 51 Federal Register 21438.
13. 45 Federal Register 50544.
14. 40 CFR §720.50 (2000).

Appendix 7.1

15 USC §§2601–2629 as Amended through 1996

(Reproduced from:
www4.law.cornell.edu/uscode/15/ch53.html)

- United States Code
    - ○ TITLE 15—COMMERCE AND TRADE
        - ■ CHAPTER 53—TOXIC SUBSTANCES CONTROL
            - ■ SUBCHAPTER I—CONTROL OF TOXIC SUB-STANCES
            - ■

Sec. 2601. Findings, policy, and intent

- (a) Findings
  The Congress finds that—
    - ○ (1) human beings and the environment are being exposed each year to a large number of chemical substances and mixtures;
    - ○ (2) among the many chemical substances and mixtures which are constantly being developed and produced, there are some whose manufacture, processing, distribution in commerce, use, or disposal may present an unreasonable risk of injury to health or the environment; and
    (3) the effective regulation of interstate commerce in such chemical substances and mixtures also necessitates the regulation of intrastate commerce in such chemical substances and mixtures.

- (b) Policy
  It is the policy of the United States that—

○ (1) adequate data should be developed with respect to the effect of chemical substances and mixtures on health and the environment and that the development of such data should be the responsibility of those who manufacture and those who process such chemical substances and mixtures;

○ (2) adequate authority should exist to regulate chemical substances and mixtures which present an unreasonable risk of injury to health or the environment, and to take action with respect to chemical substances and mixtures which are imminent hazards; and

(3) authority over chemical substances and mixtures should be exercised in such a manner as not to impede unduly or create unnecessary economic barriers to technological innovation while fulfilling the primary purpose of this chapter to assure that such innovation and commerce in such chemical substances and mixtures do not present an unreasonable risk of injury to health or the environment.

- (c) Intent of Congress

  It is the intent of Congress that the Administrator shall carry out this chapter in a reasonable and prudent manner, and that the Administrator shall consider the environmental, economic, and social impact of any action the Administrator takes or proposes to take under this chapter.

Sec. 2602. Definitions

As used in this chapter:

- (1) The term "Administrator" means the Administrator of the Environmental Protection Agency.

- (2)

  ○ (A) Except as provided in subparagraph (B), the term "chemical substance" means any organic or inorganic substance of a particular molecular identity, including—

    ▪ (i) any combination of such substances occurring in whole or in part as a result of a chemical reaction or occurring in nature and

    (ii) any element or uncombined radical.

  ○ (B) Such term does not include—

    ▪ (i) any mixture,

    ▪ (ii) any pesticide (as defined in the Federal Insecticide, Fungicide, and Rodenticide Act (7 U.S.C. 136 et seq.)) when manufactured, processed, or distributed in commerce for use as a pesticide,

- ▪ (iii) tobacco or any tobacco product,
- ▪ (iv) any source material, special nuclear material, or by-product material (as such terms are defined in the Atomic Energy Act of 1954 (42 U.S.C. 2011 et seq.) and regulations issued under such Act),
- ▪ (v) any article the sale of which is subject to the tax imposed by section 4181 of the Internal Revenue Code of 1986 (26 U.S.C. 4181) (determined without regard to any exemptions from such tax provided by section 4182 or 4221 or any other provision of such Code), and
  (vi) any food, food additive, drug, cosmetic, or device (as such terms are defined in section 201 of the Federal Food, Drug, and Cosmetic Act (21 U.S.C. 321)) when manufactured, processed, or distributed in commerce for use as a food, food additive, drug, cosmetic, or device. The term "food" as used in clause (vi) of this subparagraph includes poultry and poultry products (as defined in sections 4(e) and 4(f) of the Poultry Products Inspection Act (21 U.S.C. 453(e) and (f))), meat and meat food products (as defined in section 1(j) of the Federal Meat Inspection Act (21 U.S.C. 601(j))), and eggs and egg products (as defined in section 4 of the Egg Products Inspection Act (21 U.S.C. 1033)).

- (3) The term "commerce" means trade, traffic, transportation, or other commerce (A) between a place in a State and any place outside of such State, or (B) which affects trade, traffic, transportation, or commerce described in clause (A).

- (4) The terms "distribute in commerce" and "distribution in commerce" when used to describe an action taken with respect to a chemical substance or mixture or article containing a substance or mixture mean to sell, or the sale of, the substance, mixture, or article in commerce; to introduce or deliver for introduction into commerce, or the introduction or delivery for introduction into commerce of, the substance, mixture, or article; or to hold, or the holding of, the substance, mixture, or article after its introduction into commerce.

- (5) The term "environment" includes water, air, and land and the interrelationship which exists among and between water, air, and land and all living things.

- (6) The term "health and safety study" means any study of any effect of a chemical substance or mixture on health or the environment or on both, including underlying data and epidemiological studies, studies of occupational exposure to a chemical substance or mixture, toxicological, clinical, and ecological studies of a chemical substance or mixture, and any test performed pursuant to this chapter.

- (7) The term "manufacture" means to import into the customs territory of the United States (as defined in general note 2 of the Harmonized Tariff Schedule of the United States), produce, or manufacture.

- (8) The term "mixture" means any combination of two or more chemical substances if the combination does not occur in nature and is not, in whole or in part, the result of a chemical reaction; except that such term does include any combination which occurs, in whole or in part, as a result of a chemical reaction if none of the chemical substances comprising the combination is a new chemical substance and if the combination could have been manufactured for commercial purposes without a chemical reaction at the time the chemical substances comprising the combination were combined.

- (9) The term "new chemical substance" means any chemical substance which is not included in the chemical substance list compiled and published under section 2607(b) of this title.

- (10) The term "process" means the preparation of a chemical substance or mixture, after its manufacture, for distribution in commerce—
    - (A) in the same form or physical state as, or in a different form or physical state from, that in which it was received by the person so preparing such substance or mixture, or
    - (B) as part of an article containing the chemical substance or mixture.

- (11) The term "processor" means any person who processes a chemical substance or mixture.

- (12) The term "standards for the development of test data" means a prescription of—
    - (A) the—
        - (i) health and environmental effects, and
        (ii) information relating to toxicity, persistence, and other characteristics which affect health and the environment, for which test data for a chemical substance or mixture are to be developed and any analysis that is to be performed on such data, and
        (B) to the extent necessary to assure that data respecting such effects and characteristics are reliable and adequate—
        - (i) the manner in which such data are to be developed,

          (ii) the specification of any test protocol or methodology to be employed in the development of such data, and
          (iii) such other requirements as are necessary to provide such assurance.

- (13) The term "State" means any State of the United States, the District of Columbia, the Commonwealth of Puerto Rico, the Virgin Islands, Guam, the Canal Zone, American Samoa, the Northern Mariana Islands, or any other territory or possession of the United States.

- (14) The term "United States," when used in the geographic sense, means all of the States.

Sec. 2603. Testing of chemical substances and mixtures

- (a) Testing requirements
If the Administrator finds that—

- (1)
  - (A)
    - (i) the manufacture, distribution in commerce, processing, use, or disposal of a chemical substance or mixture, or that any combination of such activities, may present an unreasonable risk of injury to health or the environment,
    - (ii) there are insufficient data and experience upon which the effects of such manufacture, distribution in commerce, processing, use, or disposal of such substance or mixture or of any combination of such activities on health or the environment can reasonably be determined or predicted, and
    - (iii) testing of such substance or mixture with respect to such effects is necessary to develop such data; or
  - (B)
    - (i) a chemical substance or mixture is or will be produced in substantial quantities, and (I) it enters or may reasonably be anticipated to enter the environment in substantial quantities or
  - (II) there is or may be significant or substantial human exposure to such substance or mixture,
    - (ii) there are insufficient data and experience upon which the effects of the manufacture, distribution in commerce, processing, use, or disposal of such substance or mixture or of any combination of such activities on health or the environment can reasonably be determined or predicted, and

(iii) testing of such substance or mixture with respect to such effects is necessary to develop such data; and

(2) in the case of a mixture, the effects which the mixture's manufacture, distribution in commerce, processing, use, or disposal or any combination of such activities may have on health or the environment may not be reasonably and more efficiently determined or predicted by testing the chemical substances which comprise the mixture; the Administrator shall by rule require that testing be conducted on such substance or mixture to develop data with respect to the health and environmental effects for which there is an insufficiency of data and experience and which are relevant to a determination that the manufacture, distribution in commerce, processing, use, or disposal of such substance or mixture, or that any combination of such activities, does or does not present an unreasonable risk of injury to health or the environment.

- (b) Testing requirement rule
  - (1) A rule under subsection (a) of this section shall include—
    - (A) identification of the chemical substance or mixture for which testing is required under the rule,
    - (B) standards for the development of test data for such substance or mixture, and

    (C) with respect to chemical substances which are not new chemical substances and to mixtures, a specification of the period (which period may not be of unreasonable duration) within which the persons required to conduct the testing shall submit to the Administrator data developed in accordance with the standards referred to in subparagraph (B). In determining the standards and period to be included, pursuant to subparagraphs (B) and (C), in a rule under subsection (a) of this section, the Administrator's considerations shall include the relative costs of the various test protocols and methodologies which may be required under the rule and the reasonably foreseeable availability of the facilities and personnel needed to perform the testing required under the rule. Any such rule may require the submission to the Administrator of preliminary data during the period prescribed under subparagraph (C).
  - (2)
    - (A) The health and environmental effects for which standards for the development of test data may be prescribed include carcinogenesis, mutagenesis, teratogenesis, behavioral

disorders, cumulative or synergistic effects, and any other effect which may present an unreasonable risk of injury to health or the environment. The characteristics of chemical substances and mixtures for which such standards may be prescribed include persistence, acute toxicity, subacute toxicity, chronic toxicity, and any other characteristic which may present such a risk. The methodologies that may be prescribed in such standards include epidemiologic studies, serial or hierarchical tests, in vitro tests, and whole animal tests, except that before prescribing epidemiologic studies of employees, the Administrator shall consult with the Director of the National Institute for Occupational Safety and Health.

- (B) From time to time, but not less than once each 12 months, the Administrator shall review the adequacy of the standards for development of data prescribed in rules under subsection (a) of this section and shall, if necessary, institute proceedings to make appropriate revisions of such standards.

○ (3)

- (A) A rule under subsection (a) of this section respecting a chemical substance or mixture shall require the persons described in subparagraph (B) to conduct tests and submit data to the Administrator on such substance or mixture, except that the Administrator may permit two or more of such persons to designate one such person or a qualified third party to conduct such tests and submit such data on behalf of the persons making the designation.
- (B) The following persons shall be required to conduct tests and submit data on a chemical substance or mixture subject to a rule under subsection (a) of this section:
    - (i) Each person who manufactures or intends to manufacture such substance or mixture if the Administrator makes a finding described in subsection (a)(1)(A)(ii) or (a)(1)(B)(ii) of this section with respect to the manufacture of such substance or mixture.
    - (ii) Each person who processes or intends to process such substance or mixture if the Administrator makes a finding described in subsection (a)(1)(A)(ii) or (a)(1)(B)(ii) of this section with respect to the processing of such substance or mixture.
    - (iii) Each person who manufactures or processes or intends to manufacture or process such substance or mixture if the Administrator makes a finding described in subsection (a)(1)(A)(ii) or (a)(1)(B)(ii) of

this section with respect to the distribution in commerce, use, or disposal of such substance or mixture.

○ (4) Any rule under subsection (a) of this section requiring the testing of and submission of data for a particular chemical substance or mixture shall expire at the end of the reimbursement period (as defined in subsection (c)(3)(B) of this section) which is applicable to test data for such substance or mixture unless the Administrator repeals the rule before such date; and a rule under subsection (a) of this section requiring the testing of and submission of data for a category of chemical substances or mixtures shall expire with respect to a chemical substance or mixture included in the category at the end of the reimbursement period (as so defined) which is applicable to test data for such substance or mixture unless the Administrator before such date repeals the application of the rule to such substance or mixture or repeals the rule.

○ (5) Rules issued under subsection (a) of this section (and any substantive amendment thereto or repeal thereof) shall be promulgated pursuant to section 553 of title 5 except that (A) the Administrator shall give interested persons an opportunity for the oral presentation of data, views, or arguments, in addition to an opportunity to make written submissions; (B) a transcript shall be made of any oral presentation; and (C) the Administrator shall make and publish with the rule the findings described in paragraph (1)(A) or (1)(B) of subsection (a) of this section and, in the case of a rule respecting a mixture, the finding described in paragraph (2) of such subsection.

- (c) Exemption
  ○ (1) Any person required by a rule under subsection (a) of this section to conduct tests and submit data on a chemical substance or mixture may apply to the Administrator (in such form and manner as the Administrator shall prescribe) for an exemption from such requirement.
  ○ (2) If, upon receipt of an application under paragraph (1), the Administrator determines that—
    - (A) the chemical substance or mixture with respect to which such application was submitted is equivalent to a chemical substance or mixture for which data has been submitted to the Administrator in accordance with a rule under subsection (a) of this section or for which data is being developed pursuant to such a rule, and

    (B) submission of data by the applicant on such substance or mixture would be duplicative of data which has been submitted to the Administrator in accordance with such rule or which is being developed pursuant to such rule, the Ad-

ministrator shall exempt, in accordance with paragraph (3) or

- () the chemical substance or mixture with respect to which such substance or mixture under the rule with respect to which such application was submitted.

- (3)

- (A) If the exemption under paragraph (2) of any person from the requirement to conduct tests and submit test data on a chemical substance or mixture is granted on the basis of the existence of previously submitted test data and if such exemption is granted during the reimbursement period for such test data (as prescribed by subparagraph (B)), then (unless such person and the persons referred to in clauses (i) and (ii) agree on the amount and method of reimbursement) the Administrator shall order the person granted the exemption to provide fair and equitable reimbursement (in an amount determined under rules of the Administrator)—

  - (i) to the person who previously submitted such test data, for a portion of the costs incurred by such person in complying with the requirement to submit such data, and
  (ii) to any other person who has been required under this subparagraph to contribute with respect to such costs, for a portion of the amount such person was required to contribute. In promulgating rules for the determination of fair and equitable reimbursement to the persons described in clauses (i) and (ii) for costs incurred with respect to a chemical substance or mixture, the Administrator shall, after consultation with the Attorney General and the Federal Trade Commission, consider all relevant factors, including the effect on the competitive position of the person required to provide reimbursement in relation to the person to be reimbursed and the share of the market for such substance or mixture of the person required to provide reimbursement in relation to the share of such market of the persons to be reimbursed. An order under this subparagraph shall, for purposes of judicial review, be considered final agency action.

- (B) For purposes of subparagraph (A), the reimbursement period for any test data for a chemical substance or mixture is a period—
    - (i) beginning on the date such data is submitted in accordance with a rule promulgated under subsection (a) of this section, and
    (ii) ending—
- (I) five years after the date referred to in clause (i), or
- (II) at the expiration of a period which begins on the date referred to in clause (i) and which is equal to the period which the Administrator determines was necessary to develop such data,
whichever is later.

- (4)

    - (A) If the exemption under paragraph (2) of any person from the requirement to conduct tests and submit test data on a chemical substance or mixture is granted on the basis of the fact that test data is being developed by one or more persons pursuant to a rule promulgated under subsection (a) of this section, then (unless such person and the persons referred to in clauses (i) and (ii) agree on the amount and method of reimbursement) the Administrator shall order the person granted the exemption to provide fair and equitable reimbursement (in an amount determined under rules of the Administrator)—
        - (i) to each such person who is developing such test data, for a portion of the costs incurred by each such person in complying with such rule, and
        (ii) to any other person who has been required under this subparagraph to contribute with respect to the costs of complying with such rule, for a portion of the amount such person was required to contribute. In promulgating rules for the determination of fair and equitable reimbursement to the persons described in clauses (i) and (ii) for costs incurred with respect to a chemical substance or mixture, the Administrator shall, after consultation with the Attorney General and the Federal Trade Commission, consider the factors described in the second sentence of paragraph (3)(A). An order under this subparagraph shall, for pur-

poses of judicial review, be considered final agency action.

- (B) If any exemption is granted under paragraph (2) on the basis of the fact that one or more persons are developing test data pursuant to a rule promulgated under subsection (a) of this section and if after such exemption is granted the Administrator determines that no such person has complied with such rule, the Administrator shall (i) after providing written notice to the person who holds such exemption and an opportunity for a hearing, by order terminate such exemption, and (ii) notify in writing such person of the requirements of the rule with respect to which such exemption was granted.

○ (d) Notice

Upon the receipt of any test data pursuant to a rule under subsection (a) of this section, the Administrator shall publish a notice of the receipt of such data in the Federal Register within 15 days of its receipt. Subject to section 2613 of this title, each such notice shall (1) identify the chemical substance or mixture for which data have been received; (2) list the uses or intended uses of such substance or mixture and the information required by the applicable standards for the development of test data; and (3) describe the nature of the test data developed. Except as otherwise provided in section 2613 of this title, such data shall be made available by the Administrator for examination by any person.

○ (e) Priority list

○ (1)

- (A) There is established a committee to make recommendations to the Administrator respecting the chemical substances and mixtures to which the Administrator should give priority consideration for the promulgation of a rule under subsection (a) of this section. In making such a recommendation with respect to any chemical substance or mixture, the committee shall consider all relevant factors, including—
  - (i) the quantities in which the substance or mixture is or will be manufactured,
  - (ii) the quantities in which the substance or mixture enters or will enter the environment,
  - (iii) the number of individuals who are or will be exposed to the substance or mixture in their places of employment and the duration of such exposure,
  - (iv) the extent to which human beings are or will be exposed to the substance or mixture,
  - (v) the extent to which the substance or mixture is

closely related to a chemical substance or mixture which is known to present an unreasonable risk of injury to health or the environment,

- (vi) the existence of data concerning the effects of the substance or mixture on health or the environment,
- (vii) the extent to which testing of the substance or mixture may result in the development of data upon which the effects of the substance or mixture on health or the environment can reasonably be determined or predicted, and

(viii) the reasonably foreseeable availability of facilities and personnel for performing testing on the substance or mixture. The recommendations of the committee shall be in the form of a list of chemical substances and mixtures which shall be set forth, either by individual substance or mixture or by groups of substances or mixtures, in the order in which the committee determines the Administrator should take action under subsection (a) of this section with respect to the substances and mixtures. In establishing such list, the committee shall give priority attention to those chemical substances and mixtures which are known to cause or contribute to or which are suspected of causing or contributing to cancer, gene mutations, or birth defects. The committee shall designate chemical substances and mixtures on the list with respect to which the committee determines the Administrator should, within 12 months of the date on which such substances and mixtures are first designated, initiate a proceeding under subsection (a) of this section. The total number of chemical substances and mixtures on the list which are designated under the preceding sentence may not, at any time, exceed 50.

- (B) As soon as practicable but not later than nine months after January 1, 1977, the committee shall publish in the Federal Register and transmit to the Administrator the list and designations required by subparagraph (A) together with the reasons for the committee's inclusion of each chemical substance or mixture on the list. At least every six months after the date of the transmission to the Administrator of the list pursuant to the preceding sentence, the committee shall make such previsions in the list as it determines to be necessary and shall transmit them to the Administrator together with the committee's reasons for the revisions. Upon receipt of any such revision, the Administrator shall publish in the Federal Register the list with such revision, the reasons for such revision, and the designations made under subparagraph (A). The Administrator shall provide reasonable opportunity to any interested person to file with the Administrator written comments on the committee's list, any revision of such list by the committee, and designations made by the committee, and shall make such comments available to the public. Within the 12-month period beginning on the date of the first inclusion on the list of a chemical substance or mixture designated by the committee under subparagraph (A) the Administrator shall with respect to such chemical substance or mixture either initiate a rulemaking proceeding under subsection (a) of this section or if such a proceeding is not initiated within such period, publish in the Federal Register the Administrator's reason for not initiating such a proceeding.

○ (2)

- (A) The committee established by paragraph (1)(A) shall consist of eight members as follows:
    - (i) One member appointed by the Administrator from the Environmental Protection Agency.
    - (ii) One member appointed by the Secretary of Labor from officers or employees of the Department of Labor engaged in the Secretary's activities under the Occupational Safety and Health Act of 1970 (29 U.S.C. 651 et seq.).
    - (iii) One member appointed by the Chairman of the Council on Environmental Quality from the Council or its officers or employees.
    - (iv) One member appointed by the Director of the National Institute for Occupational Safety and Health from officers or employees of the Institute.
    - (v) One member appointed by the Director of the National Institute of Environmental Health Sciences from officers or employees of the Institute.
    - (vi) One member appointed by the Director of the National Cancer Institute from officers or employees of the Institute.
    - (vii) One member appointed by the Director of the National Science Foundation from officers or employees of the Foundation.
    - (viii) One member appointed by the Secretary of Commerce from officers or employees of the Department of Commerce.

- (B)
  - (i) An appointed member may designate an individual to serve on the committee on the member's behalf. Such a designation may be made only with the approval of the applicable appointing authority and only if the individual is from the entity from which the member was appointed.
  - (ii) No individual may serve as a member of the committee for more than four years in the aggregate. If any member of the committee leaves the entity from which the member was appointed, such member may not continue as a member of the committee, and the member's position shall be considered to be vacant. A vacancy in the committee shall be filled in the same manner in which the original appointment was made.
  - (iii) Initial appointments to the committee shall be made not later than the 60th day after January 1, 1977. Not later than the 90th day after such date the members of the committee shall hold a meeting for the selection of a chairperson from among their number.
- (C)
  - (i) No member of the committee, or designee of such member, shall accept employment or compensation from any person subject to any requirement of this chapter or of any rule promulgated or order issued thereunder, for a period of at least 12 months after termination of service on the committee.
  - (ii) No person, while serving as a member of the committee, or designee of such member, may own any stocks or bonds, or have any pecuniary interest, of substantial value in any person engaged in the manufacture, processing, or distribution in commerce of any chemical substance or mixture subject to any requirement of this chapter or of any rule promulgated or order issued thereunder.
  - (iii) The Administrator, acting through attorneys of the Environmental Protection Agency, or the Attorney General may bring an action in the appropriate district court of the United States to restrain any violation of this subparagraph.
- (D) The Administrator shall provide the committee such administrative support services as may be necessary to enable the committee to carry out its function under this subsection.

- (f) Required actions
  Upon the receipt of—
    ○ (1) any test data required to be submitted under this chapter, or
    ○ (2) any other information available to the Administrator, which indicates to the Administrator that there may be a reasonable basis to conclude that a chemical substance or mixture presents or will present a significant risk of serious or widespread harm to human beings from cancer, gene mutations, or birth defects, the Administrator shall, within the 180-day period beginning on the date of the receipt of such data or information, initiate appropriate action under section 2604, 2605, or 2606 of this title to prevent or reduce to a sufficient extent such risk or publish in the Federal Register a finding that such risk is not unreasonable. For good cause shown the Administrator may extend such period for an additional period of not more than 90 days. The Administrator shall publish in the Federal Register notice of any such extension and the reasons therefor. A finding by the Administrator that a risk is not unreasonable shall be considered agency action for purposes of judicial review under chapter 7 of title 5. This subsection shall not take effect until two years after January 1, 1977.

- (g) Petition for standards for the development of test data
  A person intending to manufacture or process a chemical substance for which notice is required under section 2604(a) of this title and who is not required under a rule under subsection (a) of this section to conduct tests and submit data on such substance may petition the Administrator to prescribe standards for the development of test data for such substance. The Administrator shall by order either grant or deny any such petition within 60 days of its receipt. If the petition is granted, the Administrator shall prescribe such standards for such substance within 75 days of the date the petition is granted. If the petition is denied, the Administrator shall publish, subject to section 2613 of this title, in the Federal Register the reasons for such denial.

Sec. 2604. Manufacturing and processing notices

- (a) In general
    ○ (1) Except as provided in subsection (h) of this section, no person may—
        ▪ (A) manufacture a new chemical substance on or after the 30th day after the date on which the Administrator first publishes the list required by section 2607(b) of this title, or
        ▪ (B) manufacture or process any chemical substance for a use which the Administrator has determined, in accordance

with paragraph (2), is a significant new use, unless such person submits to the Administrator, at least 90 days before such manufacture or processing, a notice, in accordance with subsection (d) of this section, of such person's intention to manufacture or process such substance and such person complies with any applicable requirement of subsection (b) of this section.

○ (2) A determination by the Administrator that a use of a chemical substance is a significant new use with respect to which notification is required under paragraph (1) shall be made by a rule promulgated after a consideration of all relevant factors, including—

- (A) the projected volume of manufacturing and processing of a chemical substance,
- (B) the extent to which a use changes the type or form of exposure of human beings or the environment to a chemical substance,
- (C) the extent to which a use increases the magnitude and duration of exposure of human beings or the environment to a chemical substance, and

(D) the reasonably anticipated manner and methods of manufacturing, processing, distribution in commerce, and disposal of a chemical substance.

- (b) Submission of test data

- (1)

○ (A) If (i) a person is required by subsection (a)(1) of this section to submit a notice to the Administrator before beginning the manufacture or processing of a chemical substance, and (ii) such person is required to submit test data for such substance pursuant to a rule promulgated under section 2603 of this title before the submission of such notice, such person shall submit to the Administrator such data in accordance with such rule at the time notice is submitted in accordance with subsection (a)(1) of this section.

○ (B) If—

- (i) a person is required by subsection (a)(1) of this section to submit a notice to the Administrator, and

(ii) such person has been granted an exemption under section 2603(c) of this title from the requirements of a rule promulgated under section 2603 of this title before the submission of such notice, such person may not, before the expiration of the 90-day period which begins on the date of the submission in accordance with such rule of the test data the submission or development of which was the basis for the exemption, manufacture such substance if such person is subject to subsection (a)(1)(A) of this section or manufacture or process such substance for a significant new use if the person is subject to subsection (a)(1)(B) of this section.

- (2)
    - (A) If a person—
        - (i) is required by subsection (a)(1) of this section to submit a notice to the Administrator before beginning the manufacture or processing of a chemical substance listed under paragraph (4), and
        (ii) is not required by a rule promulgated under section 2603 of this title before the submission of such notice to submit test data for such substance, such person shall submit to the Administrator data prescribed by subparagraph (B) at the time notice is submitted in accordance with subsection (a)(1) of this section.
    - (B) Data submitted pursuant to subparagraph (A) shall be data which the person submitting the data believes show that—
        - (i) in the case of a substance with respect to which notice is required under subsection (a)(1)(A) of this section, the manufacture, processing, distribution in commerce, use, and disposal of the chemical substance or any combination of such activities will not present an unreasonable risk of injury to health or the environment, or
        - (ii) in the case of a chemical substance with respect to which notice is required under subsection (a)(1)(B) of this section, the intended significant new use of the chemical substance will not present an unreasonable risk of injury to health or the environment.

- (3) Data submitted under paragraph (1) or (2) shall be made available, subject to section 2613 of this title, for examination by interested persons.

- (4)
    - (A)
        - (i) The Administrator may, by rule, compile and keep current a list of chemical substances with respect to which the Administrator finds that the manufacture, processing, distribution in commerce, use, or disposal, or any combination of such activities, presents or may present an unreasonable risk of injury to health or the environment.
        - (ii) In making a finding under clause (i) that the manufacture, processing, distribution in commerce, use, or disposal of a chemical substance or any combination of such activities presents or may present an unreasonable risk of injury

to health or the environment, the Administrator shall consider all relevant factors, including—

- (I) the effects of the chemical substance on health and the magnitude of human exposure to such substance; and
  (II) the effects of the chemical substance on the environment and the magnitude of environmental exposure to such substance.
- (B) The Administrator shall, in prescribing a rule under subparagraph (A) which lists any chemical substance, identify those uses, if any, which the Administrator determines, by rule under subsection (a)(2) of this section, would constitute a significant new use of such substance.
- (C) Any rule under subparagraph (A), and any substantive amendment or repeal of such a rule, shall be promulgated pursuant to the procedures specified in section 553 of title 5, except that
  - (i) the Administrator shall give interested persons an opportunity for the oral presentation of data, views, or arguments, in addition to an opportunity to make written submissions, (ii) a transcript shall be kept of any oral presentation, and (iii) the Administrator shall make and publish with the rule the finding described in subparagraph (A).

- (c) Extension of notice period
The Administrator may for good cause extend for additional periods (not to exceed in the aggregate 90 days) the period, prescribed by subsection (a) or (b) of this section before which the manufacturing or processing of a chemical substance subject to such subsection may begin. Subject to section 2613 of this title, such an extension and the reasons therefor shall be published in the Federal Register and shall constitute a final agency action subject to judicial review.

- (d) Content of notice; publications in the Federal Register
  - (1) The notice required by subsection (a) of this section shall include—
    - (A) insofar as known to the person submitting the notice or insofar as reasonably ascertainable, the information described in subparagraphs (A), (B), (C), (D), (F), and (G) of section 2607(a)(2) of this title, and
    (B) in such form and manner as the Administrator may prescribe, any test data in the possession or control of the person giving such notice which are related to the effect of any manufacture, processing, distribution in commerce, use, or disposal of such substance or any article containing such substance, or of any combination of such activities, on health or the environment, and
    (C) a description of any other data concerning the envi-

ronmental and health effects of such substance, insofar as known to the person making the notice or insofar as reasonably ascertainable. Such a notice shall be made available, subject to section 2613 of this title, for examination by interested persons.

- (2) Subject to section 2613 of this title, not later than five days (excluding Sundays and legal holidays) after the date of the receipt of a notice under subsection (a) of this section or of data under subsection (b) of this section, the Administrator shall publish in the Federal Register a notice which—
    - (A) identifies the chemical substance for which notice or data has been received;
    - (B) lists the uses or intended uses of such substance; and

    (C) in the case of the receipt of data under subsection (b) of this section, describes the nature of the tests performed on such substance and any data which was developed pursuant to subsection (b) of this section or a rule under section 2603 of this title. A notice under this paragraph respecting a chemical substance shall identify the chemical substance by generic class unless the Administrator determines that more specific identification is required in the public interest.

- (3) At the beginning of each month the Administrator shall publish a list in the Federal Register of (A) each chemical substance for which notice has been received under subsection (a) of this section and for which the notification period prescribed by subsection (a), (b), or (c) of this section has not expired, and (B) each chemical substance for which such notification period has expired since the last publication in the Federal Register of such list.

- (e) Regulation pending development of information

- (1)
    - (A) If the Administrator determines that—
        - (i) the information available to the Administrator is insufficient to permit a reasoned evaluation of the health and environmental effects of a chemical substance with respect to which notice is required by subsection (a) of this section; and

        (ii)(I) in the absence of sufficient information to permit the Administrator to make such an evaluation, the manufacture, processing, distribution in commerce, use, or disposal of such substance, or any combination of such activities, may present an unreasonable risk of injury to health or the environment, or
    - (II) such substance is or will be produced in substantial quantities,

and such substance either enters or may reasonably be anticipated to enter the environment in substantial quantities or there is or may be significant or substantial human exposure to the substance, the Administrator may issue a proposed order, to take effect on the expiration of the notification period applicable to the manufacturing or processing of such substance under subsection (a), (b), or (c) of this section, to prohibit or limit the manufacture, processing, distribution in commerce, use, or disposal of such substance or to prohibit or limit any combination of such activities.

- (B) A proposed order may not be issued under subparagraph (A) respecting a chemical substance (i) later than 45 days before the expiration of the notification period applicable to the manufacture or processing of such substance under subsection (a), (b), or (c) of this section, and (ii) unless the Administrator has, on or before the issuance of the proposed order, notified, in writing, each manufacturer or processor, as the case may be, of such substance of the determination which underlies such order.

- (C) If a manufacturer or processor of a chemical substance to be subject to a proposed order issued under subparagraph (A) files with the Administrator (within the 30-day period beginning on the date such manufacturer or processor received the notice required by subparagraph (B)(ii)) objections specifying with particularity the provisions of the order deemed objectionable and stating the grounds therefor, the proposed order shall not take effect.

- (2)

  - (A)

    - (i) Except as provided in clause (ii), if with respect to a chemical substance with respect to which notice is required by subsection (a) of this section, the Administrator makes the determination described in paragraph (1)(A) and if—

  - (I) the Administrator does not issue a proposed order under paragraph (1) respecting such substance, or

  - (II) the Administrator issues such an order respecting such substance but such order does not take effect because objections were filed under paragraph (1)(C) with respect to it, the Administrator, through attorneys of the Environmental Protection Agency, shall apply to the United States District Court for the District of Columbia or the United States district court for the judicial district in which the manufacturer or processor, as the case may be, of such substance is found, resides, or transacts business for an injunction to prohibit or limit the manufacture, processing, distribution in commerce, use, or disposal of such substance (or to prohibit or limit any combination of such activities).

- (ii) If the Administrator issues a proposed order under paragraph (1)(A) respecting a chemical substance but such order does not take effect because objections have been filed under paragraph (1)(C) with respect to it, the Administrator is not required to apply for an injunction under clause (i) respecting such substance if the Administrator determines, on the basis of such objections, that the determinations under paragraph (1)(A) may not be made.

- (B) A district court of the United States which receives an application under subparagraph (A)(i) for an injunction respecting a chemical substance shall issue such injunction if the court finds that—

  - (i) the information available to the Administrator is insufficient to permit a reasoned evaluation of the health and environmental effects of a chemical substance with respect to which notice is required by subsection (a) of this section; and (ii)(I) in the absence of sufficient information to permit the Administrator to make such an evaluation, the manufacture, processing, distribution in commerce, use, or disposal of such substance, or any combination of such activities, may present an unreasonable risk of injury to health or the environment, or

- (II) such substance is or will be produced in substantial quantities, and such substance either enters or may reasonably be anticipated to enter the environment in substantial quantities or there is or may be significant or substantial human exposure to the substance.

- (C) Pending the completion of a proceeding for the issuance of an injunction under subparagraph (B) respecting a chemical substance, the court may, upon application of the Administrator made through attorneys of the Environmental Protection Agency, issue a temporary restraining order or a preliminary injunction to prohibit the manufacture, processing, distribution in commerce, use, or disposal of such a substance (or any combination of such activities) if the court finds that the notification period applicable under subsection (a), (b), or (c) of this section to the manufacturing or processing of such substance may expire before such proceeding can be completed.

- (D) After the submission to the Administrator of test data sufficient to evaluate the health and environmental effects of a chemical substance subject to an injunction issued under subparagraph (B) and the evaluation of such data by the Administrator, the district court of the United States which issued such injunction shall, upon petition dissolve the injunction unless the Administrator has initiated a proceeding for the issuance of a rule under section 2605(a) of this title respecting the substance. If such a proceeding has been initiated, such court shall continue the injunction in effect until the effective date of the rule promulgated in such proceeding or, if such proceeding is terminated without the promulgation of a rule, upon the termination of the proceeding, whichever occurs first.

- (f) Protection against unreasonable risks
  - (1) If the Administrator finds that there is a reasonable basis to conclude that the manufacture, processing, distribution in commerce, use, or disposal of a chemical substance with respect to which notice is required by subsection (a) of this section, or that any combination of such activities, presents or will present an unreasonable risk of injury to health or environment before a rule promulgated under section 2605 of this title can protect against such risk, the Administrator shall, before the expiration of the notification period applicable under subsection (a), (b), or (c) of this section to the manufacturing or processing of such substance, take the action authorized by paragraph (2) or (3) to the extent necessary to protect against such risk.
  - (2) The Administrator may issue a proposed rule under section 2605(a) of this title to apply to a chemical substance with respect to which a finding was made under paragraph (1)—
    - (A) a requirement limiting the amount of such substance which may be manufactured, processed, or distributed in commerce,
    - (B) a requirement described in paragraph (2), (3), (4), (5), (6), or (7) of section 2605(a) of this title, or
      - (C) any combination of the requirements referred to in subparagraph (B). Such a proposed rule shall be effective upon its publication in the Federal Register. Section 2605(d)(2)(B) of this title shall apply with respect to such rule.
    - (3)
      - (A) The Administrator may—
        - (i) issue a proposed order to prohibit the man-

ufacture, processing, or distribution in commerce of a substance with respect to which a finding was made under paragraph (1), or

- (ii) apply, through attorneys of the Environmental Protection Agency, to the United States District Court for the District of Columbia or the United States district court for the judicial district in which the manufacturer, or processor, as the case may be, of such substance, is found, resides, or transacts business for an injunction to prohibit the manufacture, processing, or distribution in commerce of such substance. A proposed order issued under clause (i) respecting a chemical substance shall take effect on the expiration of the notification period applicable under subsection (a), (b), or (c) of this section to the manufacture or processing of such substance.

- (B) If the district court of the United States to which an application has been made under subparagraph (A)(ii) finds that there is a reasonable basis to conclude that the manufacture, processing, distribution in commerce, use, or disposal of the chemical substance with respect to which such application was made, or that any combination of such activities, presents or will present an unreasonable risk of injury to health or the environment before a rule promulgated under section 2605 of this title can protect against such risk, the court shall issue an injunction to prohibit the manufacture, processing, or distribution in commerce of such substance or to prohibit any combination of such activities.

- (C) The provisions of subparagraphs (B) and (C) of subsection (e)(1) of this section shall apply with respect to an order issued under clause (i) of subparagraph (A); and the provisions of subparagraph (C) of subsection (e)(2) of this section shall apply with respect to an injunction issued under subparagraph (B).

- (D) If the Administrator issues an order pursuant to subparagraph (A)(i) respecting a chemical substance and objections are filed in accordance with subsection (e)(1)(C) of this section, the Administrator shall seek an injunction under subparagraph (A)(ii) respecting such substance unless the Administrator determines, on the basis of such objections, that such substance does not or will not present an unreasonable risk of injury to health or the environment.

○ (g) Statement of reasons for not taking action

If the Administrator has not initiated any action under this section or section 2605 or 2606 of this title to prohibit or limit the manufacture, processing, distribution in commerce, use, or disposal of a chemical substance, with respect to which notification or data is required by subsection (a)(1)(B) or (b) of this section, before the expiration of the notification period applicable to the manufacturing or processing of such substance, the Administrator shall publish a statement of the Administrator's reasons for not initiating such action. Such a statement shall be published in the Federal Register before the expiration of such period. Publication of such statement in accordance with the preceding sentence is not a prerequisite to the manufacturing or processing of the substance with respect to which the statement is to be published.

○ (h) Exemptions

■ (1) The Administrator may, upon application, exempt any person from any requirement of subsection (a) or (b) of this section to permit such person to manufacture or process a chemical substance for test marketing purposes—

■ (A) upon a showing by such person satisfactory to the Administrator that the manufacture, processing, distribution in commerce, use, and disposal of such substance, and that any combination of such activities, for such purposes will not present any unreasonable risk of injury to health or the environment, and

(B) under such restrictions as the Administrator considers appropriate.

■ (2)

■ (A) The Administrator may, upon application, exempt any person from the requirement of subsection (b)(2) of this section to submit data for a chemical substance. If, upon receipt of an application under the preceding sentence, the Administrator determines that—

○ (i) the chemical substance with respect to which such application was submitted is equivalent to a chemical substance for which data has been submitted to the Administrator as required by subsection (b)(2) of this section, and

(ii) submission of data by the applicant on such substance would be duplicative of data which has been submitted to the Administrator in accordance with such subsection, the Administrator shall exempt the applicant from the requirement to submit such data on such substance. No exemption which is granted under this subparagraph with respect to the submission of data for a chemical substance may take effect before the beginning of the reimbursement period applicable to such data.

- (B) If the Administrator exempts any person, under subparagraph (A), from submitting data required under subsection (b)(2) of this section for a chemical substance because of the existence of previously submitted data and if such exemption is granted during the reimbursement period for such data, then (unless such person and the persons referred to in clauses (i) and (ii) agree on the amount and method of reimbursement) the Administrator shall order the person granted the exemption to provide fair and equitable reimbursement (in an amount determined under rules of the Administrator)—

○ (i) to the person who previously submitted the data on which the exemption was based, for a portion of the costs incurred by such person in complying with the requirement under subsection (b)(2) of this section to submit such data, and

(ii) to any other person who has been required under this subparagraph to contribute with respect to such costs, for a portion of the amount such person was required to contribute. In promulgating rules for the determination of fair and equitable reimbursement to the persons described in clauses (i) and (ii) for costs incurred with respect to a chemical substance, the Administrator shall, after consultation with the Attorney General and the Federal Trade Commission, consider all relevant factors, including the effect on the competitive position of the person required to provide reimbursement in relation to the persons to be reimbursed and the share of the market for such substance of the person required to provide reimbursement in relation to the share of such market of the persons to be reimbursed. For purposes of judicial review, an order under this subparagraph shall be considered final agency action.

- (C) For purposes of this paragraph, the reimbursement period for any previously submitted data for a chemical substance is a period—

○ (i) beginning on the date of the termination of the prohibition, imposed under this section, on the manufacture or processing of such substance by the person who submitted such data to the Administrator, and

(ii) ending—

- (I) five years after the date referred to in clause (i), or
- (II) at the expiration of a period which begins on the date referred to in clause (i) and is equal to the period which the Administrator determines was necessary to develop such data, whichever is later.

- (3) The requirements of subsections (a) and (b) of this section do not apply with respect to the manufacturing or processing of any chemical substance which is manufactured or processed, or proposed to be manufactured or processed, only in small quantities (as defined by the Administrator by rule) solely for purposes of—
- (A) scientific experimentation or analysis, or
- (B) chemical research on, or analysis of such substance or another substance, including such research or analysis for the development of a product, if all persons engaged in such experimentation, research, or analysis for a manufacturer or processor are notified (in such form and manner as the Administrator may prescribe) of any risk to health which the manufacturer, processor, or the Administrator has reason to believe may be associated with such chemical substance.
  - (4) The Administrator may, upon application and by rule, exempt the manufacturer of any new chemical substance from all or part of the requirements of this section if the Administrator determines that the manufacture, processing, distribution in commerce, use, or disposal of such chemical substance, or that any combination of such activities, will not present an unreasonable risk of injury to health or the environment. A rule promulgated under this paragraph (and any substantive amendment to, or repeal of, such a rule) shall be promulgated in accordance with paragraphs (2) and (3) of section 2605(c) of this title.
  - (5) The Administrator may, upon application, make the requirements of subsections (a) and (b) of this section inapplicable with respect to the manufacturing or processing of any chemical substance (A) which exists temporarily as a result of a chemical reaction in the manufacturing or processing of a mixture or another chemical substance, and (B) to which there is no, and will not be, human or environmental exposure.
  - (6) Immediately upon receipt of an application under paragraph (1) or (5) the Administrator shall publish in the Federal Register notice of the receipt of such application. The Administrator shall give interested persons an opportunity to comment upon any such application and shall, within 45 days of its receipt, either approve or deny the application. The Administrator shall publish in the Federal Register notice of the approval or denial of such an application.

○ (i) "Manufacture" and "process" defined
For purposes of this section, the terms "manufacture" and "process" mean manufacturing or processing for commercial purposes.

Sec. 2605. Regulation of hazardous chemical substances and mixtures

• (a) Scope of regulation
If the Administrator finds that there is a reasonable basis to conclude that the manufacture, processing, distribution in commerce, use, or disposal of a chemical substance or mixture, or that any combination of such activities, presents or will present an unreasonable risk of injury to health or the environment, the Administrator shall by rule apply one or more of the following requirements to such substance or mixture to the extent necessary to protect adequately against such risk using the least burdensome requirements:

○ (1) A requirement (A) prohibiting the manufacturing, processing, or distribution in commerce of such substance or mixture, or (B) limiting the amount of such substance or mixture which may be manufactured, processed, or distributed in commerce.

○ (2) A requirement—

▪ (A) prohibiting the manufacture, processing, or distribution in commerce of such substance or mixture for (i) a particular use or (ii) a particular use in a concentration in excess of a level specified by the Administrator in the rule imposing the requirement, or

▪ (B) limiting the amount of such substance or mixture which may be manufactured, processed, or distributed in commerce for

▪ (i) a particular use or (ii) a particular use in a concentration in excess of a level specified by the Administrator in the rule imposing the requirement.

○ (3) A requirement that such substance or mixture or any article containing such substance or mixture be marked with or accompanied by clear and adequate warnings and instructions with respect to its use, distribution in commerce, or disposal or with respect to any combination of such activities. The form and content of such warnings and instructions shall be prescribed by the Administrator.

○ (4) A requirement that manufacturers and processors of such substance or mixture make and retain records of the processes used to manufacture or process such substance or mixture and monitor or conduct tests which are reasonable and necessary to assure compliance with the requirements of any rule applicable under this subsection.

○ (5) A requirement prohibiting or otherwise regulating any manner or method of commercial use of such substance or mixture.
○ (6)
- (A) A requirement prohibiting or otherwise regulating any manner or method of disposal of such substance or mixture, or of any article containing such substance or mixture, by its manufacturer or processor or by any other person who uses, or disposes of, it for commercial purposes.
- (B) A requirement under subparagraph (A) may not require any person to take any action which would be in violation of any law or requirement of, or in effect for, a State or political subdivision, and shall require each person subject to it to notify each State and political subdivision in which a required disposal may occur of such disposal.
○ (7) A requirement directing manufacturers or processors of such substance or mixture (A) to give notice of such unreasonable risk of injury to distributors in commerce of such substance or mixture and, to the extent reasonably ascertainable, to other persons in possession of such substance or mixture or exposed to such substance or mixture, (B) to give public notice of such risk of injury, and (C) to replace or repurchase such substance or mixture as elected by the person to which the requirement is directed. Any requirement (or combination of requirements) imposed under this subsection may be limited in application to specified geographic areas.

- (b) Quality control
If the Administrator has a reasonable basis to conclude that a particular manufacturer or processor is manufacturing or processing a chemical substance or mixture in a manner which unintentionally causes the chemical substance or mixture to present or which will cause it to present an unreasonable risk of injury to health or the environment—
    ○ (1) the Administrator may by order require such manufacturer or processor to submit a description of the relevant quality control procedures followed in the manufacturing or processing of such chemical substance or mixture; and
    (2) if the Administrator determines—
    - (A) that such quality control procedures are inadequate to prevent the chemical substance or mixture from presenting such risk of injury, the Administrator may order the manufacturer or processor to revise such quality control procedures to the extent necessary to remedy such inadequacy; or
    - (B) that the use of such quality control procedures has resulted in the distribution in commerce of chemical substances or mixtures which present an unreasonable risk of injury to health or the environment, the Administrator may order the manufacturer or processor to (i) give notice of such risk to

processors or distributors in commerce of any such substance or mixture, or to both, and, to the extent reasonably ascertainable, to any other person in possession of or exposed to any such substance, (ii) to give public notice of such risk, and (iii) to provide such replacement or repurchase of any such substance or mixture as is necessary to adequately protect health or the environment. A determination under subparagraph (A) or (B) of paragraph (2) shall be made on the record after opportunity for hearing in accordance with section 554 of title 5. Any manufacturer or processor subject to a requirement to replace or repurchase a chemical substance or mixture may elect either to replace or repurchase the substance or mixture and shall take either such action in the manner prescribed by the Administrator.

- (c) Promulgation of subsection (a) rules
    - (1) In promulgating any rule under subsection (a) of this section with respect to a chemical substance or mixture, the Administrator shall consider and publish a statement with respect to—
        - (A) the effects of such substance or mixture on health and the magnitude of the exposure of human beings to such substance or mixture,
        - (B) the effects of such substance or mixture on the environment and the magnitude of the exposure of the environment to such substance or mixture,
        - (C) the benefits of such substance or mixture for various uses and the availability of substitutes for such uses, and
        (D) the reasonably ascertainable economic consequences of the rule, after consideration of the effect on the national economy, small business, technological innovation, the environment, and public health. If the Administrator determines that a risk of injury to health or the environment could be eliminated or reduced to a sufficient extent by actions taken under another Federal law (or laws) administered in whole or in part by the Administrator, the Administrator may not promulgate a rule under subsection (a) of this section to protect against such risk of injury unless the Administrator finds, in the Administrator's discretion, that it is in the public interest to protect against such risk under this chapter. In making such a finding the Administrator shall consider
            - (i) all relevant aspects of the risk, as determined by the Administrator in the Administrator's discretion, (ii) a comparison of the estimated costs of complying with actions taken under this chapter and under such law (or laws), and (iii) the relative efficiency of actions

under this chapter and under such law (or laws) to protect against such risk of injury.

○ (2) When prescribing a rule under subsection (a) the Administrator shall proceed in accordance with section 553 of title 5 (without regard to any reference in such section to sections 556 and 557 of such title), and shall also (A) publish a notice of proposed rulemaking stating with particularity the reason for the proposed rule; (B) allow interested persons to submit written data, views, and arguments, and make all such submissions publicly available; (C) provide an opportunity for an informal hearing in accordance with paragraph (3); (D) promulgate, if appropriate, a final rule based on the matter in the rulemaking record (as defined in section 2618(a) of this title), and (E) make and publish with the rule the finding described in subsection (a) of this section.

○ (3) Informal hearings required by paragraph (2)(C) shall be conducted by the Administrator in accordance with the following requirements:

■ (A) Subject to subparagraph (B), an interested person is entitled—

■ (i) to present such person's position orally or by documentary submissions (or both), and

(ii) if the Administrator determines that there are disputed issues of material fact it is necessary to resolve, to present such rebuttal submissions and to conduct (or have conducted under subparagraph (B)(ii)) such cross-examination of persons as the Administrator determines (I) to be appropriate, and (II) to be required for a full and true disclosure with respect to such issues.

■ (B) The Administrator may prescribe such rules and make such rulings concerning procedures in such hearings to avoid unnecessary costs or delay. Such rules or rulings may include

■ (i) the imposition of reasonable time limits on each interested person's oral presentations, and (ii) requirements that any cross-examination to which a person may be entitled under subparagraph (A) be conducted by the Administrator on behalf of that person in such manner as the Administrator determines (I) to be appropriate, and (II) to be required for a full and true disclosure with respect to disputed issues of material fact.

■ (C)

■ (i) Except as provided in clause (ii), if a group of persons each of whom under subparagraphs (A) and (B) would be entitled to conduct (or have conducted) cross-examination and who are determined by the Administrator to have the same or similar interests in the proceeding cannot agree upon a single representative of such interests for purposes of cross-examination, the Administrator may make rules and rulings

- (I) limiting the representation of such interest for such purposes, and (II) governing the manner in which such cross-examination shall be limited.

    - (ii) When any person who is a member of a group with respect to which the Administrator has made a determination under clause (i) is unable to agree upon group representation with the other members of the group, then such person shall not be denied under the authority of clause (i) the opportunity to conduct (or have conducted) cross-examination as to issues affecting the person's particular interests if (I) the person satisfies the Administrator that the person has made a reasonable and good faith effort to reach agreement upon group representation with the other members of the group and (II) the Administrator determines that there are substantial and relevant issues which are not adequately presented by the group representative.

- (D) A verbatim transcript shall be taken of any oral presentation made, and cross-examination conducted in any informal hearing under this subsection. Such transcript shall be available to the public.

○ (4)

- (A) The Administrator may, pursuant to rules prescribed by the Administrator, provide compensation for reasonable attorneys' fees, expert witness fees, and other costs of participating in a rulemaking proceeding for the promulgation of a rule under subsection (a) of this section to any person—

    - (i) who represents an interest which would substantially contribute to a fair determination of the issues to be resolved in the proceeding, and
    (ii) if—

- (I) the economic interest of such person is small in comparison to the costs of effective participation in the proceeding by such person, or

- (II) such person demonstrates to the satisfaction of the Administrator that such person does not have sufficient resources adequately to participate in the proceeding without compensation under this subparagraph. In determining for purposes of clause (i) if an interest will substantially contribute to a fair determination of the issues to be resolved in a proceeding, the Administrator shall take into account the number and complexity of such issues and the extent to which representation of such interest will contribute to widespread public participation in the proceeding and representation of a fair balance of interests for the resolution of such issues.

- (B) In determining whether compensation should be provided to a person under subparagraph (A) and the amount of such compensation, the Administrator shall take into account the financial burden which will be incurred by such person in participating in the rulemaking proceeding. The Administrator shall take such action as may be necessary to ensure that the aggregate amount of compensation paid under this paragraph in any fiscal year to all persons who, in rulemaking proceedings in which they receive compensation, are persons who either—
  - (i) would be regulated by the proposed rule, or
  - (ii) represent persons who would be so regulated, may not exceed 25 per centum of the aggregate amount paid as compensation under this paragraph to all persons in such fiscal year.

○ (5) Paragraph (1), (2), (3), and (4) of this subsection apply to the promulgation of a rule repealing, or making a substantive amendment to, a rule promulgated under subsection (a) of this section.

- (d) Effective date
  ○ (1) The Administrator shall specify in any rule under subsection (a) of this section the date on which it shall take effect, which date shall be as soon as feasible.
  ○ (2)
    - (A) The Administrator may declare a proposed rule under subsection (a) of this section to be effective upon its publication in the Federal Register and until the effective date of final action taken, in accordance with subparagraph (B), respecting such rule if—
      - (i) the Administrator determines that—
    - (I) the manufacture, processing, distribution in commerce, use, or disposal of the chemical substance or mixture subject to such proposed rule or any combination of such activities is likely to result in an unreasonable risk of serious or widespread injury to health or the environment before such effective date; and
    - (II) making such proposed rule so effective is necessary to protect the public interest; and

(ii) in the case of a proposed rule to prohibit the manufacture, processing, or distribution of a chemical substance or mixture because of the risk determined under clause (i)(I), a court has in an action under section 2606 of this title granted relief with respect to such risk associated with such substance or mixture. Such a proposed rule which is made so effective shall not, for purposes of judicial review, be considered final agency action.

- (B) If the Administrator makes a proposed rule effective upon its publication in the Federal Register, the Administrator shall, as expeditiously as possible, give interested persons prompt notice of such action, provide reasonable opportunity, in accordance with paragraphs (2) and (3) of subsection (c) of this section, for a hearing on such rule, and either promulgate such rule (as proposed or with modifications) or revoke it; and if such a hearing is requested, the Administrator shall commence the hearing within five days from the date such request is made unless the Administrator and the person making the request agree upon a later date for the hearing to begin, and after the hearing is concluded the Administrator shall, within ten days of the conclusion of the hearing, either promulgate such rule (as proposed or with modifications) or revoke it.

- (e) Polychlorinated biphenyls
  - (1) Within six months after January 1, 1977, the Administrator shall promulgate rules to—
    - (A) prescribe methods for the disposal of polychlorinated biphenyls, and
    (B) require polychlorinated biphenyls to be marked with clear and adequate warnings, and instructions with respect to their processing, distribution in commerce, use, or disposal or with respect to any combination of such activities. Requirements prescribed by rules under this paragraph shall be consistent with the requirements of paragraphs (2) and (3).
  - (2)
    - (A) Except as provided under subparagraph (B), effective one year after January 1, 1977, no person may manufacture, process, or distribute in commerce or use any polychlorinated biphenyl in any manner other than in a totally enclosed manner.
    - (B) The Administrator may by rule authorize the manufacture, processing, distribution in commerce or use (or any

combination of such activities) of any polychlorinated biphenyl in a manner other than in a totally enclosed manner if the Administrator finds that such manufacture, processing, distribution in commerce, or use (or combination of such activities) will not present an unreasonable risk of injury to health or the environment.

- (C) For the purposes of this paragraph, the term "totally enclosed manner" means any manner which will ensure that any exposure of human beings or the environment to a polychlorinated biphenyl will be insignificant as determined by the Administrator by rule.

○ (3)

- (A) Except as provided in subparagraphs (B) and (C)—
    - (i) no person may manufacture any polychlorinated biphenyl after two years after January 1, 1977, and
    (ii) no person may process or distribute in commerce any polychlorinated biphenyl after two and one-half years after such date.
- (B) Any person may petition the Administrator for an exemption from the requirements of subparagraph (A), and the Administrator may grant by rule such an exemption if the Administrator finds that—
    - (i) an unreasonable risk of injury to health or environment would not result, and
    (ii) good faith efforts have been made to develop a chemical substance which does not present an unreasonable risk of injury to health or the environment and which may be substituted for such polychlorinated biphenyl. An exemption granted under this subparagraph shall be subject to such terms and conditions as the Administrator may prescribe and shall be in effect for such period (but not more than one year from the date it is granted) as the Administrator may prescribe.
- (C) Subparagraph (A) shall not apply to the distribution in commerce of any polychlorinated biphenyl if such polychlorinated biphenyl was sold for purposes other than resale before two and one half years after October 11, 1976.

○ (4) Any rule under paragraph (1), (2)(B), or (3)(B) shall be promulgated in accordance with paragraphs (2), (3), and (4) of subsection (c) of this section.

○ (5) This subsection does not limit the authority of the Administrator, under any other provision of this chapter or any other Federal law, to take action respecting any polychlorinated biphenyl.

Sec. 2606. Imminent hazards

- (a) Actions authorized and required
  - (1) The Administrator may commence a civil action in an appropriate district court of the United States—
    - (A) for seizure of an imminently hazardous chemical substance or mixture or any article containing such a substance or mixture,
    - (B) for relief (as authorized by subsection (b) of this section) against any person who manufactures, processes, distributes in commerce, or uses, or disposes of, an imminently hazardous chemical substance or mixture or any article containing such a substance or mixture, or
    - (C) for both such seizure and relief. A civil action may be commenced under this paragraph notwithstanding the existence of a rule under section 2603 of this title, 2604 of this title, 2605 of this title, or subchapter IV of this chapter or an order under section 2604 of this title or subchapter IV of this chapter, and notwithstanding the pendency of any administrative or judicial proceeding under any provision of this chapter.
  - (2) If the Administrator has not made a rule under section 2605(a) of this title immediately effective (as authorized by section 2605(d)(2)(A)(i) of this title) with respect to an imminently hazardous chemical substance or mixture, the Administrator shall commence in a district court of the United States with respect to such substance or mixture or article containing such substance or mixture a civil action described in subparagraph (A), (B), or (C) of paragraph (1).

- (b) Relief authorized
  - (1) The district court of the United States in which an action under subsection (a) of this section is brought shall have jurisdiction to grant such temporary or permanent relief as may be necessary to protect health or the environment from the unreasonable risk associated with the chemical substance, mixture, or article involved in such action.
  - (2) In the case of an action under subsection (a) of this section brought against a person who manufactures, processes, or distributes in commerce a chemical substance or mixture or an article containing a chemical substance or mixture, the relief authorized by paragraph (1) may include the issuance of a mandatory order requiring (A) in the case of purchasers of such substance, mixture, or article known to the defendant, notification to such purchasers of

the risk associated with it; (B) public notice of such risk; (C) recall; (D) the replacement or repurchase of such substance, mixture, or article; or (E) any combination of the actions described in the preceding clauses.

- ○ (3) In the case of an action under subsection (a) of this section against a chemical substance, mixture, or article, such substance, mixture, or article may be proceeded against by process of libel for its seizure and condemnation. Proceedings in such an action shall conform as nearly as possible to proceedings in rem in admiralty.

- (c) Venue and consolidation

- (1)

  - ○ (A) An action under subsection (a) of this section against a person who manufactures, processes, or distributes a chemical substance or mixture or an article containing a chemical substance or mixture may be brought in the United States District Court for the District of Columbia, or for any judicial district in which any of the defendants is found, resides, or transacts business; and process in such an action may be served on a defendant in any other district in which such defendant resides or may be found. An action under subsection (a) of this section against a chemical substance, mixture, or article may be brought in any United States district court within the jurisdiction of which the substance, mixture, or article is found.
  - ○ (B) In determining the judicial district in which an action may be brought under subsection (a) of this section in instances in which such action may be brought in more than one judicial district, the Administrator shall take into account the convenience of the parties.
  - ○ (C) Subpoenas requiring attendance of witnesses in an action brought under subsection (a) of this section may be served in any judicial district.

- (2) Whenever proceedings under subsection (a) of this section involving identical chemical substances, mixtures, or articles are pending in courts in two or more judicial districts, they shall be consolidated for trial by order of any such court upon application reasonably made by any party in interest, upon notice to all parties in interest.

- (d) Action under section 2605
  Where appropriate, concurrently with the filing of an action under subsection (a) of this section or as soon thereafter as may be practicable, the Administrator shall initiate a proceeding for the promulgation of a rule under section 2605(a) of this title.

- (e) Representation

Notwithstanding any other provision of law, in any action under subsection (a) of this section, the Administrator may direct attorneys of the Environmental Protection Agency to appear and represent the Administrator in such an action.

- (f) "Imminently hazardous chemical substance or mixture" defined

For the purposes of subsection (a) of this section, the term "imminently hazardous chemical substance or mixture" means a chemical substance or mixture which presents an imminent and unreasonable risk of serious or widespread injury to health or the environment. Such a risk to health or the environment shall be considered imminent if it is shown that the manufacture, processing, distribution in commerce, use, or disposal of the chemical substance or mixture, or that any combination of such activities, is likely to result in such injury to health or the environment before a final rule under section 2605 of this title can protect against such risk.

Sec. 2607. Reporting and retention of information

- (a) Reports
  - (1) The Administrator shall promulgate rules under which—
    - (A) each person (other than a small manufacturer or processor) who manufactures or processes or proposes to manufacture or process a chemical substance (other than a chemical substance described in subparagraph (B)(ii)) shall maintain such records, and shall submit to the Administrator such reports, as the Administrator may reasonably require, and

      (B) each person (other than a small manufacturer or processor) who manufactures or processes or proposes to manufacture or process—
      - (i) a mixture, or
      - (ii) a chemical substance in small quantities (as defined by the Administrator by rule) solely for purposes of scientific experimentation or analysis or chemical research on, or analysis of, such substance or another substance, including any such research or analysis for the development of a product, shall maintain records and submit to the Administrator reports but only to the extent the Administrator determines the maintenance of records or submission of reports, or both, is necessary for the effective enforcement of this chapter. The Administrator may not require in a rule promulgated under this paragraph the maintenance of

records or the submission of reports with respect to changes in the proportions of the components of a mixture unless the Administrator finds that the maintenance of such records or the submission of such reports, or both, is necessary for the effective enforcement of this chapter. For purposes of the compilation of the list of chemical substances required under subsection (b) of this section, the Administrator shall promulgate rules pursuant to this subsection not later than 180 days after January 1, 1977.

○ (2) The Administrator may require under paragraph (1) maintenance of records and reporting with respect to the following insofar as known to the person making the report or insofar as reasonably ascertainable:

- (A) The common or trade name, the chemical identity, and the molecular structure of each chemical substance or mixture for which such a report is required.

- (B) The categories or proposed categories of use of each such substance or mixture.

- (C) The total amount of each such substance and mixture manufactured or processed, reasonable estimates of the total amount to be manufactured or processed, the amount manufactured or processed for each of its categories of use, and reasonable estimates of the amount to be manufactured or processed for each of its categories of use or proposed categories of use.

- (D) A description of the byproducts resulting from the manufacture, processing, use, or disposal of each such substance or mixture.

- (E) All existing data concerning the environmental and health effects of such substance or mixture.

- (F) The number of individuals exposed, and reasonable estimates of the number who will be exposed, to such substance or mixture in their places of employment and the duration of such exposure.

- (G) In the initial report under paragraph (1) on such substance or mixture, the manner or method of its disposal, and in any subsequent report on such substance or mixture, any change in such manner or method. To the extent feasible, the Administrator shall not require under paragraph (1), any reporting which is unnecessary or duplicative.

○ (3)

- (A)

    - (i) The Administrator may by rule require a small manufacturer or processor of a chemical substance to submit to the Administrator such information respecting the chemical substance as the Administrator may require for publication of the first list of chemical substances required by subsection (b) of this section.

- (ii) The Administrator may by rule require a small manufacturer or processor of a chemical substance or mixture—

- (I) subject to a rule proposed or promulgated under section 2603, 2604(b)(4), or 2605 of this title, or an order in effect under section 2604(e) of this title, or

- (II) with respect to which relief has been granted pursuant to a civil action brought under section 2604 or 2606 of this title, to maintain such records on such substance or mixture, and to submit to the Administrator such reports on such substance or mixture, as the Administrator may reasonably require. A rule under this clause requiring reporting may require reporting with respect to the matters referred to in paragraph (2).

- (B) The Administrator, after consultation with the Administrator of the Small Business Administration, shall by rule prescribe standards for determining the manufacturers and processors which qualify as small manufacturers and processors for purposes of this paragraph and paragraph (1).

- (b) Inventory
  - (1) The Administrator shall compile, keep current, and publish a list of each chemical substance which is manufactured or processed in the United States. Such list shall at least include each chemical substance which any person reports, under section 2604 of this title or subsection (a) of this section, is manufactured or processed in the United States. Such list may not include any chemical substance which was not manufactured or processed in the United States within three years before the effective date of the rules promulgated pursuant to the last sentence of subsection (a)(1) of this section. In the case of a chemical substance for which a notice is submitted in accordance with section 2604 of this title, such chemical substance shall be included in such list as of the earliest date (as determined by the Administrator) on which such substance was manufactured or processed in the United States. The Administrator shall first publish such a list not later than 315 days after January 1, 1977. The Administrator shall not include in such list any chemical substance which is manufactured or processed only in small quantities (as defined by the Administrator by rule) solely for purposes of scientific experimentation or analysis or chemical research on, or analysis of,

such substance or another substance, including such research or analysis for the development of a product.

○ (2) To the extent consistent with the purposes of this chapter, the Administrator may, in lieu of listing, pursuant to paragraph (1), a chemical substance individually, list a category of chemical substances in which such substance is included.

- (c) Records

Any person who manufactures, processes, or distributes in commerce any chemical substance or mixture shall maintain records of significant adverse reactions to health or the environment, as determined by the Administrator by rule, alleged to have been caused by the substance or mixture. Records of such adverse reactions to the health of employees shall be retained for a period of 30 years from the date such reactions were first reported to or known by the person maintaining such records. Any other record of such adverse reactions shall be retained for a period of five years from the date the information contained in the record was first reported to or known by the person maintaining the record. Records required to be maintained under this subsection shall include records of consumer allegations of personal injury or harm to health, reports of occupational disease or injury, and reports or complaints of injury to the environment submitted to the manufacturer, processor, or distributor in commerce from any source. Upon request of any duly designated representative of the Administrator, each person who is required to maintain records under this subsection shall permit the inspection of such records and shall submit copies of such records.

- (d) Health and safety studies

The Administrator shall promulgate rules under which the Administrator shall require any person who manufactures, processes, or distributes in commerce or who proposes to manufacture, process, or distribute in commerce any chemical substance or mixture (or with respect to paragraph (2), any person who has possession of a study) to submit to the Administrator—

○ (1) lists of health and safety studies (A) conducted or initiated by or for such person with respect to such substance or mixture at any time, (B) known to such person, or (C) reasonably ascertainable by such person, except that the Administrator may exclude certain types or categories of studies from the requirements of this subsection if the Administrator finds that submission of lists of such studies are unnecessary to carry out the purposes of this chapter; and

(2) copies of any study contained on a list submitted pursuant to paragraph (1) or otherwise known by such person.

- (e) Notice to Administrator of substantial risks
Any person who manufactures, processes, or distributes in commerce as chemical substance or mixture and who obtains information which reasonably supports the conclusion that such substance or mixture presents a substantial risk of injury to health or the environment shall immediately inform the Administrator of such information unless such person has actual knowledge that the Administrator has been adequately informed of such information.

- (f) "Manufacture" and "process" defined
For purposes of this section, the terms "manufacture" and "process" mean manufacture or process for commercial purposes.

Sec. 2608. Relationship to other Federal laws

- (a) Laws not administered by the Administrator
  - (1) If the Administrator has reasonable basis to conclude that the manufacture, processing, distribution in commerce, use, or disposal of a chemical substance or mixture, or that any combination of such activities, presents or will present an unreasonable risk of injury to health or the environment and determines, in the Administrator's discretion, that such risk may be prevented or reduced to a sufficient extent by action taken under a Federal law not administered by the Administrator, the Administrator shall submit to the agency which administers such law a report which describes such risk and includes in such description a specification of the activity or combination of activities which the Administrator has reason to believe so presents such risk. Such report shall also request such agency—
  - (A)
    - (i) to determine if the risk described in such report may be prevented or reduced to a sufficient extent by action taken under such law, and
    (ii) if the agency determines that such risk may be so prevented or reduced, to issue an order declaring whether or not the activity or combination of activities specified in the description of such risk presents such risk; and
    (B) to respond to the Administrator with respect to the matters described in subparagraph (A). Any report of the Administrator shall include a detailed statement of the information on which it is based and shall be published in the Federal Register. The agency receiving a request under such a report shall make the requested determination, issue the requested order, and make the requested response within

such time as the Administrator specifies in the request, but such time specified may not be less than 90 days from the date the request was made. The response of an agency shall be accompanied by a detailed statement of the findings and conclusions of the agency and shall be published in the Federal Register.

- (2) If the Administrator makes a report under paragraph (1) with respect to a chemical substance or mixture and the agency to which such report was made either—
  - (A) issues an order declaring that the activity or combination of activities specified in the description of the risk described in the report does not present the risk described in the report, or
  - (B) initiates, within 90 days of the publication in the Federal Register of the response of the agency under paragraph (1), action under the law (or laws) administered by such agency to protect against such risk associated with such activity or combination of activities, the Administrator may not take any action under section 2605 or 2606 of this title with respect to such risk.

- (3) If the Administrator has initiated action under section 2605 or 2606 of this title with respect to a risk associated with a chemical substance or mixture which was the subject of a report made to an agency under paragraph (1), such agency shall before taking action under the law (or laws) administered by it to protect against such risk consult with the Administrator for the purpose of avoiding duplication of Federal action against such risk.

- (b) Laws administered by the Administrator
  The Administrator shall coordinate actions taken under this chapter with actions taken under other Federal laws administered in whole or in part by the Administrator. If the Administrator determines that a risk to health or the environment associated with a chemical substance or mixture could be eliminated or reduced to a sufficient extent by actions taken under the authorities contained in such other Federal laws, the Administrator shall use such authorities to protect against such risk unless the Administrator determines, in the Administrator's discretion, that it is in the public interest to protect against such risk by actions taken under this chapter. This subsection shall not be construed to relieve the Administrator of any requirement imposed on the Administrator by such other Federal laws.

- (c) Occupational safety and health
  In exercising any authority under this chapter, the Administrator shall not, for purposes of section 653(b)(1) of title 29, be deemed to be exercis-

ing statutory authority to prescribe or enforce standards or regulations affecting occupational safety and health.

- (d) Coordination
  In administering this chapter, the Administrator shall consult and coordinate with the Secretary of Health and Human Services and the heads of any other appropriate Federal executive department or agency, any relevant independent regulatory agency, and any other appropriate instrumentality of the Federal Government for the purpose of achieving the maximum enforcement of this chapter while imposing the least burdens of duplicative requirements on those subject to the chapter and for other purposes. The Administrator shall, in the report required by section 2629 of this title, report annually to the Congress on actions taken to coordinate with such other Federal departments, agencies, or instrumentalities, and on actions taken to coordinate the authority under this chapter with the authority granted under other Acts referred to in subsection (b) of this section.

Sec. 2609. Research, development, collection, dissemination, and utilization of data

- (a) Authority
  The Administrator shall, in consultation and cooperation with the Secretary of Health and Human Services and with other heads of appropriate departments and agencies, conduct such research, development, and monitoring as is necessary to carry out the purposes of this chapter. The Administrator may enter into contracts and may make grants for research, development, and monitoring under this subsection. Contracts may be entered into under this subsection without regard to section 3324(a) and (b) of title 31 and section 5 of title 41.

- (b) Data systems
  - (1) The Administrator shall establish, administer, and be responsible for the continuing activities of an interagency committee which shall design, establish, and coordinate an efficient and effective system, within the Environmental Protection Agency, for the collection, dissemination to other Federal departments and agencies, and use of data submitted to the Administrator under this chapter.
  - (2)
    - (A) The Administrator shall, in consultation and cooperation with the Secretary of Health and Human Services and other heads of appropriate departments and agencies design, establish, and coordinate an efficient and effective system

for the retrieval of toxicological and other scientific data which could be useful to the Administrator in carrying out the purposes of this chapter. Systematized retrieval shall be developed for use by all Federal and other departments and agencies with responsibilities in the area of regulation or study of chemical substances and mixtures and their effect on health or the environment.

- (B) The Administrator, in consultation and cooperation with the Secretary of Health and Human Services, may make grants and enter into contracts for the development of a data retrieval system described in subparagraph (A). Contracts may be entered into under this subparagraph without regard to section 3324(a) and (b) of title 31 and section 5 of title 41.

- (c) Screening techniques

The Administrator shall coordinate, with the Assistant Secretary for Health of the Department of Health and Human Services, research undertaken by the Administrator and directed toward the development of rapid, reliable, and economical screening techniques for carcinogenic, mutagenic, teratogenic, and ecological effects of chemical substances and mixtures.

- (d) Monitoring

The Administrator shall, in consultation and cooperation with the Secretary of Health and Human Services, establish and be responsible for research aimed at the development, in cooperation with local, State, and Federal agencies, of monitoring techniques and instruments which may be used in the detection of toxic chemical substances and mixtures and which are reliable, economical, and capable of being implemented under a wide variety of conditions.

- (e) Basic research

The Administrator shall, in consultation and cooperation with the Secretary of Health and Human Services, establish research programs to develop the fundamental scientific basis of the screening and monitoring techniques described in subsections (c) and (d) of this section, the bounds of the reliability of such techniques, and the opportunities for their improvement.

- (f) Training

The Administrator shall establish and promote programs and workshops to train or facilitate the training of Federal laboratory and technical personnel in existing or newly developed screening and monitoring techniques.

- (g) Exchange of research and development results
  The Administrator shall, in consultation with the Secretary of Health and Human Services and other heads of appropriate departments and agencies, establish and coordinate a system for exchange among Federal, State, and local authorities of research and development results respecting toxic chemical substances and mixtures, including a system to facilitate and promote the development of standard data format and analysis and consistent testing procedures.

Sec. 2610. Inspections and subpoenas

- (a) In general
  For purposes of administering this chapter, the Administrator, and any duly designated representative of the Administrator, may inspect any establishment, facility, or other premises in which chemical substances, mixtures, or products subject to subchapter IV of this chapter are manufactured, processed, stored, or held before or after their distribution in commerce and any conveyance being used to transport chemical substances, mixtures, such products, or such articles in connection with distribution in commerce. Such an inspection may only be made upon the presentation of appropriate credentials and of a written notice to the owner, operator, or agent in charge of the premises or conveyance to be inspected. A separate notice shall be given for each such inspection, but a notice shall not be required for each entry made during the period covered by the inspection. Each such inspection shall be commenced and completed with reasonable promptness and shall be conducted at reasonable times, within reasonable limits, and in a reasonable manner.

- (b) Scope
  - (1) Except as provided in paragraph (2), an inspection conducted under subsection (a) of this section shall extend to all things within the premises or conveyance inspected (including records, files, papers, processes, controls, and facilities) bearing on whether the requirements of this chapter applicable to the chemical substances, mixtures, or products subject to subchapter IV of this chapter within such premises or conveyance have been complied with.
  - (2) No inspection under subsection (a) of this section shall extend to—
    - (A) financial data,
    - (B) sales data (other than shipment data),
    - (C) pricing data,
    - (D) personnel data, or
    - (E) research data (other than data required by this chapter or

under a rule promulgated thereunder), unless the nature and extent of such data are described with reasonable specificity in the written notice required by subsection (a) of this section for such inspection.

- (c) Subpoenas
In carrying out this chapter, the Administrator may by subpoena require the attendance and testimony of witnesses and the production of reports, papers, documents, answers to questions, and other information that the Administrator deems necessary. Witnesses shall be paid the same fees and mileage that are paid witnesses in the courts of the United States. In the event of contumacy, failure, or refusal of any person to obey any such subpoena, any district court of the United States in which venue is proper shall have jurisdiction to order any such person to comply with such subpoena. Any failure to obey such an order of the court is punishable by the court as a contempt thereof.

Sec. 2611. Exports

- (a) In general
  - (1) Except as provided in paragraph (2) and subsection (b) of this section, this chapter (other than section 2607 of this title) shall not apply to any chemical substance, mixture, or to an article containing a chemical substance or mixture, if—
    - (A) it can be shown that such substance, mixture, or article is being manufactured, processed, or distributed in commerce for export from the United States, unless such substance, mixture, or article was, in fact, manufactured, processed, or distributed in commerce, for use in the United States, and
    (B) such substance, mixture, or article (when distributed in commerce), or any container in which it is enclosed (when so distributed), bears a stamp or label stating that such substance, mixture, or article is intended for export.
  - (2) Paragraph (1) shall not apply to any chemical substance, mixture, or article if the Administrator finds that the substance, mixture, or article will present an unreasonable risk of injury to health within the United States or to the environment of the United States. The Administrator may require, under section 2603 of this title, testing of any chemical substance or mixture exempted from this chapter by paragraph (1) for the purpose of determining whether or not such substance or mixture presents an unreasonable risk of injury to health within the United States or to the environment of the United States.

- (b) Notice
  - ○ (1) If any person exports or intends to export to a foreign country a chemical substance or mixture for which the submission of data is required under section 2603 or 2604(b) of this title, such person shall notify the Administrator of such exportation or intent to export and the Administrator shall furnish to the government of such country notice of the availability of the data submitted to the Administrator under such section for such substance or mixture.
  - ○ (2) If any person exports or intends to export to a foreign country a chemical substance or mixture for which an order has been issued under section 2604 of this title or a rule has been proposed or promulgated under section 2604 or 2605 of this title, or with respect to which an action is pending, or relief has been granted under section 2604 or 2606 of this title, such person shall notify the Administrator of such exportation or intent to export and the Administrator shall furnish to the government of such country notice of such rule, order, action, or relief.

Sec. 2612. Entry into customs territory of the United States

- (a) In general
  - ○ (1) The Secretary of the Treasury shall refuse entry into the customs territory of the United States (as defined in general note 2 of the Harmonized Tariff Schedule of the United States) of any chemical substance, mixture, or article containing a chemical substance or mixture offered for such entry if—
    - ▪ (A) it fails to comply with any rule in effect under this chapter, or
    - ▪ (B) it is offered for entry in violation of section 2604 of this title, 2605 of this title, or subchapter IV of this chapter, a rule or order under section 2604 of this title, 2605 of this title, or subchapter IV of this chapter, or an order issued in a civil action brought under section 2604 of this title, 2606 of this title or subchapter IV of this chapter.
  - ○ (2) If a chemical substance, mixture, or article is refused entry under paragraph (1), the Secretary of the Treasury shall notify the consignee of such entry refusal, shall not release it to the consignee, and shall cause its disposal or storage (under such rules as the Secretary of the Treasury may prescribe) if it has not been exported by the consignee within 90 days from the date of receipt of notice of such refusal, except that the Secretary of the Treasury may, pending a review by the Administrator of the entry refusal, release to the consignee such substance, mixture, or article on execution of bond for the amount of the full invoice of such substance, mixture, or

article (as such value is set forth in the customs entry), together with the duty thereon. On failure to return such substance, mixture, or article for any cause to the custody of the Secretary of the Treasury when demanded, such consignee shall be liable to the United States for liquidated damages equal to the full amount of such bond. All charges for storage, cartage, and labor on and for disposal of substances, mixtures, or articles which are refused entry or release under this section shall be paid by the owner or consignee, and in default of such payment shall constitute a lien against any future entry made by such owner or consignee.

- (b) Rules

The Secretary of the Treasury, after consultation with the Administrator, shall issue rules for the administration of subsection (a) of this section.

Sec. 2613. Disclosure of data

- (a) In general

Except as provided by subsection (b) of this section, any information reported to, or otherwise obtained by, the Administrator (or any representative of the Administrator) under this chapter, which is exempt from disclosure pursuant to subsection (a) of section 552 of title 5 by reason of subsection (b)(4) of such section, shall, notwithstanding the provisions of any other section of this chapter, not be disclosed by the Administrator or by any officer or employee of the United States, except that such information—

  - (1) shall be disclosed to any officer or employee of the United States—
    - (A) in connection with the official duties of such officer or employee under any law for the protection of health or the environment, or
    - (B) for specific law enforcement purposes;
  - (2) shall be disclosed to contractors with the United States and employees of such contractors if in the opinion of the Administrator such disclosure is necessary for the satisfactory performance by the contractor of a contract with the United States entered into on or after October 11, 1976, for the performance of work in connection with this chapter and under such conditions as the Administrator may specify;
  - (3) shall be disclosed if the Administrator determines it necessary to protect health or the environment against an unreasonable risk of injury to health or the environment; or
  - (4) may be disclosed when relevant in any proceeding under this chapter, except that disclosure in such a proceeding shall be made

in such manner as to preserve confidentiality to the extent practicable without impairing the proceeding. In any proceeding under section 552(a) of title 5 to obtain information the disclosure of which has been denied because of the provisions of this subsection, the Administrator may not rely on section 552(b)(3) of such title to sustain the Administrator's action.

- (b) Data from health and safety studies
  - ○ (1) Subsection (a) does not prohibit the disclosure of—
    - ▪ (A) any health and safety study which is submitted under this chapter with respect to—
      - ▪ (i) any chemical substance or mixture which, on the date on which such study is to be disclosed has been offered for commercial distribution, or
      - ▪ (ii) any chemical substance or mixture for which testing is required under section 2603 of this title or for which notification is required under section 2604 of this title, and (B) any data reported to, or otherwise obtained by, the Administrator from a health and safety study which relates to a chemical substance or mixture described in clause (i) or (ii) of subparagraph (A). This paragraph does not authorize the release of any data which discloses processes used in the manufacturing or processing of a chemical substance or mixture or, in the case of a mixture, the release of data disclosing the portion of the mixture comprised by any of the chemical substances in the mixture.
  - ○ (2) If a request is made to the Administrator under subsection (a) of section 552 of title 5 for information which is described in the first sentence of paragraph (1) and which is not information described in the second sentence of such paragraph, the Administrator may not deny such request on the basis of subsection (b)(4) of such section.

- (c) Designation and release of confidential data
  - ○ (1) In submitting data under this chapter, a manufacturer, processor, or distributor in commerce may (A) designate the data which such person believes is entitled to confidential treatment under subsection (a) of this section, and (B) submit such designated data separately from other data submitted under this chapter. A designation under this paragraph shall be made in writing and in such manner as the Administrator may prescribe.
  - ○ (2)
    - ▪ (A) Except as provided by subparagraph (B), if the Administrator proposes to release for inspection data which has

been designated under paragraph (1)(A), the Administrator shall notify, in writing and by certified mail, the manufacturer, processor, or distributor in commerce who submitted such data of the intent to release such data. If the release of such data is to be made pursuant to a request made under section 552(a) of title 5, such notice shall be given immediately upon approval of such request by the Administrator. The Administrator may not release such data until the expiration of 30 days after the manufacturer, processor, or distributor in commerce submitting such data has received the notice required by this subparagraph.

- (B)
    - (i) Subparagraph (A) shall not apply to the release of information under paragraph (1), (2), (3), or (4) of subsection (a) of this section, except that the Administrator may not release data under paragraph (3) of subsection (a) of this section unless the Administrator has notified each manufacturer, processor, and distributor in commerce who submitted such data of such release. Such notice shall be made in writing by certified mail at least 15 days before the release of such data, except that if the Administrator determines that the release of such data is necessary to protect against an imminent, unreasonable risk of injury to health or the environment, such notice may be made by such means as the Administrator determines will provide notice at least 24 hours before such release is made.
    - (ii) Subparagraph (A) shall not apply to the release of information described in subsection (b)(1) of this section other than information described in the second sentence of such subsection.

- (d) Criminal penalty for wrongful disclosure
    - (1) Any officer or employee of the United States or former officer or employee of the United States, who by virtue of such employment or official position has obtained possession of, or has access to, material the disclosure of which is prohibited by subsection (a) of this section, and who knowing that disclosure of such material is prohibited by such subsection, willfully discloses the material in any manner to any person not entitled to receive it, shall be guilty of a misdemeanor and fined not more than $5,000 or imprisoned for not more than one year, or both. Section 1905 of title 18 does not apply with respect to the publishing, divulging, disclosure, or making known of, or making available, information reported or otherwise obtained under this chapter.

○ (2) For the purposes of paragraph (1), any contractor with the United States who is furnished information as authorized by subsection (a)(2) of this section, and any employee of any such contractor, shall be considered to be an employee of the United States.

- (e) Access by Congress
Notwithstanding any limitation contained in this section or any other provision of law, all information reported to or otherwise obtained by the Administrator (or any representative of the Administrator) under this chapter shall be made available, upon written request of any duly authorized committee of the Congress, to such committee.

Sec. 2614. Prohibited acts

It shall be unlawful for any person to—
- (1) fail or refuse to comply with (A) any rule promulgated or order issued under section 2603 of this title, (B) any requirement prescribed by section 2604 or 2605 of this title, (C) any rule promulgated or order issued under section 2604 or 2605 of this title, or (D) any requirement of subchapter II of this chapter or any rule promulgated or order issued under subchapter II of this chapter;

- (2) use for commercial purposes a chemical substance or mixture which such person knew or had reason to know was manufactured, processed, or distributed in commerce in violation of section 2604 or 2605 of this title, a rule or order under section 2604 or 2605 of this title, or an order issued in action brought under section 2604 or 2606 of this title;

- (3) fail or refuse to (A) establish or maintain records, (B) submit reports, notices, or other information, or (C) permit access to or copying of records, as required by this chapter or a rule thereunder; or

- (4) fail or refuse to permit entry or inspection as required by section 2610 of this title.

Sec. 2615. Penalties

- (a) Civil
  ○ (1) Any person who violates a provision of section 2614 or 2689 of this title shall be liable to the United States for a civil penalty in an amount not to exceed $25,000 for each such violation. Each day such a violation continues shall, for purposes of this subsection, constitute a separate violation of section 2614 or 2689 of this title.

○ (2)

- (A) A civil penalty for a violation of section section 2614 or 2689 of this title shall be assessed by the Administrator by an order made on the record after opportunity (provided in accordance with this subparagraph) for a hearing in accordance with section 554 of title 5. Before issuing such an order, the Administrator shall give written notice to the person to be assessed a civil penalty under such order of the Administrator's proposal to issue such order and provide such person an opportunity to request, within 15 days of the date the notice is received by such person, such a hearing on the order.
- (B) In determining the amount of a civil penalty, the Administrator shall take into account the nature, circumstances, extent, and gravity of the violation or violations and, with respect to the violator, ability to pay, effect on ability to continue to do business, any history of prior such violations, the degree of culpability, and such other matters as justice may require.
- (C) The Administrator may compromise, modify, or remit, with or without conditions, any civil penalty which may be imposed under this subsection. The amount of such penalty, when finally determined, or the amount agreed upon in compromise, may be deducted from any sums owing by the United States to the person charged.

○ (3) Any person who requested in accordance with paragraph (2)(A) a hearing respecting the assessment of a civil penalty and who is aggrieved by an order assessing a civil penalty may file a petition for judicial review of such order with the United States Court of Appeals for the District of Columbia Circuit or for any other circuit in which such person resides or transacts business. Such a petition may only be filed within the 30-day period beginning on the date the order making such assessment was issued.

○ (4) If any person fails to pay an assessment of a civil penalty—

- (A) after the order making the assessment has become a final order and if such person does not file a petition for judicial review of the order in accordance with paragraph (3), or
- (B) after a court in an action brought under paragraph (3) has entered a final judgment in favor of the Administrator, the Attorney General shall recover the amount assessed (plus interest at currently prevailing rates from the date of the expiration of the 30-day period referred to in paragraph (3) or the date of such final judgment, as the case may be) in an action brought in any appropriate district court of the United States. In such an action, the validity, amount, and appropriateness of such penalty shall not be subject to review.

- (b) Criminal

Any person who knowingly or willfully violates any provision of section 2614 or 2689 of this title, shall, in addition to or in lieu of any civil penalty which may be imposed under subsection (a) of this section for such violation, be subject, upon conviction, to a fine of not more than $25,000 for each day of violation, or to imprisonment for not more than one year, or both.

Sec. 2616. Specific enforcement and seizure

- (a) Specific enforcement
    - (1) The district courts of the United States shall have jurisdiction over civil actions to—
        - (A) restrain any violation of section 2614 or 2689 of this title,
        - (B) restrain any person from taking any action prohibited by section 2604 of this title, 2605 of this title, or subchapter IV of this chapter, or by a rule or order under section 2604 of this title, 2605 of this title, or subchapter IV of this chapter,
        - (C) compel the taking of any action required by or under this chapter, or
        - (D) direct any manufacturer or processor of a chemical substance, mixture, or product subject to subchapter IV of this chapter manufactured or processed in violation of section 2604 of this title, 2605 of this title, or subchapter IV of this chapter, or a rule or order under section 2604 of this title, 2605 of this title, or subchapter IV of this chapter, and distributed in commerce, (i) to give notice of such fact to distributors in commerce of such substance, mixture, or product and, to the extent reasonably ascertainable, to other persons in possession of such substance, mixture, or product or exposed to such substance, mixture, or product, (ii) to give public notice of such risk of injury, and (iii) to either replace or repurchase such substance, mixture, or product, whichever the person to which the requirement is directed elects.
    - (2) A civil action described in paragraph (1) may be brought—
        - (A) in the case of a civil action described in subparagraph (A) of such paragraph, in the United States district court for the judicial district wherein any act, omission, or transaction constituting a violation of section 2614 of this title occurred or wherein the defendant is found or transacts business, or
        - (B) in the case of any other civil action described in such paragraph, in the United States district court for the judicial district wherein the defendant is found or transacts business.

In any such civil action process may be served on a defendant in any judicial district in which a defendant resides or may be found. Subpoenas requiring attendance of witnesses in any such action may be served in any judicial district.

- (b) Seizure
Any chemical substance, mixture, or product subject to subchapter IV of this chapter which was manufactured, processed, or distributed in commerce in violation of this chapter or any rule promulgated or order issued under this chapter or any article containing such a substance or mixture shall be liable to be proceeded against, by process of libel, for the seizure and condemnation of such substance, mixture, product, or article, in any district court of the United States within the jurisdiction of which such substance, mixture, product, or article is found. Such proceedings shall conform as nearly as possible to proceedings in rem in admiralty.

Sec. 2617. Preemption

- (a) Effect on State law
    - (1) Except as provided in paragraph (2), nothing in this chapter shall affect the authority of any State or political subdivision of a State to establish or continue in effect regulation of any chemical substance, mixture, or article containing a chemical substance or mixture.
    - (2) Except as provided in subsection (b) of this section—
        - (A) if the Administrator requires by a rule promulgated under section 2603 of this title the testing of a chemical substance or mixture, no State or political subdivision may, after the effective date of such rule, establish or continue in effect a requirement for the testing of such substance or mixture for purposes similar to those for which testing is required under such rule; and
        (B) if the Administrator prescribes a rule or order under section 2604 or 2605 of this title (other than a rule imposing a requirement described in subsection (a)(6) of section 2605 of this title) which is applicable to a chemical substance or mixture, and which is designed to protect against a risk of injury to health or the environment associated with such substance or mixture, no State or political subdivision of a State may, after the effective date of such requirement, establish or continue in effect, any requirement which is applicable to such substance or mixture, or an article containing such substance or mixture, and which is designed to protect against such risk unless such requirement (i) is identical to the requirement

prescribed by the Administrator, (ii) is adopted under the authority of the Clean Air Act (42 U.S.C. 7401 et seq.) or any other Federal law, or (iii) prohibits the use of such substance or mixture in such State or political subdivision (other than its use in the manufacture or processing of other substances or mixtures).

- (b) Exemption

Upon application of a State or political subdivision of a State the Administrator may by rule exempt from subsection (a)(2) of this section, under such conditions as may be prescribed in such rule, a requirement of such State or political subdivision designed to protect against a risk of injury to health or the environment associated with a chemical substance, mixture, or article containing a chemical substance or mixture if—

- (1) compliance with the requirement would not cause the manufacturing, processing, distribution in commerce, or use of the substance, mixture, or article to be in violation of the applicable requirement under this chapter described in subsection (a)(2) of this section, and

(2) the State or political subdivision requirement (A) provides a significantly higher degree of protection from such risk than the requirement under this chapter described in subsection (a)(2) of this section and (B) does not, through difficulties in marketing, distribution, or other factors, unduly burden interstate commerce.

Sec. 2618. Judicial review

- (a) In general

- (1)

- (A) Not later than 60 days after the date of the promulgation of a rule under section 2603(a), 2604(a)(2), 2604(b)(4), 2605(a), 2605(e), or 2607 of this title, or under subchapter II or IV of this chapter, any person may file a petition for judicial review of such rule with the United States Court of Appeals for the District of Columbia Circuit or for the circuit in which such person resides or in which such person's principal place of business is located. Courts of appeals of the United States shall have exclusive jurisdiction of any action to obtain judicial review (other than in an enforcement proceeding) of such a rule if any district court of the United States would have had jurisdiction of such action but for this subparagraph.

- (B) Courts of appeals of the United States shall have exclusive jurisdiction of any action to obtain judicial review (other than in an enforcement proceeding) of an order issued under subparagraph (A) or (B) of section 2605(b)(1) of this title if any district court of the United States would have had jurisdiction of such action but for this subparagraph.

- (2) Copies of any petition filed under paragraph (1)(A) shall be transmitted forthwith to the Administrator and to the Attorney General by the clerk of the court with which such petition was filed. The provisions of section 2112 of title 28 shall apply to the filing of the rulemaking record of proceedings on which the Administrator based the rule being reviewed under this section and to the transfer of proceedings between United States courts of appeals.

- (3) For purposes of this section, the term "rulemaking record" means—
  - (A) the rule being reviewed under this section;
  - (B) in the case of a rule under section 2603(a) of this title, the finding required by such section, in the case of a rule under section 2604(b)(4) of this title, the finding required by such section, in the case of a rule under section 2605(a) of this title the finding required by section 2604(f) or 2605(a) of this title, as the case may be, in the case of a rule under section 2605(a) of this title, the statement required by section 2605(c)(1) of this title, and in the case of a rule under section 2605(e) of this title, the findings required by paragraph (2)(B) or (3)(B) of such section, as the case may be, and in the case of a rule under subchapter IV of this chapter, the finding required for the issuance of such a rule;
  - (C) any transcript required to be made of oral presentations made in proceedings for the promulgation of such rule;
  - (D) any written submission of interested parties respecting the promulgation of such rule; and

    (E) any other information which the Administrator considers to be relevant to such rule and which the Administrator identified, on or before the date of the promulgation of such rule, in a notice published in the Federal Register.

- (b) Additional submissions and presentations; modifications
  If in an action under this section to review a rule the petitioner or the Administrator applies to the court for leave to make additional oral submissions or written presentations respecting such rule and shows to the satisfaction of the court that such submissions and presentations would be material and that there were reasonable grounds for the submissions and failure to make such submissions and presentations in the proceeding before the Administrator, the court may order the Administrator to provide additional opportunity to make such submissions and presentations. The Administrator may modify or set aside the rule being reviewed or make a

new rule by reason of the additional submissions and presentations and shall file such modified or new rule with the return of such submissions and presentations. The court shall thereafter review such new or modified rule.

- (c) Standard of review

- (1)
  - ○ (A) Upon the filing of a petition under subsection (a)(1) of this section for judicial review of a rule, the court shall have jurisdiction (i) to grant appropriate relief, including interim relief, as provided in chapter 7 of title 5, and (ii) except as otherwise provided in subparagraph (B), to review such rule in accordance with chapter 7 of title 5.
  - ○ (B) Section 706 of title 5 shall apply to review of a rule under this section, except that—
    - ▪ (i) in the case of review of a rule under section 2603(a), 2604(b)(4), 2605(a), or 2605(e) of this title, the standard for review prescribed by paragraph (2)(E) of such section 706 shall not apply and the court shall hold unlawful and set aside such rule if the court finds that the rule is not supported by substantial evidence in the rulemaking record (as defined in subsection (a)(3) of this section) taken as a whole;
    - ▪ (ii) in the case of review of a rule under section 2605(a) of this title, the court shall hold unlawful and set aside such rule if it finds that—
  - ○ (I) a determination by the Administrator under section 2605(c)(3) of this title that the petitioner seeking review of such rule is not entitled to conduct (or have conducted) cross-examination or to present rebuttal submissions, or
  - ○ (II) a rule of, or ruling by, the Administrator under section 2605(c)(3) of this title limiting such petitioner's cross-examination or oral presentations, has precluded disclosure of disputed material facts which was necessary to a fair determination by the Administrator of the rulemaking proceeding taken as a whole; and section 706(2)(D) shall not apply with respect to a determination, rule, or ruling referred to in subclause (I) or (II); and
  - (iii) the court may not review the contents and adequacy of—
  - ○ (I) any statement required to be made pursuant to section 2605(c)(1) of this title, or
  - ○ (II) any statement of basis and purpose required by section 553(c) of title 5 to be incorporated in the rule except as part of a review of the rulemaking record taken as a whole. The term "evidence" as used in clause (i) means any matter in the rulemaking record.
  - ○ (C) A determination, rule, or ruling of the Administrator described

in subparagraph (B)(ii) may be reviewed only in an action under this section and only in accordance with such subparagraph.

- (2) The judgment of the court affirming or setting aside, in whole or in part, any rule reviewed in accordance with this section shall be final, subject to review by the Supreme Court of the United States upon certiorari or certification, as provided in section 1254 of title 28.

- (d) Fees and costs
The decision of the court in an action commenced under subsection (a) of this section, or of the Supreme Court of the United States on review of such a decision, may include an award of costs of suit and reasonable fees for attorneys and expert witnesses if the court determines that such an award is appropriate.

- (e) Other remedies
The remedies as provided in this section shall be in addition to and not in lieu of any other remedies provided by law.

Sec. 2619. Citizens' civil actions

- (a) In general
Except as provided in subsection (b) of this section, any person may commence a civil action—
  - (1) against any person (including (A) the United States, and (B) any other governmental instrumentality or agency to the extent permitted by the eleventh amendment to the Constitution) who is alleged to be in violation of this chapter or any rule promulgated under section 2603, 2604, or 2605 of this title, or subchapter II or IV of this chapter, or order issued under section 2604 of this title or subchapter II or IV of this chapter to restrain such violation, or
  - (2) against the Administrator to compel the Administrator to perform any act or duty under this chapter which is not discretionary. Any civil action under paragraph (1) shall be brought in the United States district court for the district in which the alleged violation occurred or in which the defendant resides or in which the defendant's principal place of business is located. Any action brought under paragraph (2) shall be brought in the United States District Court for the District of Columbia, or the United States district court for the judicial district in which the plaintiff is domiciled. The district courts of the United States shall have jurisdiction over suits brought under this section, without regard to the amount in controversy or the citizenship of the parties. In any civil action under this subsection process may be served on a defendant in any judicial

district in which the defendant resides or may be found and sub-poenas for witnesses may be served in any judicial district.

- (b) Limitation

  No civil action may be commenced—
  - ○ (1) under subsection (a)(1) of this section to restrain a violation of this chapter or rule or order under this chapter—
    - ▪ (A) before the expiration of 60 days after the plaintiff has given notice of such violation (i) to the Administrator, and (ii) to the person who is alleged to have committed such violation, or
    - ▪ (B) if the Administrator has commenced and is diligently prosecuting a proceeding for the issuance of an order under section 2615(a)(2) of this title to require compliance with this chapter or with such rule or order or if the Attorney General has commenced and is diligently prosecuting a civil action in a court of the United States to require compliance with this chapter or with such rule or order, but if such proceeding or civil action is commenced after the giving of notice, any person giving such notice may intervene as a matter of right in such proceeding or action; or
  - ○ (2) under subsection (a)(2) of this section before the expiration of 60 days after the plaintiff has given notice to the Administrator of the alleged failure of the Administrator to perform an act or duty which is the basis for such action or, in the case of an action under such subsection for the failure of the Administrator to file an action under section 2606 of this title, before the expiration of ten days after such notification. Notice under this subsection shall be given in such manner as the Administrator shall prescribe by rule.

- (c) General
  - ○ (1) In any action under this section, the Administrator, if not a party, may intervene as a matter of right.
  - ○ (2) The court, in issuing any final order in any action brought pursuant to subsection (a) of this section, may award costs of suit and reasonable fees for attorneys and expert witnesses if the court determines that such an award is appropriate. Any court, in issuing its decision in an action brought to review such an order, may award costs of suit and reasonable fees for attorneys if the court determines that such an award is appropriate.
  - ○ (3) Nothing in this section shall restrict any right which any person (or class of persons) may have under any statute or common law to seek enforcement of this chapter or any rule or order under this chapter or to seek any other relief.

- (d) Consolidation

When two or more civil actions brought under subsection (a) of this section involving the same defendant and the same issues or violations are pending in two or more judicial districts, such pending actions, upon application of such defendants to such actions which is made to a court in which any such action is brought, may, if such court in its discretion so decides, be consolidated for trial by order (issued after giving all parties reasonable notice and opportunity to be heard) of such court and tried in—

  ○ (1) any district which is selected by such defendant and in which one of such actions is pending,
  ○ (2) a district which is agreed upon by stipulation between all the parties to such actions and in which one of such actions is pending, or
  ○ (3) a district which is selected by the court and in which one of such actions is pending. The court issuing such an order shall give prompt notification of the order to the other courts in which the civil actions consolidated under the order are pending.

Sec. 2620. Citizens' petitions

- (a) In general

Any person may petition the Administrator to initiate a proceeding for the issuance, amendment, or repeal of a rule under section 2603, 2605, or 2607 of this title or an order under section 2604(e) or 2605(b)(2) of this title.

- (b) Procedures

  ○ (1) Such petition shall be filed in the principal office of the Administrator and shall set forth the facts which it is claimed establish that it is necessary to issue, amend, or repeal a rule under section 2603, 2605, or 2607 of this title or an order under section 2604(e), 2605(b)(1)(A), or 2605(b)(1)(B) of this title.
  ○ (2) The Administrator may hold a public hearing or may conduct such investigation or proceeding as the Administrator deems appropriate in order to determine whether or not such petition should be granted.
  ○ (3) Within 90 days after filing of a petition described in paragraph (1), the Administrator shall either grant or deny the petition. If the Administrator grants such petition, the Administrator shall promptly commence an appropriate proceeding in accordance with section 2603, 2604, 2605, or 2607 of this title. If the Administrator denies such petition, the Administrator shall publish in the Federal Register the Administrator's reasons for such denial.

○ (4)

- (A) If the Administrator denies a petition filed under this section (or if the Administrator fails to grant or deny such petition within the 90-day period) the petitioner may commence a civil action in a district court of the United States to compel the Administrator to initiate a rulemaking proceeding as requested in the petition. Any such action shall be filed within 60 days after the Administrator's denial of the petition or, if the Administrator fails to grant or deny the petition within 90 days after filing the petition, within 60 days after the expiration of the 90-day period.
- (B) In an action under subparagraph (A) respecting a petition to initiate a proceeding to issue a rule under section 2603, 2605, or 2607 of this title or an order under section 2604(e) or 2605(b)(2) of this title, the petitioner shall be provided an opportunity to have such petition considered by the court in a de novo proceeding. If the petitioner demonstrates to the satisfaction of the court by a preponderance of the evidence that—
    - (i) in the case of a petition to initiate a proceeding for the issuance of a rule under section 2603 of this title or an order under section 2604(e) of this title—
- (I) information available to the Administrator is insufficient to permit a reasoned evaluation of the health and environmental effects of the chemical substance to be subject to such rule or order; and
(II) in the absence of such information, the substance may present an unreasonable risk to health or the environment, or the substance is or will be produced in substantial quantities and it enters or may reasonably be anticipated to enter the environment in substantial quantities or there is or may be significant or substantial human exposure to it; or
    - (ii) in the case of a petition to initiate a proceeding for the issuance of a rule under section 2605 or 2607 of this title or an order under section 2605(b)(2) of this title, there is a reasonable basis to conclude that the issuance of such a rule or order is necessary to protect health or the environment against an unreasonable risk of injury to health or the environment.[1]

[1] So in original. The period probably should be a semicolon. The court shall order the Administrator to initiate the action requested by the petitioner. If the court finds that the extent of the risk to health or the environment alleged by the petitioner is less than the extent of risks to health or the environment with respect to which the Administrator is taking action under this chapter and there are insufficient resources available to the Administrator to take the action requested by the petitioner, the court may permit the Administrator to defer initiating the action requested by the petitioner until such time as the court prescribes.

- (C) The court in issuing any final order in any action brought pursuant to subparagraph (A) may award costs of suit and reasonable fees for attorneys and expert witnesses if the court determines that such an award is appropriate. Any court, in issuing its decision in an action brought to review such an order, may award costs of suit and reasonable fees for attorneys if the court determines that such an award is appropriate.

○ (5) The remedies under this section shall be in addition to, and not in lieu of, other remedies provided by law.

Sec. 2621. National defense waiver

The Administrator shall waive compliance with any provision of this chapter upon a request and determination by the President that the requested waiver is necessary in the interest of national defense. The Administrator shall maintain a written record of the basis upon which such waiver was granted and make such record available for in camera examination when relevant in a judicial proceeding under this chapter. Upon the issuance of such a waiver, the Administrator shall publish in the Federal Register a notice that the waiver was granted for national defense purposes, unless, upon the request of the President, the Administrator determines to omit such publication because the publication itself would be contrary to the interests of national defense, in which event the Administrator shall submit notice thereof to the Armed Services Committees of the Senate and the House of Representatives.

Sec. 2622. Employee protection

- (a) In general
  No employer may discharge any employee or otherwise discriminate against any employee with respect to the employee's compensation, terms, conditions, or privileges of employment because the employee (or any person acting pursuant to a request of the employee) has—
    ○ (1) commenced, caused to be commenced, or is about to commence or cause to be commenced a proceeding under this chapter;
    ○ (2) testified or is about to testify in any such proceeding; or
    ○ (3) assisted or participated or is about to assist or participate in any manner in such a proceeding or in any other action to carry out the purposes of this chapter.

- (b) Remedy
    ○ (1) Any employee who believes that the employee has been discharged or otherwise discriminated against by any person in violation of subsection (a) of this section may, within 30 days after such

alleged violation occurs, file (or have any person file on the employee's behalf) a complaint with the Secretary of Labor (hereinafter in this section referred to as the "Secretary") alleging such discharge or discrimination. Upon receipt of such a complaint, the Secretary shall notify the person named in the complaint of the filing of the complaint.

- (2)
  - (A) Upon receipt of a complaint filed under paragraph (1), the Secretary shall conduct an investigation of the violation alleged in the complaint. Within 30 days of the receipt of such complaint, the Secretary shall complete such investigation and shall notify in writing the complainant (and any person acting on behalf of the complainant) and the person alleged to have committed such violation of the results of the investigation conducted pursuant to this paragraph. Within ninety days of the receipt of such complaint the Secretary shall, unless the proceeding on the complaint is terminated by the Secretary on the basis of a settlement entered into by the Secretary and the person alleged to have committed such violation, issue an order either providing the relief prescribed by subparagraph (B) or denying the complaint. An order of the Secretary shall be made on the record after notice and opportunity for agency hearing. The Secretary may not enter into a settlement terminating a proceeding on a complaint without the participation and consent of the complainant.
  - (B) If in response to a complaint filed under paragraph (1) the Secretary determines that a violation of subsection (a) of this section has occurred, the Secretary shall order (i) the person who committed such violation to take affirmative action to abate the violation, (ii) such person to reinstate the complainant to the complainant's former position together with the compensation (including back pay), terms, conditions, and privileges of the complainant's employment, (iii) compensatory damages, and (iv) where appropriate, exemplary damages. If such an order is issued, the Secretary, at the request of the complainant, shall assess against the person against whom the order is issued a sum equal to the aggregate amount of all costs and expenses (including attorneys' fees) reasonably incurred, as determined by the Secretary, by the complainant for, or in connection with, the bringing of the complaint upon which the order was issued.

- (c) Review
  - (1) Any employee or employer adversely affected or aggrieved by an order issued under subsection (b) of this section may obtain review of

the order in the United States Court of Appeals for the circuit in which the violation, with respect to which the order was issued, allegedly occurred. The petition for review must be filed within sixty days from the issuance of the Secretary's order. Review shall conform to chapter 7 of title 5.

- ○ (2) An order of the Secretary, with respect to which review could have been obtained under paragraph (1), shall not be subject to judicial review in any criminal or other civil proceeding.

- (d) Enforcement
Whenever a person has failed to comply with an order issued under subsection (b)(2) of this section, the Secretary shall file a civil action in the United States district court for the district in which the violation was found to occur to enforce such order. In actions brought under this subsection, the district courts shall have jurisdiction to grant all appropriate relief, including injunctive relief and compensatory and exemplary damages.

- (e) Exclusion
Subsection (a) of this section shall not apply with respect to any employee who, acting without direction from the employee's employer (or any agent of the employer), deliberately causes a violation of any requirement of this chapter.

Sec. 2623. Employment effects

- (a) In general
The Administrator shall evaluate on a continuing basis the potential effects on employment (including reductions in employment or loss of employment from threatened plant closures) of—
    - ○ (1) the issuance of a rule or order under section 2603, 2604, or 2605 of this title, or
    - ○ (2) a requirement of section 2604 or 2605 of this title.

- (b) Investigations
    - ○ (1) Any employee (or any representative of an employee) may request the Administrator to make an investigation of—
        - ▪ (A) a discharge or layoff or threatened discharge or layoff of the employee, or
        - ▪ (B) adverse or threatened adverse effects on the employee's employment, allegedly resulting from a rule or order under section 2603, 2604, or 2605 of this title or a requirement of section 2604 or 2605 of this title. Any such request shall be made in writing, shall set forth with reasonable particularity

the grounds for the request, and shall be signed by the employee, or representative of such employee, making the request.

○ (2)

■ (A) Upon receipt of a request made in accordance with paragraph (1) the Administrator shall (i) conduct the investigation requested, and (ii) if requested by any interested person, hold public hearings on any matter involved in the investigation unless the Administrator, by order issued within 45 days of the date such hearings are requested, denies the request for the hearings because the Administrator determines there are no reasonable grounds for holding such hearings. If the Administrator makes such a determination, the Administrator shall notify in writing the person requesting the hearing of the determination and the reasons therefor and shall publish the determination and the reasons therefor in the Federal Register.

■ (B) If public hearings are to be held on any matter involved in an investigation conducted under this subsection—

■ (i) at least five days' notice shall be provided the person making the request for the investigation and any person identified in such request,

■ (ii) such hearings shall be held in accordance with section 2605(c)(3) of this title, and

(iii) each employee who made or for whom was made a request for such hearings and the employer of such employee shall be required to present information respecting the applicable matter referred to in paragraph (1)(A) or (1)(B) together with the basis for such information.

○ (3) Upon completion of an investigation under paragraph (2), the Administrator shall make findings of fact, shall make such recommendations as the Administrator deems appropriate, and shall make available to the public such findings and recommendations.

○ (4) This section shall not be construed to require the Administrator to amend or repeal any rule or order in effect under this chapter.

Sec. 2624. Studies

● (a) Indemnification study

The Administrator shall conduct a study of all Federal laws administered by the Administrator for the purpose of determining whether and under what conditions, if any, indemnification should be accorded any person as a result of any action taken by the Administrator under any such law. The study shall—

- (1) include an estimate of the probable cost of any indemnification programs which may be recommended;
- (2) include an examination of all viable means of financing the cost of any recommended indemnification; and

(3) be completed and submitted to Congress within two years from the effective date of enactment of this chapter. The General Accounting Office shall review the adequacy of the study submitted to Congress pursuant to paragraph (3) and shall report the results of its review to the Congress within six months of the date such study is submitted to Congress.

- (b) Classification, storage, and retrieval study

The Council on Environmental Quality, in consultation with the Administrator, the Secretary of Health and Human Services, the Secretary of Commerce, and the heads of other appropriate Federal departments or agencies, shall coordinate a study of the feasibility of establishing (1) a standard classification system for chemical substances and related substances, and (2) a standard means for storing and for obtaining rapid access to information respecting such substances. A report on such study shall be completed and submitted to Congress not later than 18 months after the effective date of enactment of this chapter.

Sec. 2625. Administration

- (a) Cooperation of Federal agencies

Upon request by the Administrator, each Federal department and agency is authorized—

- (1) to make its services, personnel, and facilities available (with or without reimbursement) to the Administrator to assist the Administrator in the administration of this chapter; and

(2) to furnish to the Administrator such information, data, estimates, and statistics, and to allow the Administrator access to all information in its possession as the Administrator may reasonably determine to be necessary for the administration of this chapter.

- (b) Fees

- (1) The Administrator may, by rule, require the payment of a reasonable fee from any person required to submit data under section 2603 or 2604 of this title to defray the cost of administering this chapter. Such rules shall not provide for any fee in excess of $2,500 or, in the case of a small business concern, any fee in excess of $100. In setting a fee under this paragraph, the Administrator shall take into account the ability to pay of the person required to submit the data and the cost to the Administrator of reviewing such data. Such rules may provide for sharing such a fee in any case in which the expenses of testing are shared under section 2603 or 2604 of this title.

○ (2) The Administrator, after consultation with the Administrator of the Small Business Administration, shall by rule prescribe standards for determining the persons which qualify as small business concerns for purposes of paragraph (1).

- (c) Action with respect to categories
  ○ (1) Any action authorized or required to be taken by the Administrator under any provision of this chapter with respect to a chemical substance or mixture may be taken by the Administrator in accordance with that provision with respect to a category of chemical substances or mixtures. Whenever the Administrator takes action under a provision of this chapter with respect to a category of chemical substances or mixtures, any reference in this chapter to a chemical substance or mixture (insofar as it relates to such action) shall be deemed to be a reference to each chemical substance or mixture in such category.
  ○ (2) For purposes of paragraph (1):
    ▪ (A) The term "category of chemical substances" means a group of chemical substances the members of which are similar in molecular structure, in physical, chemical, or biological properties, in use, or in mode of entrance into the human body or into the environment, or the members of which are in some other way suitable for classification as such for purposes of this chapter, except that such term does not mean a group of chemical substances which are grouped together solely on the basis of their being new chemical substances.
    ▪ (B) The term "category of mixtures" means a group of mixtures the members of which are similar in molecular structure, in physical, chemical, or biological properties, in use, or in the mode of entrance into the human body or into the environment, or the members of which are in some other way suitable for classification as such for purposes of this chapter.

- (d) Assistance office
The Administrator shall establish in the Environmental Protection Agency an identifiable office to provide technical and other nonfinancial assistance to manufacturers and processors of chemical substances and mixtures respecting the requirements of this chapter applicable to such manufacturers and processors, the policy of the Agency respecting the application of such requirements to such manufacturers and processors,

and the means and methods by which such manufacturers and processors may comply with such requirements.

- (e) Financial disclosures
  - ○ (1) Except as provided under paragraph (3), each officer or employee of the Environmental Protection Agency and the Department of Health and Human Services who—
    - ▪ (A) performs any function or duty under this chapter, and (B) has any known financial interest (i) in any person subject to this chapter or any rule or order in effect under this chapter, or (ii) in any person who applies for or receives any grant or contract under this chapter, shall, on February 1, 1978, and on February 1 of each year thereafter, file with the Administrator or the Secretary of Health and Human Services (hereinafter in this subsection referred to as the "Secretary"), as appropriate, a written statement concerning all such interests held by such officer or employee during the preceding calendar year. Such statement shall be made available to the public.
  - ○ (2) The Administrator and the Secretary shall—
    - ▪ (A) act within 90 days of January 1, 1977—
      - ▪ (i) to define the term "known financial interests" for purposes of paragraph (1), and
        (ii) to establish the methods by which the requirement to file written statements specified in paragraph (1) will be monitored and enforced, including appropriate provisions for review by the Administrator and the Secretary of such statements; and
        (B) report to the Congress on June 1, 1978, and on June 1 of each year thereafter with respect to such statements and the actions taken in regard thereto during the preceding calendar year.
  - ○ (3) The Administrator may by rule identify specific positions with the Environmental Protection Agency, and the Secretary may by rule identify specific positions with the Department of Health and Human Services, which are of a nonregulatory or nonpolicymaking nature, and the Administrator and the Secretary may by rule provide that officers or employees occupying such positions shall be exempt from the requirements of paragraph (1).
  - ○ (4) This subsection does not supersede any requirement of chapter 11 of title 18.
  - ○ (5) Any officer or employee who is subject to, and knowingly violates, this subsection or any rule issued thereunder, shall be fined not more than $2,500 or imprisoned not more than one year, or both.

- (f) Statement of basis and purpose
Any final order issued under this chapter shall be accompanied by a statement of its basis and purpose. The contents and adequacy of any such statement shall not be subject to judicial review in any respect.

- (g) Assistant Administrator
    - (1) The President, by and with the advice and consent of the Senate, shall appoint an Assistant Administrator for Toxic Substances of the Environmental Protection Agency. Such Assistant Administrator shall be a qualified individual who is, by reason of background and experience, especially qualified to direct a program concerning the effects of chemicals on human health and the environment. Such Assistant Administrator shall be responsible for
        - (A) the collection of data, (B) the preparation of studies, (C) the making of recommendations to the Administrator for regulatory and other actions to carry out the purposes and to facilitate the administration of this chapter, and (D) such other functions as the Administrator may assign or delegate.
    - (2) The Assistant Administrator to be appointed under paragraph (1) shall be in addition to the Assistant Administrators of the Environmental Protection Agency authorized by section 1(d) of Reorganization Plan No. 3 of 1970.

Sec. 2626. Development and evaluation of test methods

- (a) In general
The Secretary of Health and Human Services, in consultation with the Administrator and acting through the Assistant Secretary for Health, may conduct, and make grants to public and nonprofit private entities and enter into contracts with public and private entities for, projects for the development and evaluation of inexpensive and efficient methods (1) for determining and evaluating the health and environmental effects of chemical substances and mixtures, and their toxicity, persistence, and other characteristics which affect health and the environment, and (2) which may be used for the development of test data to meet the requirements of rules promulgated under section 2603 of this title. The Administrator shall consider such methods in prescribing under section 2603 of this title standards for the development of test data.

- (b) Approval by Secretary
No grant may be made or contract entered into under subsection (a) of this section unless an application therefor has been submitted to and approved by the Secretary. Such an application shall be submitted in such form and manner and contain such information as the Secretary may re-

quire. The Secretary may apply such conditions to grants and contracts under subsection (a) of this section as the Secretary determines are necessary to carry out the purposes of such subsection. Contracts may be entered into under such subsection without regard to section 3324(a) and (b) of title 31 and section 5 of title 41.

Sec. 2627. State programs

- (a) In general
  For the purpose of complementing (but not reducing) the authority of, or actions taken by, the Administrator under this chapter, the Administrator may make grants to States for the establishment and operation of programs to prevent or eliminate unreasonable risks within the States to health or the environment which are associated with a chemical substance or mixture and with respect to which the Administrator is unable or is not likely to take action under this chapter for their prevention or elimination. The amount of a grant under this subsection shall be determined by the Administrator, except that no grant for any State program may exceed 75 per centum of the establishment and operation costs (as determined by the Administrator) of such program during the period for which the grant is made.

- (b) Approval by Administrator
  - (1) No grant may be made under subsection (a) of this section unless an application therefor is submitted to and approved by the Administrator. Such an application shall be submitted in such form and manner as the Administrator may require and shall—
    - (A) set forth the need of the applicant for a grant under subsection (a) of this section,
    - (B) identify the agency or agencies of the State which shall establish or operate, or both, the program for which the application is submitted,
    - (C) describe the actions proposed to be taken under such program,
    - (D) contain or be supported by assurances satisfactory to the Administrator that such program shall, to the extent feasible, be integrated with other programs of the applicant for environmental and public health protection,
    - (E) provide for the making of such reports and evaluations as the Administrator may require, and
      (F) contain such other information as the Administrator may prescribe.
  - (2) The Administrator may approve an application submitted in accordance with paragraph (1) only if the applicant has established

to the satisfaction of the Administrator a priority need, as determined under rules of the Administrator, for the grant for which the application has been submitted. Such rules shall take into consideration the seriousness of the health effects in a State which are associated with chemical substances or mixtures, including cancer, birth defects, and gene mutations, the extent of the exposure in a State of human beings and the environment to chemical substances and mixtures, and the extent to which chemical substances and mixtures are manufactured, processed, used, and disposed of in a State.

- (c) Annual reports
Not later than six months after the end of each of the fiscal years 1979, 1980, and 1981, the Administrator shall submit to the Congress a report respecting the programs assisted by grants under subsection (a) of this section in the preceding fiscal year and the extent to which the Administrator has disseminated information respecting such programs.

- (d) Authorization
For the purpose of making grants under subsection (a) of this section, there are authorized to be appropriated $1,500,000 for each of the fiscal years 1982 and 1983. Sums appropriated under this subsection shall remain available until expended.

Sec. 2628. Authorization of appropriations

There are authorized to be appropriated to the Administrator for purposes of carrying out this chapter (other than sections 2626 and 2627 of this title and subsections (a) and (c) through (g) of section 2609 of this title) $58,646,000 for the fiscal year 1982 and $62,000,000 for the fiscal year 1983. No part of the funds appropriated under this section may be used to construct any research laboratories.

Sec. 2629. Annual report

The Administrator shall prepare and submit to the President and the Congress on or before January 1, 1978, and on or before January 1 of each succeeding year a comprehensive report on the administration of this chapter during the preceding fiscal year. Such reports shall include—

- (1) a list of the testing required under section 2603 of this title during the year for which the report is made and an estimate of the costs incurred during such year by the persons required to perform such tests;

- (2) the number of notices received during such year under section 2604 of this title, the number of such notices received during such year under such section for chemical substances subject to a rule, and a summary of any action taken during such year under section 2604(g) of this title;

- (3) a list of rules issued during such year under section 2605 of this title;

- (4) a list, with a brief statement of the issues, of completed or pending judicial actions under this chapter and administrative actions under section 2615 of this title during such year;

- (5) a summary of major problems encountered in the administration of this chapter; and
  (6) such recommendations for additional legislation as the Administrator deems necessary to carry out the purposes of this chapter.

Appendix 7.2

---

19 CFR 141.20

(Reproduced from:

www.access.gpo.gov/nara/cfr/waisidx_01/
19cfrv2_01.html)

Part 141—Entry of Merchandise

§141.20 Actual owner's declaration and superseding bond of actual owner

(a) *Filing*—(1) *Declaration of owner.* A consignee in whose name an entry summary for consumption, warehouse, or temporary importation under bond is filed, or in whose name a rewarehouse entry or a manufacturing warehouse entry is made, and who desires, under the provisions of section 485(d), Tariff Act of 1930, as amended (19 U.S.C. 1485(d)), to be relieved from statutory liability for the payment of increased and additional duties shall declare at the time of the filing of the entry summary or entry documentation, as provided in §141.19(a), that he is not the actual owner of the merchandise, furnish the name and address of the owner, and file with the port director within 90 days from the time of entry (see §141.68) a declaration of the actual owner of the merchandise acknowledging that the actual owner will pay all additional and increased duties. The declaration of owner shall be filed on Customs Form 3347.

(2) *Bond of actual owner.* If the consignee desires to be relieved from contractual liability for the payment of increased and additional duties voluntarily assumed by him under the single-entry bond which he filed in connection with the entry documentation and/or entry summary, or under his continuous bond against which the entry and/or entry summary is charged, he shall file a bond of the actual owner on Customs Form 301, containing the bond conditions set forth in §113.62 of this chapter, with the port director within 90 days from the time of entry.

(b) *Appropriate party to execute and file.* Neither the declaration of the actual owner nor the bond of the actual owner shall be accepted unless executed by the actual owner or his duly authorized agent, and filed by the nominal consignee or his duly authorized agent.

(c) *Nonresident actual owner.* If the actual owner is a nonresident, the actual owner's declaration shall not be accepted as compliance with section 485(d), Tariff Act of 1930, as amended (19 U.S.C. 1485(d)), unless there is filed therewith the owner's bond on Customs Form 301, containing the bond conditions set forth in §113.62 of this chapter, with a resident corporate surety.

(d) *Filing of declaration of owner for purposes other than relief from liability.* Nothing in this section shall be construed to prevent the nominal consignee from filing the actual owner's declaration without the superseding bond for purposes other than relief from statutory liability for the payment of increased and additional duties under the provisions of section 485(d), Tariff Act of 1930, as amended (19 U.S.C. 1485(d)).

[T.D. 73–175, 38 FR 17447, July 2, 1973, as amended by T.D. 74–212, 39 FR 28420, Aug. 7, 1974; T.D. 79–221, 44 FR 46816, Aug. 9, 1979; T.D. 84–213, 49 FR 41184, Oct. 19, 1984]

19 CFR 144.21 through 144.28

(Reproduced from:
*www.access.gpo.gov/nara/cfr/waisidx_01/
19cfrv2_01.html*)

Subpart C—Transfer of Right to Withdraw Merchandise from Warehouse

§144.21 Conditions for transfer

Under the provisions of section 557(b) Tariff Act of 1930, as amended (19 U.S.C. 1557(b)), the right to withdraw all or part of merchandise entered for warehouse may be transferred by appropriate endorsement on the withdrawal form, provided that the transferee files a bond on Customs Form 301, containing the bond conditions set forth in §113.62 of this chapter. Upon the deposit of the endorsed form, properly executed, and the transferee's bond with the Customs officer designated to receive such form and bond, the transferor and his sureties shall be relieved from all undischarged liability.

[T.D. 73–175, 38 FR 17464, July 2, 1973, as amended by T.D. 84–213, 49 FR 41185, Oct. 19, 1984; 49 FR 44867, Nov. 9, 1984]

§144.22 Endorsement of transfer on withdrawal form

Transfer of the right to withdraw merchandise entered for warehouse shall be established by an appropriate endorsement on the withdrawal form by the person primarily liable for payment of duties before the transfer is completed, i.e., the person who made the warehouse or rewarehouse entry or a transferee of the withdrawal right of such person. Endorsement shall be made on whichever of the following withdrawal forms is applicable:

(a) Customs Form 7501 for:

(i) A duty paid warehouse withdrawal for consumption;

(ii) Withdrawal with no duty payment (diplomatic use);

(iii) Merchandise to be withdrawn as vessel or aircraft supplies and equipment under §10.60 of this chapter or other conditionally free merchandise;

(b) Customs Form 7512 for merchandise to be withdrawn for transportion, exportation, or transportation and exportation; or

[T.D. 82–204, 47 FR 49376, Nov. 1, 1982, as amended by T.D. 95–81, 60 FR 52295, Oct. 6, 1995]

§144.23   Endorsement in blank

If the transferor wishes to do so, he may endorse the withdrawal form to authorize the right to withdraw the merchandise specified thereon but leave the space for the name of the transferee blank. A holder of a withdrawal form so endorsed and otherwise fully executed may insert his own name in the blank space, deposit such form and his transferee's bond with the Customs officer designated to receive such form and bond, and thereby establish his right to withdraw the merchandise.

§144.24   Transferee's bond

The transferee's bond shall be on Customs Form 301 and contain the bond conditions set forth in §113.62 of this chapter.

[T.D. 84–213, 49 FR 41185, Oct. 19, 1984]

§144.25   Deposit of forms

Either the transferor or the transferee may deposit the endorsed withdrawal form and transferee's bond with the Customs officer designated to receive such form and bond.

§144.26   Further transfer

The right of a transferee to withdraw the merchandise may not be revoked by the transferor but may be retransferred by the transferee.

§144.27   Withdrawal from warehouse by transferee

At any time within the warehousing period, a transferee who has established his right to withdraw merchandise may withdraw all or part of the merchandise covered by the transfer by filing any authorized kind of withdrawal from warehouse in accordance with subpart D of this part.

§144.28 Protest by transferee

(a) *Entries on or after January 12, 1971.* A transferee of merchandise entered for warehouse on or after January 12, 1971, shall have the right to file a protest under section 514, Tariff Act of 1930, as amended (19 U.S.C. 1514), to the same extent that such right would have been available to the transferor.

(b) *Entries prior to January 12, 1971.* A transferee of merchandise entered for warehouse prior to January 12, 1971, shall have no right to file a protest, except under the conditions set forth in section 557(b), Tariff Act of 1930, as amended (19 U.S.C. 1557(b)), prior to the amendments made thereto by Pub. L. 91–685, effective January 12, 1971 (T.D. 71–55).

40 CFR §§721.90 and 721.91

(Reproduced from:

*www.access.gpo.gov/nara/cfr/waisidx_00/
40cfrv23_00.html*)

Part 721—Significant New Uses of Chemical Substances

§721.90   Release to water

Whenever a substance is identified in subpart E of this part as being subject to this section, a significant new use of the substance is:

(a) Any predictable or purposeful release of a manufacturing stream associated with any use of the substance, from any site:

(1) Into the waters of the United States.

(2) Into the waters of the United States without application of one or more of the following treatment technologies as specified in subpart E of this part either by the discharger or, in the case of a release through publicly-owned treatment works, by a combination of treatment by the discharger and the publicly-owned treatment works:

(i) Chemical precipitation and settling.

(ii) Biological treatment (activated sludge or equivalent) plus clarification.

(iii) Steam stripping.

(iv) Resin or activated carbon adsorption.

(v) Chemical destruction or conversion.

(vi) Primary wastewater treatment.

(3) Into the waters of the United States without primary wastewater treatment, and secondary wastewater treatment as defined in 40 CFR part 133.

(4) Into the waters of the United States if the quotient from the following formula:

$$\frac{\text{number of kilograms/day/site released}}{\text{receiving stream flow (million liters/day)}} \times 1000 = \text{N parts per billion}$$

exceeds the level specified in subpart E of this part when calculated using the methods described in §721.91. In lieu of calculating the above quotient, monitoring or alternative calculations may be used to predict the surface water concentration which will result from the intended release of the substance, if the monitoring procedures or calculations have been approved for such purpose by EPA. EPA will review and act on written requests to approve monitoring procedures or alternative calculations within 90 days after such requests are received. EPA will inform submitters of the disposition of such requests in writing, and will explain the reasons therefor when they are denied.

(b) Any predictable or purposeful release of a process stream containing the substance associated with any use of the substance from any site:

(1) Into the waters of the United States.

(2) Into the waters of the United States without application of one or more of the following treatment technologies as specified in subpart E of this part either by the discharger or, in the case of a release through publicly-owned treatment works, by a combination of treatment by the discharger and the publicly-owned treatment works:

(i) Chemical precipitation and settling.

(ii) Biological treatment (activated sludge or equivalent) plus clarification.

(iii) Steam stripping.

(iv) Resin or activated carbon adsorption.

(v) Chemical destruction or conversion.

(vi) Primary wastewater treatment.

(3) Into the waters of the United States without primary wastewater treatment, and secondary wastewater treatment as defined in 40 CFR part 133.

(4) Into the waters of the United States if the quotient from the following formula:

$$\frac{\text{number of kilograms/day/site released}}{\text{receiving stream flow (million liters/day)}} \times 1000 = \text{N parts per billion}$$

exceeds the level specified in subpart E of this part when calculated using the methods described in §721.91. In lieu of calculating the above quotient, monitoring or alternative calculations may be used to predict the surface water concentration which will result from the intended release of the substance, if the monitoring procedures or calculations have been approved for such purpose by EPA. EPA will review and act on written requests to approve monitoring procedures or alternative calculations within 90 days after such requests are received. EPA will inform submitters of the disposition of such requests in writing, and will explain the reasons therefor when they are denied.

(c) Any predictable or purposeful release of a use stream containing the substance associated with any use of the substance from any site:

(1) Into the waters of the United States.

(2) Into the waters of the United States without application of one or more of the following treatment technologies as specified in subpart E of this part either by the discharger or, in the case of a release through publicly-owned treatment works, by a combination of treatment by the discharger and the publicly-owned treatment works:

(i) Chemical precipitation and settling.

(ii) Biological treatment (activated sludge or equivalent) plus clarification.

(iii) Steam stripping.

(iv) Resin or activated carbon adsorption.

(v) Chemical destruction or conversion.

(vi) Primary wastewater treatment.

(3) Into the waters of the United States without primary wastewater treatment, and secondary wastewater treatment as defined in 40 CFR part 133.

(4) Into the waters of the United States if the quotient from:

$$\frac{\text{number of kilograms/day/site released}}{\text{receiving stream flow (million liters/day)}} \times 1000 = N \text{ parts per billion}$$

exceeds the level specified in subpart E of this part, when calculated using the methods described in §721.91. In lieu of calculating the above quotient, however, monitoring or alternative calculations may be used to predict the surface water concentration expected to result from intended release of the substance, if the monitoring procedures or calculations have been approved for such purpose by EPA. EPA will review and act on written requests to approve monitoring procedures or alternative calculations within 90 days after such requests are received. EPA will inform submitters of the disposition of such requests in writing, and will explain the reasons therefor when they are denied.

§721.91 Computation of estimated surface water concentrations: Instructions

These instructions describe the use of the equation specified in §721.90(a)(4) and (b)(4) to compute estimated surface water concentrations which will result from release of a substance identified in subpart E of this part. The equation shall be computed for each site using the stream flow rate appropriate for the site according to paragraph (b) of this section, and the highest number of kilograms calculated to be released for that site on a given day according to paragraph (a) of this section. Two variables shall be considered in computing the equation, the number of kilograms released, and receiving stream flow.

(a) *Number of kilograms released.* (1) To calculate the number of kilograms of substance to be released from manufacturing, processing, or use operations, as specified in the numerator of the equation, develop a process description diagram which describes each manufacturing, processing, or use operation involving the substance. The process description must include the major unit operation steps and chemical conversions. A unit operation is a functional step in a manufacturing, processing, or use operation where substances undergo chemical changes and/or changes in location, temperature, pressure, physical state, or similar characteristics. Include steps in which the substance is formulated into mixtures, suspensions, solutions, etc.

(2) Indicate on each diagram the entry point of all feedstocks (e.g., reactants, solvents, and catalysts) used in the operation. Identify each feedstock and specify its approximate weight regardless of whether the process is continuous or batch.

(3) Identify all release points from which the substance or wastes containing the substance will be released into air, land, or water. Indicate these release points on the diagram. Do not include accidental releases or fugitive emissions.

(4) For releases identified in the diagram that are destined for water, estimate the amount of substance that will be released before the substance enters control technology. The kilograms of substance released may be estimated based on:

(i) The mass balance of the operation, i.e., totaling inputs and outputs, including wastes for each part of the process such that outputs equal inputs. The amount released to water may be the difference between the amount of the substance in the starting material (or formed in a reaction) minus the amount of waste material removed from each part of the process and not released to water and the amount of the substance in the final product.

(ii) Physical properties such as water solubility where a known volume of water being discharged is assumed to contain the substance at concentrations equal to its solubility in water. This approach is particularly useful where the waste stream results from separation of organic/water phases or filtration of the substance from an aqueous stream to be discharged.

(iii) Measurements of flow rates of the process/use stream and known concentrations of the substance in the stream.

(5) After releases of a substance to water are estimated for each operation on a site, total the releases of the substance to water from all operations at that site. The value (number of kilograms) specified in the numerator of the equation should reflect total kilograms of substance released to water per day from all operations at a single site.

(6) Use the highest expected daily release of the substance for each site.

(b) *Receiving stream flow.* (1) The receiving stream flow shall be expressed in million liters per day (MLD). The flow rate data to be used must be for the point of release on the water body that first receives release of the substance whether by direct discharge from a site, or by indirect discharge through a Publicly-Owned Treatment Works (POTW) for each site. The flow rate reported shall be the lowest 7-day average stream flow with a recurrence interval of 10 years (7-Q-10). If the 7-Q-10 flow rate is not available for the actual point of release, the stream flow rate should be used from the U.S. Geological Survey (USGS) gauging station that is nearest the point of release that is expected to have a flow rate less than or equal to the receiving stream flow at the point of release.

(2) Receiving stream flow data may be available from the National Pollutant Discharge Elimination System (NPDES) permit for the site or the POTW releasing the substance to surface water, from the NPDES permitwriting authority for the site or the POTW, or from USGS publications, such as the water-data report series.

(3) If receiving stream flow data are not available for a stream, either the value of 10 MLD or the daily flow of wastewater from the site or the POTW releasing the substance must be used as an assumed minimum stream flow. Similarly, if stream flow data are not available because the location of the point of release of the substance to surface water is a lake, estuary, bay, or ocean, then the flow rate to be used must be the daily flow of wastewater from the site or the POTW releasing the substance to surface water. Wastewater flow data may be available from the NPDES permit or NPDES authority for the site or the POTW releasing the substance to water.

Appendix 7.5

Notice of Commencement Form (EPA Form 7710-56)

(Reproduced from:
www.epa.gov/opptintr/newchems/noc1.pdf)

O.M.B. No. 2070-0012   Approval Expires 10/31/96

	Agency Use Only	Date of Receipt

&EPA U.S. Environmental Protection Agency
NOTICE OF COMMENCEMENT OF
MANUFACTURE OR IMPORT (40 CFR§720.102)

Part I - SUBMITTER IDENTIFICATION   Document Control #:

Manufacturer/Importer (in U.S.)
Name of Authorized Official   Mailing Address (number and street)

Company Name   City, State, ZIP code   CBI

Technical Contact (in U.S.)
Name   Telephone Number   CBI

Part II - Premanufacture Notice (PMN) "P" Case Number:

Part III - Check the appropriate box and provide the exact date of manufacture or importation:

☐ First Commercial Manufacture**
date: _____

☐ First Commercial Importation***
date: _____

** Date of commencement is the date of completion of non-exempt manufacture of the first amount (batch, drum, etc.)
*** For importers, the date of commencement is the date that the new chemical substance clears U.S. customs.

Part IV - Manufacturing Plant Site(s) or Importing Site(s): (Importers, provide street address of destination)

CBI

Part V - Specific Chemical Identity: (For Consolidated submissions, each substance must have a separate NOC form with the specific identity of each chemical substance.)

CBI

Part VI - Generic Chemical Name (if chemical identity is claimed CBI*):

Part VII - Substance Identity Confidentiality Status:

☐ I wish to continue to claim the substance identity confidential and the substantiation to support this claim is attached. Failure to submit the required substantiation in accordance with 40 CFR 720.85(b) will result in a waiver of your claim.

☐ I previously claimed the substance identity as confidential and hereby relinquish that claim.

☐ I did not claim the substance identity as confidential in my original PMN submission.

You must submit your completed notice no later than 30 calendar days after the first date of commercial manufacture/importation to the address shown below:

U.S. Environmental Protection Agency
OPPT Document Control Office (7407)
401 M Street, S.W.
Washington, D.C. 20460
ATTN: Notice of Commencement

Signature of authorized official   Date

Note: CBI - refers to the term "Confidential Business Information". Mark (X) in the box if the information is to be held confidential.

EPA Form 7710-56 (8-95)

Appendix 7.6

Premanufacture Notice (PMN) Form (EPA Form 7710-25)

(Reproduced from:
*www.epa.gov/opptintr/newchems/pmnpart1.pdf
and www.epa.gov/opptintr/newchems/
pmnpart2.pdf*)

Form Approved. O.M.B. No. 2070-0012. Approval Expires 10-31-96.

U. S. ENVIRONMENTAL PROTECTION AGENCY	AGENCY USE ONLY

⊕EPA PREMANUFACTURE NOTICE

FOR NEW CHEMICAL SUBSTANCES

Date of receipt

When completed send this form to

DOCUMENT CONTROL OFFICER
OFFICE OF POLLUTION PREVENTION
AND TOXIC SUBSTANCES, 7407
U.S. E.P.A. 401 M STREET, SW
WASHINGTON, D.C. 20460

Enter the total number of pages in the Premanufacture Notice ▶

Document control number   EPA case number

GENERAL INSTRUCTIONS   TS-⬚⬚⬚⬚⬚⬚

- You must provide all information requested in this form to the extent that it is known to or reasonably ascertainable by you. Make reasonable estimates if you do not have actual data.
- Before you complete this form, you should read the "Instructions Manual for Premanufacture Notification" (the Instructions Manual is available from the Toxic Substances Control Act (TSCA) Information Service by calling 202-554-1404, or faxing 202-554-5603).
- If a user fee has been remitted for this notice (40 CFR 700.45), indicate in the boxes above the TS-user fee identification number you have generated. Remember, your user fee ID number must also appear on your corresponding fee remittance, which is sent to: EPA, HQ Accounting Operations Branch (PM-266), P.O. 360399M, Pittsburgh, PA 15251-6399, Attn. TSCA User Fee.

Part I – GENERAL INFORMATION

You must provide the currently correct Chemical Abstracts (CA) Name of the new chemical substance, even if you claim the identity as confidential. You may authorize another person to submit chemical identity information for you, but your submission will not be complete and the review will not begin until EPA receives this information. A letter in support of your submission should reference your TS user fee identification number. You must submit an original and two copies of this notice including all test data . If you claimed any information as confidential, a single sanitized copy must also be submitted.

Part II – HUMAN EXPOSURE AND ENVIRONMENTAL RELEASE

If there are several manufacture, processing, or use operations to be described in Part II, sections A and B of this notice, reproduce the sections as needed.

Part III – LIST OF ATTACHMENTS

Attach additional sheets if there is not enough space to answer a question fully. Label each continuation sheet with the corresponding section heading. In Part III, list these attachments, any test data or other data and any optional information included in the notice.

OPTIONAL INFORMATION

You may include any information that you want EPA to consider in evaluating the new substance. On page 11 of this form, space has been provided for you to describe pollution prevention and recycling information you may have regarding the new substance.

So-called "binding" boxes are included throughout this form for you to indicate your willingness to be bound to certain statements you make in this notice, such as use, production volume, protective equipment . . . This option is intended to reduce delays that routinely accompany the development of consent orders or Significant New Use Rules. Except in the case of exemption applications (such as TMEA, LVE, LOREX) where certain information provided in such notification is binding on the submitter when the Agency approves the exemption application, checking a binding box in this notice does not by itself prohibit the submitter from later deviating from the information (except chemical identity) reported in the form.

CONFIDENTIALITY CLAIMS

You may claim any information in this notice as confidential. To assert a claim on the form, mark (X) the confidential box next to the information that you claim as confidential. To assert a claim in an attachment, circle or bracket the information you claim as confidential. If you claim information in the notice as confidential, you must also provide a sanitized version of the notice, (including attachments). For additional instructions on claiming information as confidential, read the Instructions Manual.

⬚ Mark (X) if any information in this notice is claimed as confidential.

TEST DATA AND OTHER DATA

You are required to submit all test data in your possession or control and to provide a description of all other data known to or reasonably ascertainable by you, if these data are related to the health and environmental effects of the manufacture, processing, distribution in commerce, use, or disposal of the new chemical substance. Standard literature citations may be submitted for data in the open scientific literature. Complete test data (written in English), not summaries of data, must be submitted if they do not appear in the open literature. You should clearly identify whether test data is on the substance or on an analog. Also, the chemical composition of the tested material should be characterized. Following are examples of test data and other data. Data should be submitted according to the requirements of §720.50 of the Premanufacture Notification Rule (40 CFR Part 720).

Test Data   (Check Below any included in this notice)

- Environmental fate data ⬚ Yes
- Other data ⬚ Yes
- Health effects data ⬚ Yes
- Risk assessments
- Environmental effects data ⬚ Yes
- Structure/activity relationships
- Physical/Chemical Properties * ⬚ Yes
- Test data not in the possession or control of the submitter

*A physical and chemical properties worksheet is located on the last page of this form.

TYPE OF NOTICE   (Check Only One)

⬚ PMN (Premanufacture Notice)

⬚ INTERMEDIATE PMN (submitted in sequence with final product PMN)

⬚ SNUN (Significant New Use Notice)

⬚ TMEA (Test Marketing Exemption Application)

⬚ LVE (Low Volume Exemption) @ 40 CFR 723.50 (c)(1)

⬚ LOREX (Low Release/Low Exposure Exemption) @ 723.50(c)(2)

⬚ LVE Modification   ⬚ LOREX Modification

IS THIS A CONSOLIDATED PMN? ⬚ Yes

# of chemicals _____
(Prenotice Communication # required, enter # on page 3)

EPA FORM 7710-25   (Rev. 5-95)   Replaces previous editions of EPA Form 7710-25

Public reporting burden for this collection of information is estimated to average 110 hours per response, including time for reviewing instructions, searching existing data sources, gathering and maintaining the data needed, and completing and reviewing the collection of information. Send comments regarding the burden estimate or any other aspect of this collection of information, including suggestions for reducing this burden, to Chief, Information Policy Branch, PM-223, U.S. Environmental Protection Agency, 401 M. St., S.W., Washington, D.C. 20460; and to the Office of Management and Budget, Paperwork Reduction Act (2070-0012), Washington, D.C. 20503.

CERTIFICATION

I certify that to the best of my knowledge and belief:

1. The company named in Part I, section A, subsection 1a of this notice form intends to manufacture or import for a commercial purpose, other than in small quantities solely for research and development, the substance identified in Part I, Section B.

2. All information provided in this notice is complete and truthful as of the date of submission.

3. I am submitting with this notice all test data in my possession or control and a description of all other data known to or reasonably ascertainable by me as required by §720.50 of the Premanufacture Notification Rule.

Additional Certification Statements:

If you are submitting a PMN, Intermediate PMN, Consolidated PMN, or SNUN, check the following **user fee** certification statement that applies:

☐ The Company named in Part I, Section A has remitted the fee of $2500 specified in 40 CFR 700.45 (b), or

☐ The Company named in Part I, Section A has remitted the fee of $1000 for an Intermediate PMN (defined @ 40 CFR 700.43) in accordance with 40 CFR 700.45(b), or

☐ The Company named in Part I, Section A is a small business concern under 40 CFR 700.43 and has remitted a fee of $100 in accordance with 40 CFR 700.45 (b).

If you are submitting a **low volume exemption (LVE)** application in accordance with 40 CFR 723.50 (c) (1) or a **Low release and low exposure exemption (LoREX)** application in accordance with 40 CFR 723.50 (c) (2), check the following certification statements:

☐ The manufacturer submitting this notice intends to manufacture or import the new chemical substance for commercial purposes, other than in small quantities solely for research and development, under the terms of 40 CFR 723.50.

☐ The manufacturer is familiar with the terms of this section and will comply with those terms; and

☐ The new chemical substance for which the notice is submitted meets all applicable exemption conditions.

☐ If this application is for an LVE in accordance with 40 CFR 723.50 (c)(1), the manufacturer intends to commence manufacture of the exempted substance for commercial purposes within 1 year of the date of the expiration of the 30 day review period.

The accuracy of the statements you make in this notice should reflect your best prediction of the anticipated facts regarding the chemical substance described herein. Any knowing and willful misinterpretation is subject to criminal penalty pursuant to 18 USC 1001.

		Confidential
Signature and title of Authorized Official (Original Signature Required)	**Date**	
Signature of agent - (if applicable)	**Date**	

Part I -- GENERAL INFORMATION

Section A -- SUBMITTER IDENTIFICATION	Confidential

Mark (X) the "Confidential" box next to any subsection you claim as confidential.

1a. Person Submitting Notice (in U.S.)

Name of authorized official	Position
Company	
Mailing address (number and street)	
City, State, ZIP Code	

b. Agent (if applicable)

Name of authorized official	Position
Company	
Mailing address (number and street)	
City, State, ZIP Code	Telephone Area Code Number

c. If you are submitting this notice as part of a joint submission, mark (X) this box. ⟶ ☐

Joint Submitter (if applicable)

Name of authorized official	Position
Company	
Mailing address (number and street)	
City, State, ZIP Code	Telephone Area Code Number

2. Technical Contact (in U.S.)

Name	Position
Company	
Mailing address (number and street)	
City, State, ZIP Code	Telephone Area Code Number

3. If you have had a prenotice communication (PC) concerning this notice and EPA assigned a PC Number to the notice, enter the number. ⟶   Mark (X) if none ⟶ ☐

4. If you previously submitted an exemption application for the chemical substance covered by this notice, enter the exemption number assigned by EPA. If you previously submitted a PMN for this substance enter the PMN number assigned by EPA (i.e. withdrawn or incomplete). ⟶   Mark (X) if none ⟶ ☐

5. If you have submitted a notice of Bona fide intent to manufacture or import for the chemical substance covered by this notice, enter the notice number assigned by EPA. ⟶   Mark (X) if none ⟶ ☐

6. Type of Notice -- Mark (X)

1. ☐ Manufacture Only
   ☐ Binding Option Mark (x)

2. ☐ Import Only
   ☐ Binding Option Mark (x)

3. ☐ Both

FORM EPA 7710-25   (Rev. 5-95)   Page 3

Part I -- GENERAL INFORMATION -- Continued

► **Section B -- CHEMICAL IDENTITY INFORMATION:** You must provide a currently correct Chemical Abstracts (CA) name of the substance based on the ninth Collective Index (9CI) of CA nomenclature rules and conventions.

Mark (X) the "Confidential" box next to any item you claim as confidential.

Complete either item 1 (Class 1 or 2 substances) or 2 (Polymers) as appropriate. Complete all other items.

If another person will submit chemical identity information for you (for either item 1 or 2), mark (X) the box at the right. Identify the name, company, and address of that person in a continuation sheet. → ☐ | Confi-dential

1. **Class 1 or 2 chemical substances** (for definitions of class 1 and class 2 substances, see the **Instructions Manual**)

   a. Class of substance -- Mark (X)   1 ☐   Class 1   **or**   2 ☐   Class 2

   b. Chemical name (Currently correct Chemical Abstracts (CA) Name that is consistent with TSCA Inventory listings for similar substances. For Class 1 substances a CA Index Name must be provided. For Class 2 substances either a CA Index Name or CA Preferred Name must be provided, whichever is appropriate based on CA 9CI nomenclature rules and conventions.).

   c. Please identify which method you used to develop or obtain the specified chemical indentity information reported in this notice: (check one).

   ☐ Method 1 (CAS Inventory Expert Service - a copy of the Identification report obtained from the CAS Inventory Expert Service must be submitted as an attachment to this notice)   ☐ Method 2 (Other Source)

   d. Molecular formula and CAS Registry Number (if a number already exists for the substance)

CAS #

   e. For a **class 1** substance, provide a complete and correct chemical structure diagram. For a **class 2** substance -- (1) List the immediate precursor substances with their respective CAS Registry Numbers. (2) Describe the nature of the reaction or process. (3) Indicate the range of composition and the typical composition (where appropriate). (4) Provide a correct representative or partial chemical structure diagram, as complete as can be known, if one can be reasonably ascertained.

☐ Mark (X) this box if you attach a continuation sheet.

Part I -- GENERAL INFORMATION -- Continued

▶ **Section B -- CHEMICAL IDENTITY INFORMATION -- Continued**

2. Polymers (For a definition of polymer, see the **Instructions Manual**.)   [Confidential]

   a. Indicate the **number-average weight** of the lowest molecular weight composition of the polymer you intend to manufacture. Indicate **maximum weight percent** of low molecular weight species (not including residual monomers, reactants, or solvents) below 500 and below 1,000 absolute molecular weight of that composition.

     Describe the methods of measurement or the basis for your estimates: GPC ☐ Other ☐ : (Specify) _____

      i) **lowest** number average molecular weight: _____

      ii) maximum weight % below 500 molecular weight: _____

      iii) maximum weight % below 1000 molecular weight: _____

   ☐ Mark (X) this box if you attach a continuation sheet.

   b. You must make separate confidentiality claims for monomer or other reactant identity, composition information, and residual information. Mark (X) the "Confidential" box next to any item you claim as confidential.

     (1) -- Provide the **specific** chemical name and CAS Registry Number (if a number exists) of each monomer or other reactant used in the manufacture of the polymer.

     (2) -- Mark (X) this column if entry in column (1) is confidential.

     (3) -- Indicate the **typical** weight percent of each monomer or other reactant in the polymer.

     (4) -- **Mark (X) the identity** column if you want a monomer or other reactant used at two weight percent or less to be listed as part of the **polymer description on the TSCA Chemical Substance Inventory.**

     (5) -- Mark (X) this column if entries in columns (3) and (4) are confidential.

     (6) -- Indicate the **maximum** weight percent of each monomer or other reactant that may be present as a residual in the polymer as manufactured for commercial purposes.

     (7) -- Mark (X) this column if entry in column (6) is confidential.

Monomer or other reactant and CAS Registry Number (1)	Confidential (2)	Typical composition (3)	Identity Mark (X) (4)	Confidential (5)	Maximum residual (6)	Confidential (7)
		%			%	
		%			%	
		%			%	
		%			%	
		%			%	
		%			%	
		%			%	

   ☐ Mark (X) this box if you attach a continuation sheet.

   c. Please identify which method you used to develop or obtain the specified chemical identity information reported in this notice. (check one).
     ☐ Method 1 (CAS Inventory Expert Service - a copy of the identification report obtained from CAS Inventory Expert Service must be submitted as an attachment to this notice)   ☐ Method 2 (other source)

   d. The currently correct Chemical Abstracts (CA) name for the polymer that is consistent with TSCA Inventory listings for similar polymers.

   e. Provide a correct representative or partial chemical structure diagram, as complete as can be known, if one can be reasonably ascertained.

   ☐ Mark (X) this box if you attach a continuation sheet.

FORM EPA 7710-25 (Rev. 5-95)   Page 5

Part I — GENERAL INFORMATION -- Continued

▶ **Section B – CHEMICAL IDENTITY INFORMATION – Continued**

3. Impurities
(a) – Identify each impurity that may be reasonably anticipated to be present in the chemical substance as manufactured for commercial purposes.
Provide the CAS Registry Number if available. If there are unidentified impurities, enter "unidentified."
(b) –Estimate the maximum weight % of each impurity. If there are unidentified impurities, estimate their total weight %.

Impurity and CAS Registry Number (a)	Maximum percent (b)	Confidential
	%	
	%	
	%	
	%	
	%	
	%	
	%	

☐ Mark (X) this box if you attach a continuation sheet.

4. Synonyms – Enter any chemical synonyms for the new chemical substance identified in subsection 1 or 2.	Confidential

☐ Mark (X) this box if you attach a continuation sheet.

5. Trade identification – List trade names for the new chemical substance identified in subsection 1 or 2.	

☐ Mark (X) this box if you attach a continuation sheet.

6. **Generic chemical name** -- If you claim chemical identity as confidential, you must provide a generic chemical name for your substance
that reveals the specific chemical identity of the new chemical substance to the maximum extent possible.
Refer to the TSCA Chemical Substance Inventory, 1985 Edition, Appendix B for guidance on developing
generic names.

☐ Mark (X) this box if you attach a continuation sheet.

7. **Byproducts** – Describe any byproducts resulting from the manufacture, processing, use, or disposal of the new chemical substance. Provide the CAS
Registry Number if available.

Byproduct (1)	CAS Registry Number (2)	Confidential

☐ Mark (X) this box if you attach a continuation sheet.

Part I -- GENERAL INFORMATION -- Continued

▶ **Section C -- PRODUCTION, IMPORT, AND USE INFORMATION:**

Mark (X) the "Confidential" box next to any item you claim as confidential.

1. **Production volume** -- Estimate the **maximum** production volume during the first 12 months of production. Also estimate the maximum production volume for any consecutive 12-month period during the first three years of production. Estimates should be on 100% new chemical substance basis. *For a Low Volume Exemption application, if you choose to have your notice reviewed at a lower production volume than 10,000 kg/yr, specify the volume and mark (x) in the binding box. If granted, you are bound to this volume..*

Maximum first 12-month production (kg/yr) (100% new chemical substance basis)	Maximum 12-month production (kg/yr) (100% new chemical substance basis)	Confidential	Binding Option Mark (x)

2. **Use Information** -- You must make separate confidentiality claims for the description of the category of use, the percent of production volume devoted to each category, the formulation of the new substance, and other use information. Mark (X) the "Confidential" Box next to any item you claim as confidential.
   a. (1) -- Describe each intended category of use of the new chemical substance by function and application.
   (2) -- Mark (X) this column if entry in column (1) is confidential business information (CBI).
   (3) -- Indicate your willingness to have the information provided in column (1) binding.
   (4) -- Estimate the percent of total production for the first three years devoted to each category of use.
   (5) -- Mark (X) this column if entry in column (4) is confidential business information (CBI).
   (6) -- Estimate the percent of the new substance as formulated in mixtures, suspensions, emulsions, solutions, or gels as manufactured for commercial purposes at sites under your control associated with each category of use.
   (7) -- Mark (X) this column if entry in column (6) is confidential business information (CBI).
   (8) -- Indicate % of product volume expected for the listed "use" sectors. Mark more than one box if appropriate. Mark (X) to indicate your willingness to have the use type provided in (8) binding.
   (9) -- Mark (X) this column if entry(ies) in column (8) is (are) confidential business information (CBI).

Category of use (1) (by function and application i.e. a dispersive dye for finishing polyester fibers)	CBI (2)	Binding Option Mark (x) (3)	Production % (4)	CBI (5)	% in Formulation (6)	CBI (7)	% of substance expected per use (8)					CBI (9)
							Site-limited	Con-* sumer	Indus-trial	Com-mercial	Binding Option	
			%		%							
			%		%							
			%		%							
			%		%							
			%		%							
			%		%							

*If you have identified a "consumer" use, please provide on a continuation sheet a detailed description of the use(s) of this chemical substance in consumer products. In addition include estimates of the concentration of the new chemical substance as expected in consumer products and describe the chemical reactions by which this substance loses its identity in the consumer product.

☐ Mark (X) this box if you attach a continuation sheet.

b. Generic use description

If you claim any category of use description in subsection 2a as confidential, enter a generic description of that category. Read the **Instructions Manual** for examples of generic use descriptions.

☐ Mark (X) this box if you attach a continuation sheet.

3. **Hazard Information** -- Include in the notice a copy of reasonable facsimile of any hazard warning statement, label, material safety data sheet, or other information which will be provided to any person who is reasonably likely to be exposed to this substance regarding protective equipment or practices for the safe handling, transport, use, or disposal of the new substance. List in part III hazard information you include.

Binding Option Mark (x)

☐ Mark (X) this box if you attach hazard information.

Part II -- HUMAN EXPOSURE AND ENVIRONMENTAL RELEASE

▶**Section A -- INDUSTRIAL SITES CONTROLLED BY THE SUBMITTER** Mark (X) the "Confidential" box next to any item you claim as confidential.

Complete section A for each type of manufacture, processing, or use operation involving the new chemical substance at industrial sites you control. Importers do not have to complete this section for operations outside the U.S. ; however, you may still have reporting requirements if there are further industrial processing or use operations after import. You must describe these operations. See instructions manual.

1. Operation description	Confidential
a. Identity -- Enter the identity of the site at which the operation will occur.	

Name

Site address (number and street)

City, County, State, ZIP Code

If the same operation will occur at more than one site, enter the number of sites. Identify the additional sites on a continuation sheet, and if any of the sites have significantly different production rates or operations, include all the information requested in this section for those sites as attachments. ⟶	# of sites	

[ ] Mark (X) this box if you attach a continuation sheet.

b. Type -- Mark (X) [ ] Manufacturing   [ ] Processing   [ ] Use

c. Amount and Duration -- Complete 1 or 2 as appropriate

1. Batch	Maximum kg/batch (100 % new chemical substance)	Hours/batch	Batches/year
2. Continuous	Maximum kg/day (100 % new chemical substance)	Hours/day	Days/year

d. Process description [ ] Mark (X) to indicate your willingness to have your process description binding.

(1) Diagram the major unit operation steps and chemical conversions. Include interim storage and transport containers (specify-e.g. 5 gallon pails, 55 gallon drum, rail car, tank truck, etc.).
(2) Provide the identity, the approximate weight (by kg/day or kg/batch on an 100% new chemical substance basis), and entry point of all starting materials and feedstocks (including reactants, solvents, and catalysts, etc.), and of all products, recycle streams, and wastes. Include cleaning chemicals (note frequency if not used daily or per batch.).
(3) Identify by number the points of release, including small or intermittant releases, to the environment of the new chemical substance.

[ ] Mark (X) this box if you attach a continuation sheet.

Part II -- HUMAN EXPOSURE AND ENVIRONMENTAL RELEASE -- Continued

▶ Section A -- INDUSTRIAL SITES CONTROLLED BY THE SUBMITTER -- Continued

2. **Occupational Exposure** -- You must make separate confidentiality claims for the description of worker activity, physical form of the new chemical substance, number of workers exposed, and duration of activity. Mark (X) the "Confidential" box next to any item you claim as confidential.
   (1) -- Describe the activities (e.g. bag dumping, tote filling, unloading drums, sampling, cleaning, etc.) in which workers may be exposed to the substance.
   (2) -- Mark (X) this column if entry in column (1) is confidential business information (CBI).
   (3) -- Describe any protective equipment and engineering controls used to protect workers.
   (4) and (6) -- Indicate you willingness to have the information provided in column (3) or (5) binding.
   (5) -- Indicate the physical form(s) of the new chemical substance (e.g. solid: crystal, granule, powder, or dust) and % new chemical substance (if part of a mixture) at the time of exposure.
   (7) -- Mark (X) this column if entry in column (5) is confidential business information (CBI).
   (8) -- Estimate the maximum number of workers involved in each activity for all sites combined.
   (9) -- Mark (X) this column if entry in column (8) is confidential business information (CBI).
   (10) and (11) -- Estimate the maximum duration of the activity for any worker in hours per day and days per year.
   (12) -- Mark (X) this column if entries in columns (10) and (11) are confidential business information (CBI).

Worker activity (e.g. bag dumping, filling drums) (1)	CBI (2)	Protective Equipment/ Engineering Controls (3)	Binding Option Mark (x) (4)	Physical form(s) (e.g.solid:powder) and % new substance (5)	Binding Option Mark (x) (6)	CBI (7)	# of Workers Exposed (8)	CBI (9)	Maximum Hrs/day (10)	duration Days/yr (11)	CBI (12)

☐ Mark (X) this box if you attach a continuation sheet.

3. **Environmental Release and Disposal** -- You must make separate confidentiality claims for the release number and the amount of the new chemical substance released and other release and disposal information. Mark (X) the "Confidential" box next to each item you claim as confidential.
   (1) -- Enter the number of each release point identified in the process description, part II, section A, subsection 1d(3).
   (2) -- Estimate the amount of the new substance released (a) directly to the environment or (b) into control technology (in kg/day or kg/batch).
   (3) -- Mark (X) this column if entries in columns (1) and (2) are confidential business information (CBI).
   (4) -- Identify the media of release i.e. stack air, fugitive air (optional-see Instruction Manual), surface water, on-site or off-site land or incineration, POTW, or other (please specify) to which the new substance will be released from that release point.
   (5) -- a. Describe control technology, if any, ,and control efficiency that will be used to limit the release of the new substance to the environment. For releases disposed of on land, characterize the disposal method and state whether it is approved for disposal of RCRA hazardous waste. On a continuation sheet , for each site describe any additional disposal methods that will be used and whether the waste is subject to secondary or tertiary on-site treatment. b. Estimate the amount released to the environment after control technology (in kg/day).
   (6) -- Mark (X) this column if entries in columns (4) and (5) are confidential business information (CBI).
   (7) -- Identify the destination(s) of releases to water. Please supply NPDES (National Pollutant Discharge Elimination System) numbers for direct dischargers or NPDES numbers of the POTW (Publicly Owned Treatment Works). Mark (X) if the POTW name or NPDES # is confidential business information (CBI).

Release Number (1)	Amount of new substance released (2a)	(2b)	CBI (3)	Media of release e.g. stack air (4)	Control technology and efficiency (you may wish to optionally attach efficiency data) (5a)	Binding Mark (x)	(5b)	CBI (6)

(7) Mark (X) the destination(s) of releases to water.	☐ POTW provide name(s) below:	CBI ☐	Navigable waterway	☐ Other - Specify	provide **NPDES #**	CBI

☐ Mark (X) this box if you attach a continuation sheet.

Part II -- HUMAN EXPOSURE AND ENVIRONMENTAL RELEASE -- Continued

▶ **Section B -- INDUSTRIAL SITES CONTROLLED BY OTHERS**

Complete section B for typical processing or use operations involving the new chemical substance at sites you do not control. Importers do not have to complete this section for operations outside the U.S.; however, you must report any processing or use activities after import. See the Instructions Manual. *Complete a separate section B for each type of processing, or use operation involving the new chemical substance.* If the same operation is performed at more than one site describe the **typical** operation common to these sites. Identify additional sites on a continuation sheet.

1. **Operation Description -** To claim information in this section as confidential, circle or bracket the specific information that you claim as confidential

   (1) -- Diagram the major unit operation steps and chemical conversions, including interim storage and transport containers (specify- e.g. 5 gallon pails, 55gallon drums, rail cars, tank trucks, etc). On the diagram, identify by letter and briefly describe each worker activity. **(2)** -- Provide the identity, the approximate weight (by kg/day or kg/batch, on an 100% new chemical substance basis), and entry point of all feedstocks (including reactants, solvents and catalysts, etc) and of all products, recycle streams, and wastes. Include cleaning chemicals (note frequency if not used daily or per batch). **(3)** -- Identify by number the points of release, including small or intermittent releases, to the environment of the new chemical substance.

   (4) Please enter the # of sites (remember to identify the locations of these sites on a continuation sheet) :

___# of sites___

☐ Mark (X) this box if you attach a continuation sheet.

2. **Worker Exposure/Environmental Release**

   (1) -- From the diagram above, provide the letter for each worker activity. Complete 2-8 for each worker activity described.
   (2) -- Estimate the number of workers exposed for all sites combined.
   (4) -- Estimate the typical duration of exposure per worker in (a) hours per day and (b) days per year.
   (6) -- Describe physical form of exposure and % new chemical substance (if in mixture), and any protective equipment and engineering controls used to protect workers.
   (7) -- Estimate the percent of the new substance as formulated when packaged or used as a final product.
   (9) -- From the process diagram above, enter the number of each release point. Complete 9-13 for each release point identified.
   (10) -- Estimate the amount of the new substance released (a) directly to the environment or (b) into control technology to the environment (in kg/day or kg/batch).
   (12) -- Describe media of release i.e. stack air, fugitive air (optional-see Instructions Manual), surface water, on-site or off-site land or incineration, POTW, or other (specify) and control technology, if any, that will be used to limit the release of the new substance to the environment.
   (14) -- Identify byproducts which may result from the operation.
   (3), (5), (8), (11), (13) and (15) - Mark (X) this column if any of the proceeding entries are confidential business information (CBI).

Letter of Activity	# of Workers Exposed	CBI	Duration of Exposure		CBI	Protective Equip. / Engineering Controls/ Physical Form and % new substance	% in Form-ulation	CBI	Release Number	Amount of New Substance Released		CBI	Media of Release & Control Technology	CBI
(1)	(2)	(3)	(4a)	(4b)	(5)	(6)	(7)	(8)	(9)	(10a)	(10b)	(11)	(12)	(13)
(14) -- Byproducts:														(15)

☐ Mark (X) this box if you attach a continuation sheet.

OPTIONAL POLLUTION PREVENTION INFORMATION

To claim information in this section as confidential circle or bracket the specific information that you claim as confidential.

In this section you may provide information not reported elsewhere in this form regarding your efforts to reduce or minimize potential risks associated with activities surrounding manufacturing, processing, use and disposal of the PMN substance. Please include new information pertinent to pollution prevention, including source reduction, recycling activities and safer processes or products available due to the new chemical substance. Source reduction includes the reduction in the amount or toxicity of chemical wastes by technological modification, process and procedure modification, product reformulation, raw materials substitution, and/or inventory control. Recycling refers to the reclamation of useful chemical components from wastes that would otherwise be treated or released as air emissions or water discharges, or land disposal. Descriptions of pollution prevention, source reduction and recycling should emphasize potential risk reduction subsequent to compliance with existing regulatory requirements and can be either quantitative or qualitative. The EPA is interested in this information to assess <u>overall net</u> reductions in toxicity or environmental releases and exposures, not the shifting of risks to other environmental media or non-environmental areas (e.g., occupational or consumer exposure). In addition, information on the relative cost or performance characteristics of the PMN substance to potential alternatives may be provided. **All information provided in this section will be taken into consideration during the review of this substance. See the revised Instructions Manual that includes a Pollution Prevention manual for guidance and examples.**

Describe the expected net benefits, such as (1) an overall reduction in risk to human health or the environment; (2) a reduction in the volume manufactured; (3) a reduction in the generation of waste materials through recycling, source reduction or other means; (4) a reduction in potential toxicity or human exposure and/or environmental release; (5) an increase in product performance, a decrease in the cost of production and/or improved operation efficiency of the new chemical substance in comparison to existing chemical substances used in similar applications; or (6) the extent to which the new chemical substance may be a substitute for an existing substance that poses a greater overall risk to human health or the environment.

☐ Mark (X) this box if you attach a continuation sheet.

FORM EPA 7710-25   (Rev. 5-95)   Page 11

Part III -- LIST OF ATTACHMENTS

Attach continuation sheets for sections of the form and test data and other data (including physical/chemical properties and structure/activity information), and optional information after this page. Clearly identify the attachment and the section of the form to which it relates, if appropriate. Number consecutively the pages of the attachments. In the column below, enter the inclusive page numbers of each attachment.

Mark (X) the "Confidential" box next to any attachment **name** you claim as confidential. Read the **Instructions Manual** for guidance on how to claim any information in an attachment as confidential. You must include with the sanitized copy of the notice form a sanitized version of any attachment in which you claim information as confidential.

Attachment name	Attachment page number(s)	Confi-dential
Material Safety Data Sheet (MSDS)		

☐ Mark (X) this box if you attach a continuation sheet. Enter the attachment name and number.

PHYSICAL AND CHEMICAL PROPERTIES WORKSHEET

To assist EPA's review of physical and chemical properties data, please complete the following worksheet for data you provide and include it in the notice. Identify the property measured, the page of the notice on which the property appears, the value of the property, the units in which the property is measured (as necessary), and whether or not the property is claimed as confidential. The physical state of the neat substance should be provided. These measured properties should be for the neat (100% pure) chemical substance. Properties that are measured for mixtures or formulations should be so noted (% PMN substance in _). You are not required to submit this worksheet; however, EPA strongly recommends that you do so, as it will simplify review and ensure that confidential information is properly protected. You should submit this worksheet as a supplement to your submission of test data. This worksheet is not a substitute for submission of test data.

Property (a)	Mark (X) if provided	Page number (b)	Value (c)	Measured or Estimate (M or E)	Confidential Mark (X) (d)
Physical state of neat substance			___ (s) ___ (l) ___ (g)		
Vapor pressure ___ °C @ Temperature			Torr		
Density/relative density			g/cm3		
Solubility ___ °C @ Temperature Solvent ___			g/L		
Solubility in water @ Temperature ___ °C			°C		
Melting temperature			°C		
Boiling/sublimation temperature @ ___ torr pressure					
Spectra					
Dissociation constant					
Particle size distribution					
Octanol/water partition coefficient					
Henry's Law constant					
Volitalization from water					
Volitalization from soil					
pH @ concentration ___					
Flammability					
Explodability					
Adsorption/coefficient					
Other - Specify					

Compliance Verification Form (CVF)

Compliance Verification Form

Product Name _____

Company Name _____

Address _____

Tel. No. _____ Fax No. _____

1. Chemical Composition: Please provide composition information in the table below (Use additional sheet if necessary) and sign at the end of item 1.

    If your product composition is company confidential in its entirety, please initial here _____, and respond to item 2 below. If your product composition is partly company confidential, please fill out the non-confidential portion of it and respond to item 2 below. Thank you.

Chemical Name	CAS Number	Weight %	Impurity? (Y/N)

Does the composition account for 100% of the product? ☐Yes ☐No
Have you identified all intentional components of your product? ☐Yes ☐No

Signature _____ Date _____

Name _____ Title _____

2. If you claimed product composition either partly or in its entirety as company confidential, please review the compliance statement below and sign. Thank you.

"I certify that all components of the product named above meet the current Inventory Requirements of the U.S. EPA's Toxic Substances Control Act (TSCA) and the use of the product is not subject to the Significant New Use Rule (SNUR)".

Signature _____ Date _____

Name _____ Title _____

Please return to: Name: Fax. No
                            Address:

PMN and NOC Instruction Manual

(Reproduced from:
www.epa.gov/opptintr/newchems/tscaman2.pdf)

This draft for the Manual is posted for submitters' convenience. The Manual has been through internal review in EPA's New Chemicals Program, but has not completed OMB review. It is being posted on our Web page to assist submitters, but will not be published until OMB review is completed.

INSTRUCTION MANUAL FOR REPORTING UNDER THE TSCA §5 NEW CHEMICALS PROGRAM

How this Manual is organized:

I. GENERAL INSTRUCTIONS FOR REPORTING UNDER THE TOXIC SUBSTANCES CONTROL ACT §5 NEW CHEMICALS PROGRAM— PREMANUFACTURE NOTICE FORM

A. **Substances Which Must Be Reported ("New" Chemicals)**
B. **Who Must Submit a Premanufacture Notice**
C. **Substances Excluded from Notification**
D. **When to Submit a Notice**
E. **Filling out and Submitting the Form**

Note: Per the USEPA, the document is provided in the draft form for the sole purpose of keeping the submitters up to date with the available information on the subject.

F. Binding Boxes
G. Test Data and Other Data
H. Confidentiality
I. Consolidated Notices
J. Submission of Information by Others
K. Consultation with EPA Concerning the Premanufacture Notice
L. Notice of Commencement (NOC) of Manufacture (or Import) (40 CFR §720.102)
M. Recordkeeping
N. Recognition of Pollution Prevention and Recycling Benefits

II. PAGE-BY-PAGE INSTRUCTIONS FOR COMPLETING THE TSCA §5 NEW CHEMICALS PROGRAM PREMANUFACTURE NOTICE FORM

1. Type of Notice (Page 1)
2. Certification (Page 2)

Part I—GENERAL INFORMATION

Section A—Submitter Identification (Page 3)
Section B—Chemical Identity Information (Pages 4–6)

1. Class 1 or Class 2 chemical substances (Page 4)
2. Polymers (Page 5)
3. Impurities (Page 6)
4. Synonyms
5. Trade identification
6. Generic chemical name
7. By-products—

Section C—Production, Import, & Use Information (Page 7)

1. Production volume
2. Use information
3. Hazard information

Part II—HUMAN EXPOSURE AND ENVIRONMENTAL RELEASE (Pages 8–10)

Section A—Industrial Sites Controlled by the Submitter (Pages 8–9)
Section B—Industrial Sites Controlled by Others (Page 10)

(OPTIONAL) POLLUTION PREVENTION AND RECYCLING INFORMATION (Page 11)—Technical Information

a. Process chemistry
b. Product substitution/source reduction
c. Process modifications
d. Operating practices
h. Alternative waste disposal methods
i. Toxicity information

Part III—LIST OF ATTACHMENTS (Page 12)
Physical and Chemical Properties Worksheet (Optional—page 13)

Appendix A: Types of test data for submission
Appendix B: Annotated version of form 7710-25 distributed internally by the 3M Co.
Appendix C: Contact persons

I. GENERAL INSTRUCTIONS FOR REPORTING UNDER THE TSCA §5 NEW CHEMICALS PROGRAM—PREMANUFACTURE NOTICE FORM

A. Substances Which Must Be Reported ("New" Chemicals)

1. General—You are responsible for determining whether a substance you intend to manufacture or import is a "new" chemical substance as defined by the Toxic Substances Control Act (TSCA, the Act) and 40 Code of Federal Regulations (CFR) §720.3. You must submit a premanufacture notice (PMN) if you intend to manufacture (import is considered manufacture) any new chemical substance which is not on the TSCA Inventory or otherwise excluded from notification, as discussed below. To submit a PMN, you must use Form EPA 7710-25. Part II of this Manual is a step-by-step guide to form 7710-25 (rev 5-95).

2. Bona fide request for a TSCA Inventory search—The specific identities of some chemical substances on the Inventory are confidential and therefore do not appear on the Inventory available to the public. Such substances are described by generic names in the Appendix to the Inventory. If a substance you intend to manufacture or import is not on the published Inventory but falls within one of the generic categories in the Appendix, you may request that EPA search the Inventory's confidential file. EPA will search the confidential file only if you can demonstrate a bona fide intent to manufacture or import the substance. This policy is to ensure that this search procedure cannot be used for, essentially, industrial espionage.

The procedure for demonstrating such a bona fide intent is codified at 40 CFR §720.25. Certain information must be submitted with a bona fide request: an infrared spectrum must be supplied unless this analysis is not suitable for the particular substance, in which case a spectrum or instrument readout from a more appropriate method must be submitted; a currently correct Chemical Abstracts (CA) Index name or CA preferred name, whichever is appropriate; a currently correct Chemical Abstracts Service (CAS) register number (CASRN) (if the substance already has a CASRN assigned to it); molecular formula and a complete or partial chemical structure diagram if known or reasonably ascertainable; and a description of R&D activities that have already been conducted (include, for example, years research conducted, end use application, toxicity data, etc.). Three additional information requirements established in the PMN Rule Amendments, published March 29, 1995 (60 Federal Register (FR) 60 pp. 16298–16351) for a TSCA Inventory search include: the most probable manufacturing site, "major intended application or use" of the substance, and the approximate date when the submitter would be likely to submit a §5 notice for the substance if it is not found in the Inventory. If the substance is being imported, a statement should include: a) how long the substance has been used outside of the U.S., b) name of the country(ies) in which the substance is being used, and c) whether the substance has been used outside the U.S. for the same use as that intended after proposed importation. No specific form is required to be used.

The address to which to send bonafide submissions (by US mail) is:

OPPT Document Control Officer
Mail Stop 7407
USEPA 401 M Street, South West
Washington, DC 20460

The room at which bonafides are accepted is only open until 4 PM, and if a courier service comes after that time it will be turned away. You need the physical address for sending a *bona fide* by courier. You can give the courier the phone number (202) 260–1768 to call if there are delivery problems:

OPPT Document Control Officer
Mail Stop 7407
TSCA Data Processing Center
G099 East Tower, Waterside Mall
USEPA 401 M Street, South West
Washington, DC 20460

After conducting its search, EPA will tell you if the substance is included on the Inventory and therefore not subject to premanufacture notification or if you must submit a PMN.

B. Who Must Submit a PMN

If you intend to manufacture or import a new chemical substance for a commercial purpose, you must submit a PMN to EPA. You must submit a notice if you intend to import a new substance in bulk form or as part of a mixture, but not if you intend to import the substance only as part of an article. The use of the term "manufacture" in this manual includes both manufacture and import. Importers must fully comply with the information requirements outlined at 40 CFR §720. However, importers are not required, under §720.50(d)(3), to submit any data which relates solely to exposure to humans or the environment outside the United States. Importers must submit non-exposure data such as data on health effects (including epidemiological studies), ecological effects, physical and chemical properties, or environmental fate characteristics and (on sites under their control within the United States) exposure information.

"Article" is defined at 40 CFR §720.3 as a manufactured item which: (1) is formed to a specific shape or design during manufacture; (2) has an end use function(s) dependent in whole or in part upon its shape or design during end use; and (3) either has no change of chemical composition during its end-use or only those changes in composition which have no commercial purpose separate from the article of which it is a part and that may occur as described in 40 CFR §710.4(d)(5) and 40 CFR §720.30(h)(5). Articles are excluded from PMN requirements. Fluids and particles do not meet the definition of an article and are therefore not excluded from inventory reporting requirements. Therefore, all particles or fluids must be reported for the purposes of TSCA unless they can be considered mixtures. Also, OPPT will consider items being imported to be "articles" only if they are manufactured in a specific shape or design for a particular end use application, and this design is maintained as an essential feature in the finished product.

PMNs for imported new chemical substances should be submitted by the principal importer. "Principal importer" is defined at 40 CFR §720.3(z). It is not necessarily the same as "Importer of Record" under customs regulations.

Generally, when you contract with another person to manufacture a new chemical substance, that person must submit the notice. However, if you request another person to manufacture a new chemical substance, and if you specify the identity and total amount of the substance to be manufactured and the basic technology and controls under which the substance will be produced, and if that person manufactures the substance exclusively for you, that person is considered a "toll manufacturer", and you must submit the notice. Information regarding human exposure and environmental release should be submitted on EPA form 7710.25 in Part II, Section A, Industrial Sites Controlled by the Submitter. EPA recognizes that in this and similar instances, the other manufacturer may have information useful to the Agency's review of the new chemical. Therefore, EPA strongly encourages joint submission in these situations.

This manual does not discuss biotechnology submission information requirements for uses of microorganisms subject to the Toxic Substances Control Act (TSCA). For specific information on submitting notices for biotechnology products please contact the Biotechnology Program in the New Chemicals Notice Management Branch:

Biotechnology Program
New Chemicals Notice Management Branch
Chemical Control Division
Mail Stop 7405
USEPA, 401 M Street, South West
Washington, DC 20460

For additional information on who must submit a notice, see 40 CFR §720.22 or consult a Prenotice Coordinator. Prenotice Coordinators (see Contact List, Appendix C) are staff in the New Chemicals Program who specialize in assisting with status questions and questions on how to properly complete the notifications.

C. Substances Excluded from Notification

1. Statutory Excluded Categories—§3(b) of the TSCA excludes certain substances from premanufacture notification. These include mixtures (individual substances comprising the mixtures are NOT exempted), substances manufactured solely for use as pesticides, food, food additives, drugs, or cosmetics; tobacco and tobacco products; nuclear source materials; firearms and ammunition; impurities; byproducts which have no commercial use; non-isolated intermediates; and new chemical substances manufactured solely for export. Statutory exclusions are covered also at 40 CFR §720.3(e) and (u) and through criteria at §§720.30(h)(3)–(h)(7).

2. Research and development (R&D) exemption—R&D includes synthesis of new chemical substances for analysis, experimentation, or research on new or existing chemical substances, including product development activities. R&D may include tests of the physical, chemical, production, and performance characteristics of a substance.

You do not have to submit a notice for a new substance manufactured or imported in small quantities solely for research and development as specified in 40 CFR §720.36. "Small quantities" are those not greater than reasonably necessary for research and development purposes. The quantity which is reasonable may vary depending on the nature of the research and development activities. It is your responsibility to determine what is reasonable in your situation. You do not have to apply for this exemption. However, you must submit a PMN 90 days before you intend to manufacture the substance for a purpose other than research and development undertaken in compliance with §720.36.

To qualify for the exemption, your research and development activities must be conducted under the supervision of a technically qualified individual. Persons who engage in R&D for, or obtain an R&D chemical from, a manufacturer must be notified of any risk to health which may be associated with the chemical. However, R&D conducted entirely in laboratories under prudent laboratory practices is exempted from the requirement for risk evaluation.

In accordance with 40 CFR §720.78, the following R&D records must be retained (also see I-M, Recordkeeping): information reviewed and evaluated to determine the need to make any notification of risk, documentation of the nature and method of risk notification, documentation of prudent lab practices, if used instead of risk notification and evaluation and, if an R&D substance is manufactured at greater than 100 kg/yr, records regarding the chemical identity of the substance to the extent known, the production volume, and the disposition of the R&D chemical substance must also be retained.

Manufacturers and importers who distribute an R&D substance to other persons must provide those persons with written notification of known hazards and of the requirement that the substance be used solely for R&D. For additional information on R&D requirements see the New Chemical Information Bulletin: Exemptions for Research and Development and Test Marketing available from the TSCA Assistance Information Service (TAIS, TSCA Hotline) (see Contact List, Appendix C).

3. Test-marketing exemptions (TME)—You may apply for an exemption from premanufacture notification if you plan to manufacture or import a new chemical substance for test-marketing. Test-marketing involves the distribution of a predetermined limited amount of a chemical substance, or of a mixture or article containing the chemical substance, to specified number of customers to explore market acceptability before general distribution. The submitter needs to show that the intended activity is not commercial production and is not appropriately considered to be research and development.

To approve a test-marketing exemption application, the Agency must make an affirmative finding that the new chemical substance will not present an unreasonable risk to health or the environment during the test-marketing activities. 40 CFR §720.38 identifies the type of information you should submit with a test-marketing exemption application. EPA must approve or deny the application within 45 days. If you do not provide sufficient information for EPA to make its determination, the Agency will deny the request. You should send applications for test-marketing exemptions to the OPPT Document Control Officer (7407). You are not required, but are encouraged, to use the PMN form for a TME application. For additional information on test marketing requirements see the New Chemical Information Bulletin: Exemptions for Research and Development and Test Marketing available from the TSCA hotline. Recordkeeping requirements for a test-marketing exemption are discussed at I-M, Recordkeeping.

4. §5(h)(4) exemptions—Under §5(h)(4) of the Act, you may apply to EPA for an exemption from some or all premanufacture notification requirements.

EPA may grant an exemption if it makes an affirmative finding that the manufacture, processing, distribution in commerce, use, or disposal of the new substance will not present an unreasonable risk to health or the environment. Unlike other exemptions, such as test marketing, EPA may only grant a §5(h)(4) exemption by rule. In the rulemaking proceeding, the applicant should provide information sufficient to show that the chemical substance will not present an unreasonable risk to health or the environment. You should send applications for §5(h)(4) exemptions to the OPPT Document Control Officer (7407). The following exemptions have been promulgated under §5(h)(4):

a. Low Volume Exemption (LVE)—Requirements for a LVE application are found at 40 CFR §723.50. This exemption is available for substances manufactured in quantities of 10,000 kg or less per year. Multiple LVEs can be issued to several manufacturers of a single substance, however, second and subsequent applications will be evaluated in the context of existing permitted exposures. Low volume substances are not added to the TSCA Inventory. The notice must include the site of manufacture and proposed use of the new chemical substance; these are legally binding upon the company. The manufacturer may also provide information on exposure controls. If provided, any controls specified in the notice are binding throughout the period of the exemption. EPA will grant the LVE if it determines that the substance will not present an unreasonable risk of injury. The review period for an LVE is 30 days, which can be extended if more information is required.

Companies must notify the Agency within 30 days of a change of the site of manufacture or application. Manufacturers (importers) must notify processors and industrial users that the substance can be used only for the uses specified in the exemption notice. Manufacturers must also notify processors and users of any exposure controls.

Recordkeeping requirements are discussed at I-M, Recordkeeping.

b. Polymer Exemption—Requirements for a Polymer Exemption can be found at 40 CFR §723.250. The exemption was first published at 60 Federal Register (FR) 60 pp. 16316–16336, and there is a useful discussion in the Preamble to that publication. This exemption is available for certain classes of polymers which are not chemically active or bioavailable. The manufacturers are not required to submit a polymer exemption notice to EPA prior to manufacture, but must notify the Agency by 31 January for new materials first manufactured in the preceding calendar year. Recordkeeping requirements are discussed at I-M, Recordkeeping.

c. Low Release and Exposure (LoREX) Exemption—Eligibility for this exemption category is independent of production volume level. Performance standards for this exemption are set out in 40 CFR §723.50(c) and include both absolute criteria (e.g., an upper limit on surface water releases) and goals (e.g., no worker exposure). The notice must include the site of manufacture and proposed use of the new chemical substance which are legally binding upon the company. The applicant's exemption notice must describe how exposures and releases of the new chemical substance compare to the criteria of 40 CFR §723.50(c). If the exemption is granted the applicant is responsible for complying with the standards and with any controls or limitations specified in this exemption notice. These requirements must be followed throughout the period of exemption. LoREX exemption notices are subject to 30-day review periods by the Agency. Recordkeeping requirements are discussed at I-M, Recordkeeping.

D. When to Submit a Notice

You must submit a PMN at least 90 days before you begin to manufacture or import a new chemical substance for a commercial purpose. You are required to submit a notice for LVE or LoREX exemptions 30 days before you begin to manufacture or import a new chemical substance, and 45 days before manufacture or importation under TME. If information additional to that provided with the application is needed, these periods can be extended through suspension requests. If your application is not denied, you will be able to initiate manufacture/import at the end of these review periods. It is prudent for submitters who think their substances may be subjected to required additional testing during PMN review to confer with the Agency before submitting and to submit further in advance of their hoped-for start dates than the minimum number of days for review, as additional testing will extend the Agency review period for the PMN. To initiate this process contact a prenotice coordinator.

E. Filling out and Submitting the Form

Complete the §5 notice form EPA 7710-25 (rev. 5-95) using a typewriter or by printing legibly in black ink. Information which is not submitted on a photocopy of the form available from EPA (e.g., an electronically generated form created by utilizing form-making software) must be in a format pre-approved by the Agency. Approval can be obtained by contacting:

OPPT Document Control Officer
Mail Stop 7407
USEPA 401 M Street, South West
Washington, DC 20460

All information provided **must** be in English (Except that open-literature reports can be submitted in their original languages: if data appears in the open scientific literature, the submitter need only provide a standard literature citation. A standard literature citation includes author, title, periodical name, date of publication, volume, and page numbers. The submitter can assist the Agency by providing a photocopy of the article, if desired. This is the only exception to the requirement that all information must be submitted in English). Provide all information requested on the notice form to the extent that you know or can reasonably ascertain it. If you do not know or cannot reasonably ascertain the information, enter "NK" ("not known"). Many submitters want to know what is meant by "reasonably ascertainable." In general, the Agency views information in the current literature, held by the submitter or a parent or subsidiary company, or held by a supplier to be reasonably ascertainable.

Some staff members in large corporations have expressed concern for their personal liability on information submission—that there can be information held by their organization which a reasonable search will not uncover. As an example, a branch office of a parent company may have called for a study of a substance and not have retained its results in the ordinary or expected record locations, or a study of a family of substances undertaken for commercialization of one of them may not be found when commercialization of another of those substances is later undertaken. If you think you are in some danger that you might not find all of your company's information about a substance on which you are preparing notification, you should document that you made a serious search for information, which should have yielded all reasonably ascertainable information, and keep a record of your search with your records of the submission. You should be able to make available to an EPA inspector records showing: that you identified where in your corporate organization (or your suppliers) the information might be, that you sent requests for information to each site where you think the information might be kept, and that you followed up with any non-responding site until you got a response. You should review all applicable information on the substance, such as the Material Safety Data Sheet (MSDS) for the existence of testing on the new chemical substance. It is helpful if there is a corporate information policy to ensure that this sort of information is available to a responsible PMN submitter.

You may submit continuation sheets for any subsection or item on the form. Head any continuation sheet with: the TS-number or PC-number, if any; submitter name; and the number of the question to which it is a supplement. Mark the appropriate box on the notice form if you attach continuation sheets. You may photocopy the notice form, sections of the form, or this manual as frequently as you need.

Form 7710-25 (rev. 5-95) is used for several different types of submissions. These instructions are designed to guide submitters for each of them. Send your completed notice to the Document Control Officer (DCO) for Office of Pollution Prevention and Toxics. If the notice is a PMN or Significant New Use Notice (SNUN), send with original signatures and two copies (an additional copy is required if any information is claimed to be confidential, and should be submitted with the claimed-confidential information deleted ("redacted copy," "sanitized copy"). No copies are required if it is an application for a Test Market Exemption (TME), Low Volume Exemption (LVE), or Low Release–Low Exposure Exemption (LoREX), except that if it contains confidential business information (CBI) you must submit a redacted copy, which will be placed for public viewing in the TSCA Nonconfidential Information Center, 202–260–7099 (open from 12m–4pm daily):

TSCA Nonconfidential Information Center
Mail Stop 7407
Room B607, Northeast Mall
401 M Street SW, Washington DC 20460

The US Mail address for the DCO appears on page 1 of the form. The room at which submissions are accepted is open from 8 am until 4 pm daily, and if a courier service arrives after that time it will be turned away. For sending a notice by courier the address should include the physical location:

OPPT Document Control Officer
Mail Stop 7407
TSCA Data Processing Center
G099 East Tower, Waterside Mall
USEPA 401 M Street, South West
Washington, DC 20460

You can give the courier the phone number (202) 260–1768 to call if there are delivery problems.

A user fee must be remitted for PMN and SNUN §5 notices in accordance with 40 CFR §700.45. You must create a unique alpha-numeric identification number ("TS-number") to identify and link your notice with the remittance fee. This six digit number must be placed on the first page of the form in the boxes that have been provided. This number must also be placed on your fee remittance which is sent to:

EPA, Washington Financial Management Center (Mail Stop 3303)
P.O. 360399M
Pittsburgh, PA 15251-6399
Attn. TSCA User Fee

EPA uses a private bank in Pittsburgh to receive these fees. The bank will accept certified checks, money orders and bank drafts only; after the bank has processed the payment, the TS-number is sent to EPA Headquarters with certification that payment has been made. EPA Headquarters then verifies that the appropriate remittance with a TS identification number corresponds to a user fee identification number on a PMN and further processing of the notice commences. However, if a problem arises in the payment procedure, (i.e., insufficient funds, improper usage of the TS-number), the notice will be given incomplete notice status in accordance with 40 CFR §720.65(c). The EPA will inform the submitter in writing if this action is taken.

40 CFR §720.65 specifies administrative procedures applicable to incomplete notices in general. The most frequent reason for a submission to be incomplete is a name which is not in conformance with the Ninth Collective Index (9CI) of Chemical Abstracts nomenclature rules and conventions (this definitive guide to CA nomenclature has been used since 1972.) If the notice is declared incomplete, the review period has not begun no matter when in the initial review period the notice is declared incomplete. Therefore, the review period begins again at day one when a complete notice is received. However, EPA can choose to restart the clock on the day the notice was declared incomplete if it determines that its review can be completed within the remaining period. This decision is made case-by-case. See 40 CFR §720.65(c).

F. Binding Boxes

The purpose of the binding option is to enable EPA (if necessary) to efficiently negotiate with a PMN submitter the development of §5(e) consent orders and promulgate Significant New Use Rules (SNURs) for those new chemical substances that the Agency determines may present an unreasonable risk if certain control actions are *not* implemented. This option is intended to reduce delays that can slow the development of consent orders absent such agreement. At one time, SNURs were limited to environmental release activities and certain industrial, commercial, or consumer activities, but now they can include other important activities, such as protection in the workplace and hazard communications.

Control measures instituted by the submitter to reduce exposures and/or releases of the substance may have a direct bearing on the Agency's conclusions regarding risk. Therefore, you may wish to indicate your willingness to be bound to certain submitted information on the form which is related to the issue of potential risk such as use, production volume, protective equipment, engineering controls, and/or process description. By indicating your willingness to make these commitments, you would be indicating an interest in future negotiations if the Agency deems them necessary. In order to make your intentions known to EPA, mark in the "Binding Option" box on the form located to the right of the appropriate information.

Should the Agency wish to discuss development of binding control measures for your PMN, you will be contacted by a Program Manager and negotiations may ensue. Therefore, indicating a willingness to be bound by the terms of your notice *does not* by itself prohibit the submitter from deviating from the information (except chemical identity) reported in the form. In the case of exemption applications (i.e., Test Market, Low Volume/Low Release, Low Exposure), however, certain statements are automatically binding on the submitter when the Agency approves the exemption applications.

G. Test Data and Other Data

You are required to provide three copies of any test data on the health and environmental effects of the new chemical substance, including data on physical/chemical properties, in your possession or control, and a description of any other health and environmental effects data on the substance known to or reasonably ascertainable by you. Data in the possession or control of either a parent company or an affiliated subsidiary located outside the U.S. are considered by the Agency to be data that should be known to or reasonably ascertainable by a submitter (see section E, above). Data must be submitted in English. Standard literature citations may be submitted for data in the open scientific literature. Complete test data (not summaries) must be submitted if they do not appear in the open literature. Incomplete reports (e.g., from ongoing studies) are exempt from full reporting. However, you must describe the nature and objective of any incomplete study, report, or test, the name and address of any laboratory developing the data; progress to date; type of data collected; significant preliminary results; and an anticipated completion date. If significant preliminary results or final results are obtained prior to the completion of the notice review period or any other additional information significant to the review of the notice becomes available to you, you must submit this information within 10 days of receipt, but no later than 5 days before the end of the review period. If information becomes available during the last 5 days of the review period, you must immediately inform EPA by telephone. Examples of the types of test data you must submit are provided in Appendix A of this manual. In addition, a Physical and Chemical Properties Worksheet now appears on the last page of the form. For additional information on health and safety studies and on submitting test data, see 40 CFR §§720.3 and 720.50 of the Premanufacture Notification Rule. Attach test data to the notice form and reference it by page number in Part III, List of Attachments.

You are not required to submit any data previously submitted to EPA with no claims of confidentiality if you identify in your submission the office or person to whom you submitted the data, the date it was submitted, and, if appropriate, a standard literature citation. If, however, you submitted data with claims of confidentiality, you must resubmit the data with the notice and any claim of confidentiality under 40 CFR §720.80. You also are not required to submit data related solely to product efficacy. This exception does not apply to information required in the notice, test data, or other data.

H. Confidentiality

1. Asserting claims—You may assert a claim of confidentiality for any information submitted to EPA. To assert confidentiality for specific information on the form (e.g., submitter identity, chemical identity, or use information), mark in the "Confidential" or Confidential Business Information (CBI) box on the form located to the right of the information. Also mark the box at the bottom of page 1 of the form if any information in the notice was claimed as confidential in the form. As noted above, a redacted copy of any §5 notice must be submitted with the notice, or it will be declared incomplete by the Agency and returned to the submitter.

To assert confidentiality claims for information in attachments to the form, provide a complete copy of the attachment that clearly indicates (e.g., by circling or bracketing) the information you wish to claim as confidential. Bracket only the specific information you claim as confidential. For example, if you submit a study which contains a physical or chemical property, and it is only that property which you wish to claim as confidential, bracket only that property. Do not simply stamp "Confidential" on the page which contains that property. You must also clearly and specifically mark any confidentiality claims you wish to make for information or correspondance subsequently submitted to EPA about your PMN, and you must provide a redacted version for the public file.

If you claim the identity of the new chemical substance or its category of use as confidential, you must provide a generic description of this information, as indicated in the appropriate sections of the form. Guidance on developing generic names is given in Part 2 of this manual in the Chemical Identity Information Section at Item 6.

To ensure that no confidential information is disclosed to the public, you must submit an additional copy of the notice form, including attachments, which does not contain confidential information. This version ("sanitized", "redacted") will be placed in the public file. It must contain all non-confidential information, including health and safety studies. A health and safety study means any study of any effect of a chemical substance or mixture on health or the environment or on both, including underlying data and epidemiological studies, studies of occupational exposure to a chemical substance or mixture, toxicological, clinical, and ecological, or other studies of a chemical substance or mixture, and any test performed under the Act.

Information from health and safety studies which can be claimed confidential is quite limited—this is discussed at 40 CFR §720.90. Chemical identity is assumed to be part of a health and safety study. 40 CFR §720.90(b)(2) discusses the claims which must be made and substantiated for chemical identity in a health study to be confidential: that disclosure would reveal manufacturing or processing information, that it would disclose the fraction of a mixture which the substance comprises, that the study could be interpreted without knowing the identity of the substance, and that disclosure would have harmful competitive effects on the submitter.

Not only is information which arises as a result of a formal, disciplined study included, but other information relating to the effects of a chemical substance or mixture on health or the environment is also included. In sum, any data that bear on the effects of a chemical substance on health or the environment would be included.) If you do not provide the sanitized copy with your submission, the submission will be incomplete and the review period will not begin. If you provide a subsequent amendment to your PMN or additional information/data, you must also provide a non-confidential version for the public file.

2. Substantiating claims—You are not required to provide substantiation of any confidentiality claim when you submit your notice. However, you must substantiate your claim of confidentiality for chemical identity at the time you submit a Notice of Commencement of Manufacture (NOC), if you want EPA to maintain your confidentiality claim after you begin manufacture. (NOC requirements are described in 40 CFR §720.102 and in this manual.) To substantiate that claim, you must provide EPA with detailed answers to the following questions which appear in the Premanufacture Notification Rule [40 CFR §720.85(b)(3)(iv)]:

(A) What harmful effects to your competitive position, if any, do you think would result if EPA publishes on the Inventory the identity of the chemical substance? How could a competitor use such information given the fact that the identity of the substance otherwise would appear on the Inventory of chemical substances with no link between the substance and your company or industry? How substantial would the harmful effects of disclosure be? What is the causal relationship between the disclosure and the harmful effects?

(B) For what period of time should confidential treatment be given? Until a specific date, the occurrence of a specific event, or permanently? Why?

(C) Has the chemical substance been patented? If so, have you granted licenses to others with respect to the patent as it applies to the chemical substance? If the chemical substance has been patented and therefore disclosed through the patent, why should it be treated as confidential for purposes of the Inventory?

(D) Has the identity of the chemical substance been kept confidential to the extent that your competitors do not know it is being manufactured or imported for a commercial purpose by anyone?

(E) Is the fact that someone is manufacturing or importing this chemical substance for commercial purposes available to the public, e.g., in technical journals or other publications; in libraries; or in State, local, or Federal agency public files?

(F) What measures have you taken to prevent undesired disclosure of the fact that you are manufacturing or importing this substance for a commercial purpose?

(G) To what extent has the fact that you are manufacturing or importing this chemical substance for a commercial purpose been disclosed to others? What precautions have you taken in regard to these disclosures? Has this information been disclosed to the public or to competitors?

(H) In what form does this particular chemical substance leave the site of manufacture, e.g., as part of a product; in an effluent or emission stream? If so, what measures have you taken to guard against discovery of its identity?

(I) If the chemical substance leaves the site of manufacture in a product that is available to either the public or your competitors, can they identify the substance by analyzing the product?

(J) For what purpose do you manufacture or import the substance?

(K) Has EPA, another Federal agency, or any Federal court made any pertinent confidentiality determinations regarding this chemical substance? If so, copies of such determinations must be included in the substantiation.

(L) If the notice includes a health and safety study concerning the new chemical substance, the submitter must also answer the questions in 720.90(b)(2).

This substantiation must accompany your NOC. You may be required to substantiate other confidentiality claims if EPA receives a Freedom of Information Act (FOIA) request on that information.

I. Consolidated Notices

If you are manufacturing two or more, but no more than six, structurally similar new substances, you may contact a Prenotice Coordinator to obtain approval to submit a single consolidated notice. A consolidated notice is suitable for chemical substances of **similar structure** with the same or similar uses and which share similar test data and other information. A consolidated notice must identify each new substance individually; you may not submit a consolidated notice for an open-ended category. A separate chemical identity page must be provided for each substance. A distinct Agency "PMN" number is assigned to each chemical. You may not submit a consolidated notice for a series of intermediates and a final product (they will not share common uses, test data, and other information).

EPA encourages you to submit consolidated notices when appropriate. You may submit a consolidated notice only after you have received prior approval from a Prenotice Coordinator. This request should concisely describe the chemical identity of each substance to be included in your consolidated notice (note: you need not use names from the Inventory Expert Service of the CAS to request approval for a consolidation, you need only to describe the chemical substances well enough that EPA personnel can determine whether they are similar enough for combined review. You must, however, use separately obtained Method I names for each substance in a consolidation when PMN is submitted). Many requestors provide this information using the chemical identity pages of the form or in a one page table format. Prenotice coordinators are committed to quickly responding to a request for a consolidated notice; however, this will not always be possible. Remember to enter your prenotice communications number in Part I, Section A (3) (page 3) of the form.

J. Submission of Information by Others

1. Submission by an agent—You may designate an agent to submit a §5 notice for you. Both you and the agent must sign the certification on the form. You are responsible for ensuring that all information known or reasonably ascertainable by you and all test data in your possession or control are submitted to EPA. For information on submissions by agents, see 40 CFR §720.40(e).

2. Joint submissions—You may also prepare and submit a PMN with another person. A joint submission may be useful where different persons have information required in the notice, including a situation when another person has information fundamental to the notice, but wishes to keep it confidential. For example, you may have information on the identity and the physical and chemical properties of the new substance and another person may know its manufacturing process and its intended use.

Each joint submitter must use a notice form and sign the certification on the form. Each person must also assert all confidentiality claims as described in 40 CFR §720.80 and Part I of this manual. However, you are not relieved of statutory notice requirements by arranging a joint submission. You are required to complete all mandatory sections of the form to the extent that you know or can reasonably ascertain the information, even if another person also submits information for a certain section. If you submit a joint notice, the review period will not begin until EPA has received all required parts of the notice. You should identify the joint submitter in your notice and identify the section(s) which the person is submitting. See 40 CFR §720.40(e) for additional information on joint submissions.

K. Consultation with EPA Concerning the Premanufacture Notice

1. Before notice submission—

a. General inquiries—General inquiries concerning the premanufacture notification program which are not related to a specific chemical or notice should be directed to the TSCA hotline (see Contact List, Appendix C). Copies of the PMN Rule, Instructions Manual, notice form, and other materials relating to the Rule are available by contacting the hotline. (These materials are also available at EPA regional offices.)

b. Specific inquiries—Specific inquiries concerning the PMN Rule, confidentiality, joint submissions, consolidated notices, etc., should be directed to the Prenotice Coordinator. You can contact the Prenotice Coordinator by telephone, facsimile, or email (see Contact List, Appendix C)

2. During notice review—Upon receipt of the notice by the OPPT Document Control Officer (DCO), the Agency will make an initial determination whether the notice is complete. The initial determination looks to see that the notice contains all the items required on pages 4 through 8 of the form and for apparent chemical identity problems. If no problems are seen, your form is initially determined to be "complete," goes on for further review, and Day 1 of the 90-day review period is assigned as the date of receipt by the DCO. Later and more detailed review can, however, discover other additional information that has not been provided, and you can be contacted to provide additional information at any time during the review period.

You will receive written notification if your notice is declared "incomplete" as described at 40 CFR §720.65. If your notice is initially complete, you will receive an acknowledgment letter telling you your notice number and the date of expiration of the review period. The Inventory is searched to ensure that the substance for which the notice is submitted is not already included on the TSCA Chemical Substance Inventory. If your chemical substance is on the Inventory, you will be notified that your substance is not subject to premanufacture notification, and that therefore you are free to begin manufacture immediately. If the substance is not on the Inventory, and if the substance is not dropped from consideration at the Agency's initial review meeting ("Focus Meeting," which takes place at approximately Day 20), a Program Manager will be assigned to coordinate the review of the notice and to be your official contact with the Agency throughout the remainder of the review period. Based on the Focus Meeting, it is possible to declare the PMN "incomplete," usually for missing test data. Also during the review period, the Program Manager may contact you for clarification of information you have provided in the notice or if the Agency identifies issues of concern. If you are not contacted prior to the expiration of the review period, you are free to commence manufacture of the substance identified in your notice after the review period has expired. You can check the status of your submission at the New Chemicals Internet site ("www.epa.gov/opptintr/newchms") after approximately day 30.

The Program Manager will also notify you before the review period expires if he/she will extend the review period under TSCA §5(c) or if regulatory action is being considered on the new substance under TSCA §5(e) or 5(f). 5(e) Consent Orders are typically issued with a follow-up SNUR subsequently promulgated. In addition, a Program Manager will contact you if the Agency plans to develop a non–5(e) SNUR (a case in which a 5(e) Consent Order does not precede the development of a SNUR) on the chemical substance identified in your notice.

L. Notice of Commencement (NOC) of Manufacture (or Import) (40 CFR §720.102)

If EPA has not taken any action to regulate the new chemical substance during the review period, you may begin manufacturing the new chemical substance upon expiration of the review period. EPA requires that you notify the Agency by using EPA form 7710–56, no later than 30 calendar days after the first day of such manufacture or import for non-exempt commercial purposes (for import, Day 1 is the date the material clears U.S. Customs).

Your NOC must be sent to the OPPT Document Control Officer (Mail Stop 7407). In your NOC, you must provide the specific chemical identity of the substance, its PMN number, the site of first manufacture or import, and the date when manufacture or import began. You must also substantiate a confidentiality claim for chemical identity in your letter, as described above at "2. Substantiating Claims" if you want EPA to maintain the claim after you begin manufacture or import. See 40 CFR §720.85(b) for further information on substantiating confidentiality claims.

M. Recordkeeping

Recordkeeping requirements for submissions under §5 of TSCA, and for exemptions from submission, are found in several different sections of the CFR.

PMN or SNUN: 40 CFR §720.78(a) requires that you retain documentation of information for a PMN or SNUN for five years from the date of commencement of manufacture. The records you must retain include (1) information supporting the information supplied on the notice form, (2) other data, as defined in 40 CFR §720.50(b), in your possession or control, (3) production volume for the first three years of production or import, and documentation to support your stated production volume, and (4) date of commencement of manufacture, and documentation to support your stated date. You are not required to develop information solely for recordkeeping purposes, but only to retain information you have obtained or developed in the course of completing your submission.

Research and development: 40 CFR §720.78(b) requires that if you manufacture a new chemical substance under the exemption for substances manufactured solely for research and development, you must retain documentation of compliance with the exemption until five years after they are developed.

Test-Marketing exemption: 40 CFR §720.78(c) requires that if you manufacture under a test-marketing exemption under TSCA, you must retain documentation of information in the application and documentation of your compliance with any restrictions imposed by EPA when it granted the application until five years after the final date of manufacture or import under the exemption.

LVE or LoREX: 40 CFR §723.50(n) requires that each manufacturer of a new substance reported under the terms of a low volume or LoREX exemption must maintain records of 1) the annual production volume of the new chemical substance under the exemption, and 2) documentation of information in the exemption notice in compliance with the terms of the exemption. Records must be retained for five years after date of their preparation.

Polymer exemption: 40 CFR §723.250(j) requires that a manufacturer of a polymer made under the terms of the polymer exemption must maintain records for five years from the date of commencement of manufacture for: the production volume for the first three years of manufacture, the date of commencement of manufacture, documentation of the information provided above, documentation of any other information provided in the notice, such as information that demonstrates that the new polymer is not specifically excluded from the exemption and the polymer meets the exemption criteria.

N. Recognition of Pollution Prevention and Recycling Benefits

During the course of its review, the Agency will be considering whether the activities surrounding the manufacture, processing, use, and disposal of the substance identified in the notice may present an unreasonable risk of injury to human health or the environment.

It is important, also, that EPA acquire information regarding any technological, risk reduction, or environmental *benefits* which may be possible if the new chemical being reported is introduced in commerce, and this information can in some cases enable the Agency to approve substances otherwise doubtful. Therefore, PMN submitters are encouraged to complete and provide the optional information on pollution prevention on page 11 of the form. By submitting information describing the positive pollution prevention aspects of your PMN substance, you may achieve two possible benefits: first, the pollution prevention information may enable EPA to regulate the substance less stringently than it would have absent the information, and second, the pollution prevention information may be chosen by the Agency for affirmative recognition as part of the EPA New Chemicals Pollution Prevention Recognition Project.

In the Agency's EPA New Chemicals Pollution Prevention Recognition Project, EPA seeks to promote safer new chemicals and processes, providing several forms of recognition, including a letter from the Director of OPPT, inclusion in a listing of recognized chemical substances on the OPPT Internet Homepage, and other positive publicity. If you want the Agency to consider your PMN substance for this recognition, you must explicitly request to be considered in your response at (OPTIONAL) POLLUTION PREVENTION AND RECYCLING INFORMATION and identify the pollution prevention merits of your PMN substance. To the extent that you think it helpful for Agency consideration, EPA strongly encourages you to (1) submit actual test data on the PMN substance to substantiate any pollution prevention claims you assert and (2) minimize claims of confidentiality claims to facilitate publicity regarding the PMN substance.

For example, the EPA will consider any information on methods used to minimize potential risks associated with the new substance through source reduction or recycling. Some of the benefits for which information may be provided are a reduction in the volume manufactured, a reduction in the generation of waste materials, a reduction in exposure and/or environmental release or increased performance and/or operation efficiency of the new chemical substance in comparison to existing chemical substances used in similar applications. Recycling activities include reclamation of useful chemical components from wastes that would otherwise be released as air emissions, water discharges or land releases during manufacture, process or use. All descriptions may be quantitative or qualitative.

The "Optional Pollution Prevention Information" page of the PMN requests optional information that will be used in the evaluation of the new chemical substance and to compare the relative risks and benefits of the substance as a substitute for substances with similar uses currently on the market. PMN submitters are encouraged to report any and all relevant information not reported elsewhere in the PMN which they believe to be important to a thorough regulatory decision. The page provides submitters with the opportunity to describe pollution prevention and risk reduction options considered by the company in regard to the submission. A useful format for presenting such information is provided at Section II of this Manual. Providing this pollution prevention information to EPA may benefit PMN submitters by reducing regulatory controls and/or testing requirements, if the pollution prevention information sufficiently mitigates EPA's concerns for the toxicity, human exposure, or environmental releases of the PMN substance. EPA considers this information in line with the strictures of the Pollution Prevention Act of 1990.

Under the Pollution Prevention Act of 1990 (PPA), Congress established a national policy that: (a) pollution should be prevented or reduced at the source whenever feasible, (b) pollution that cannot be prevented should be recycled in an environmentally safe manner whenever feasible, and, (c) disposal or other release into the environment should be employed only as a last resort and should be conducted in an environmentally safe manner.

EPA defines "pollution prevention" to mean "source reduction," as defined under the PPA, and other practices that reduce or eliminate the creation of pollutants through: (a) increased efficiency in the use of raw materials, energy, water, or other resources, or (b) protection of natural resources by conservation.

The PPA defines "source reduction" to mean any practice which: (a) reduces the amount of any hazardous substance, pollutant, or contaminant entering any waste stream or otherwise released into the environment (including fugitive emissions) prior to recycling, treatment, or disposal, and, (b) reduces the hazards to workers, public health, and the environment associated with the release of such substances, pollutants, or contaminants.

The term includes: equipment or technology modifications, process or procedure modifications, reformulation or redesign of products, substitution of raw materials, and improvements in housekeeping, maintenance, training, or inventory control. The term ". . . does not include any practice which alters the physical, chemical, or biological characteristics or the volume of a hazardous substance, pollutant, or contaminant through a process which itself is not integral to and necessary for the production of a product or the providing of a service" (Sec. 3(5)(B)). Thus, end of pipeline controls, such as thermal oxidizers, incinerators, or waste water treatment systems are not defined as "source reduction."

EPA is interested in information on how improved processes for handling individual new chemical substances may reduce potential exposures and releases of specific PMN substances. Submitters may include a discussion of Pollution Prevention/Risk Reduction measures actually selected for implementation and the rationale for the selection. Submitters are encouraged to consider and include information comparing the releases and exposures for various process options considered but not selected, anticipated reductions in releases and exposures which can be expected in the production of the PMN substance as compared to an existing chemical substance, and how the PMN substance and/or the product in which it is used may compare favorably with existing chemicals in terms of pollution prevention. Submitters may also describe other pollution prevention-related advantages, such as process modifications, increases in product life or durability, or decreased energy consumption, etc. A set of questions to address these concerns is put forward in Section II of this Manual.

EPA is also interested in information describing possible reductions in toxicity, and human exposure, as well as environmental release of a new chemical substance, as compared to those of already commercialized chemical substances for which the new substance may substitute. Such information may demonstrate that the new chemical substance is a viable safer substitute for an existing chemical substance.

Voluntary submission of pollution prevention information is not intended to negatively affect the outcome of EPA's review of the Premanufacture Notification. When risk reductions are documented, the information will be carefully considered during EPA's review of the PMN. Based on the information, EPA may reduce or eliminate anticipated exposure controls and testing requirements which would ordinarily have been imposed by the Agency. Completion of this section is also important if you wish your substance to be considered for the EPA New Chemicals Pollution Prevention Recognition Project.

Reference Material

EPA's new "Facility Pollution Prevention Guide," a successor to the 1988 "Waste Minimization Opportunity Assessment Manual," is available to help firms develop broad-based multimedia pollution prevention programs. Worksheets and other information are included to help facilities identify, assess, and implement opportunities for preventing pollution, including methods of controlling waste creation during the production process, as well as product design and redesign. Developed by EPA's Pollution Prevention Research Branch, and Office of Solid Waste, the Guide (Doc. No. EPA/600/R-92/088) can be ordered by mail from:

EPA Center for Environmental Research
Information Publications Unit
26 W. Martin Luther King Drive
Cincinnati, OH, 45268
(800) 490–9198

II. PAGE-BY-PAGE INSTRUCTIONS FOR COMPLETING EPA FORM 7710-25 (rev. 5-95) THE TSCA §5 NEW CHEMICALS PROGRAM PREMANUFACTURE NOTICE FORM

(Besides the page-by-page instructions below, we have attached as Appendix B an annotated version of the form distributed internally by the 3M Co., for its personnel, which submitters may find useful as an adjunct to these instructions.)

Page 1: Identify Type of Notice, Your Submission

Total number of pages: Give total pages as submitted, including attachments.

TS-number: The submitter chooses this number. It should be an alphanumeric. Most repeat submitters use some variant of their initials and the number of submissions they have made (Widget Corp. might pick WID001, WID002, for example). The Agency uses the number to track notification from its bank in Pittsburgh that the money has been received, and it can be useful to the submitter in identifying the submission to EPA personnel before a P-number has been assigned by the Agency. The TS-number should be unique to this submission from your company, do not give the same number to a subsequent submission.

Confidentiality claims: Check this box if ANY information in the form is claimed confidential.

Test and other data: Indicate which types of data are included with the PMN.

Type of notice: Please check the type of notice submitted.

Check "PMN" if the application is for a standard, final-product new chemical substance for placement on the TSCA Inventory.

The "Intermediate PMN" box should be checked if this notice is for a chemical substance which is an intermediate used in the production of a final product for which a separate notice is submitted simultaneously, and for which the submitter has no intention of making a separate, non-intermediate use. In addition, the intermediate PMN must identify the final product. Separate user fee identification numbers must be generated for and appear on each notice; although a single check may be remitted bearing all user fee identification numbers for a sequence of intermediate(s) and final product. "Certification," below, discusses fees paid by small manufacturers. For further information on "intermediate PMNs," see 40 CFR §§700.43 and 700.45(b)(2)(ii).

The "Significant New Use Notice" (SNUN) box should be marked for any notice that is submitted in accordance with a SNUR.

The "Test Marketing Exemption Application" (TMEA) box should be marked for any notice submitted in accordance with the criteria listed in 40 CFR §720.38. You are not required to use form 7710-25 (rev. 5-95) to submit a TMEA, but it is encouraged.

Boxes are also provided to identify your submission as an application for either a "Low Volume Exemption" (LVE) [see 40 CFR §723.50(c)(1)] or a "Low Release/Low Exposure Exemption" (LoREX) [see 40 CFR §723.50(c)(2)]. These exemptions must be requested through use of the PMN form. Modifications for earlier approved requests for either of these exemptions are requested by checking the modification box on the first page of the PMN form.

For an application to modify an LVE or LoREX exemption, a submitter is not required to provide again information which was submitted in a previously approved exemption. Each page from the application on which some information has changed, and a new signature page, must be provided. An application for an LVE for a substance which had been the subject of a prior LVE which had been submitted before the amendments of March 29, 1995, however, is a new application and the full form 7710-25 must be submitted.

If this notice is for a consolidated PMN, the number of chemicals (two or more, but no more than six) included in the notice should be entered on page one in the space provided. A separate PMN number is assigned to each chemical substance identified in a consolidated notice. Approval for a consolidated PMN notice must be obtained from the Prenotice Coordinator prior to submission. You are required to identify the Prenotice Communication number you were given when your consolidation was approved on page 3, question 3 of the form. Further information on submitting a consolidated notice is provided at Section I, Part I of this manual.

Page 2 Certification

The official named in Part 1, Section A of the form, as the person submitting the notice, must sign the certification on page 2 of the notice form. This official is responsible for the truth and accuracy of each statement in the certification. If an agent assists you in preparing the notice, the agent must also sign the certification. All signatures must be original, and in ink.

In addition, the submitter must check certain "user fee" certification statements as appropriate as required at 40 CFR §700. For a PMN, consolidated PMN or SNUN, a fee is required: if the submitter is a small business it must remit the fee identified in either 40 CFR §700.45(b)(1) (small business concerns remit a fee of $100). If the submitter is not a small business it must remit the fee identified at 40 CFR §700.45(b)(2) (all non-"small" submitters remit a fee of

$2,500 for final products, $1,000 if the submission is for an intermediate and is submitted with the application for the final product).

A small business concern is one whose total annual sales (include all sites, including those owned or controlled by a foreign or domestic parent company) are below $40 million for the fiscal year preceding the date of the submission of the applicable §5 notice (see 40 CFR §700.43).

When using the PMN form to submit a LVE or LOREX application in accordance with 40 CFR §723.50, all three of the corresponding certification statement boxes must be checked to acknowledge that you will manufacture under the terms of the exemption. In addition, a submitter of an LVE application must certify that the manufacturer intends to commence manufacture of the proposed exempted substance for commercial purposes within 1 year of the date of expiration of the 30-day review period. There is no fee for an LVE or LOREX.

Part I—GENERAL INFORMATION

Section A—Submitter Identification (Page 3)

1a. **Person submitting notice**—Enter information on the official who signed the general certification.

b. **Agent**—Complete only if you authorize an agent to assist you in preparing this notice. The agent must also sign the certification as noted above.

If you mark the "Confidential" box next to items a or b, all information in the item will be treated as confidential.

c. Mark the box if your submission is a joint submission. Identify in Part I, Section A (1)(c) the name of the joint submitter who is authorized by the U.S. submitter to provide some of the information required in the notice. For additional information on joint submissions, see Part I, Section J of this Manual. A notice will not be considered complete until all information is received by the Agency. If information from multiple parties will not be sent to the Agency in the same package, use your TS user identification number to link multiple notices. You can generate a TS-number solely to link submissions, even for a no-fee exemption. If you don't generate a TS-number you should ask a prenotice coordinater to issue a prenotice communication number for this purpose.

Mark the "Confidential" box next to item c if you wish this information to be treated as confidential.

If you authorize another person (e.g., a foreign manufacturer or supplier) to provide information, such as, confidentially held tradename chemical substance identification directly to EPA, indicate which information will be supplied by the other person. Identify that person by name, company, and address in a continuation sheet. That person's identity may be claimed as confidential. A letter of support for your notice should be provided by the joint submitter on its company letterhead. A notice will be considered incomplete until this information is provided. Whenever possible use your TS-user identification number to link this information.

2. **Technical contact**—Identify a person who can provide EPA with additional information on the new chemical substance during the notice review period. The technical contact identified should be located in the United States and be available to be reached by telephone during normal business hours. If you mark the "Confidential" box next to this subsection, all information in it will be treated as confidential.

3. Provide any prenotice communication number assigned to your prenotice inquiry. In addition, see Section II, Part E, of this manual for further information on submission of a consolidated PMN that requires a prenotice consultation.

4. Provide the exemption number assigned for any previous exemption application submitted for the chemical substance covered by this notice. It is especially important for an exemption modification request that you provide the EPA assigned exemption number from your original exemption application. Also, provide a previously assigned PMN number, if any, for the chemical substance.

5. Self-explanatory.

6. Mark to indicate whether you intend to manufacture or import the new chemical substance or both manufacture and import. Use the optional binding box to indicate your willingness to be bound to either import or domestic manufacture only.

Section B—Chemical Identity Information (Pages 4–6)

Submitters of PMN and exemption notices are required to provide the currently correct Chemical Abstract (CA) name for the substance(s) identified in the notice based on the Ninth Collective Index (9CI) of CA nomenclature rules and conventions, and consistent with listings for similar substances in the Inventory. EPA must receive complete and unambiguous identification of the new chemical substance. If the substance is not adequately identified, the submission will not meet statutory requirements and the notice review period will not begin. If a principal importer does not know the specific identity of the new substance, the submitter must contact the foreign manufacturer or supplier and have the specific chemical identity information required in the PMN provided directly to EPA. In this way, foreign manufacturers can protect confidential business information. The same holds true for U.S. manufacturers reporting chemical substances using a generic or trade name to identify a component of the new chemical substance. The submitter of the new chemical substance must have the supplier provide chemical identity information directly to EPA before the notice can be considered complete. This information may be provided in a letter of support from the supplier or as a joint submission between the two companies. A letter of support should be provided on the supplier company's letterhead. See Part I, Section J(2) of this manual on how and when to file a joint submission. Since a letter of support or a joint submission may be received separately by the Agency, an identification number such as a TS-user fee number or a Prenotice Communications number should be used to link a PMN with information from a supplier or foreign manufacturer. The identical identification number should appear on both pieces of correspondence submitted to EPA; otherwise, there can be a delay in processing the PMN.

The type of chemical identity information required in the notice depends on whether the substance is a Class 1 or Class 2 substance or a polymer. A **Class 1 chemical substance** is a substance whose composition, except for impurities, can be represented by a definite chemical structural diagram. For Class 1 substances, a name that is consistent with the nomenclature rules and conventions of the 9th Collective Index of the Chemical Abstracts Service (CAS) and with current TSCA Inventory listings must be provided. Examples of such substances are 1,3-butadiene, benzene, and sodium chloride.

A **Class 2 chemical substance** is a substance whose composition **cannot** be easily represented by a definite chemical structural diagram. Such a substance is generally derived from natural sources or complex reactions. Its composition may be complex, difficult to characterize, and variable. For Class 2 substances and polymers, a CA Index Name or CA Preferred Name must be provided. In addition, for a Class 2 substance, the notice must identify the immediate chemical precursors and reactants by specific chemical name and Chemical Abstracts Service Registry Number (CASRN), if the number is available. Trade names or generic names of chemical precursors or reactants are not acceptable as substitutes for specific chemical names. Unacceptable names would include, e.g., "chlorinated naphthalene," "glycerol monoester of hydrogenated cottonseed oil acids," or a "reaction product of x, y, z."

A **polymer** is a substance composed of molecules characterized by the regular or irregular repetition of one or more types of identical monomeric units. In most cases, the number of monomeric units is quite large and not precisely known.

If the substance is clearly a Class 1 or 2 substance, then items a–d of Question 1 on Page 4 must be properly completed. If the substance has been named as a polymer (whether the Exemption Rule requirements are met, or not), then items a–c of Question 2 on Page 5 must be addressed and answered completely. If you are uncertain whether the chemical substance is a Class 1 or 2 substance or a polymer, then both Questions 1 and 2 on Pages 4 and 5 should be completed in full, or contact the Prenotice Coordinator for further assistance. If the variability of the composition of a reaction product is too complex to be described as distinct individual components, then reaction product nomenclature is employed. If, however, components of a reaction product can be readily identified and will always be present in the reaction product, then the components should be specifically identified and may be listed individually on the TSCA Inventory. All reaction products may be reported as Class 2 substances.

Submitters may obtain the correct chemical identity of the PMN substance through the Chemical Abstracts Service (CAS) Inventory Expert Service (so-called Method 1) *or* from any other source (so-called Method 2). Consolidations MUST be submitted with Method 1 names. A notice submitter must identify in the chemical identity section which method they used to report the substance's identity. For Method 1, a copy of the CAS report must be attached to the notice. Submitters who choose to develop their own chemical identity are cautioned that the Agency will consider submissions incomplete and thus delay their review if incorrect nomenclature is received from a source other than CAS. If the Inventory Expert Service has been used, the Agency will work with the IES to agree on a name, and the review period will not be affected. Use of CAS services other than the Inventory Expert Service, including CAS' Registry Service, will also be considered Method 2. In all cases, each chemical substance in a consolidated submission must be identified by Method I.

1. Class 1 or Class 2 chemical substances (Page 4)

a. Mark the appropriate class.

b. Enter the specific chemical name of the new chemical substance.

For a Class 1 substance, the name must be a clear description of a unique substance. In describing the chemical substance, the EPA requires Chemical Abstracts Service (CAS) chemical nomenclature be used for identification purposes when it is available. There is a separate box in question one, item c on page 4 for entry of the CAS number. The Agency encourages submitters to have contact with CAS prior to submission in order to obtain concise chemical identity information. Use the CAS standard rules of chemical nomenclature to identify the new substance. Identify the positions of attachment of chemical groups or of unsaturation, if any, by using locants. The chemical name should contain all of the information known about the details of the structure and should permit the drawing of an unambiguous chemical structural diagram.

The chemical name of a Class 2 substance must describe the chemical substance as completely as possible. In some cases, the name may be similar to the names used to describe Class 1 compounds, but it should indicate the substance's multiple components. For example, "polychlorinated biphenyl" indicates a composition that has multiple components varying both in the number and the placement of the substituent chlorine atoms. In other cases, the best possible name may only identify the substance as a reaction product of specified reactants, for example, "anhydrosorbitol monoester of hydrogenated castor oil acids."

c. Provide a molecular formula that gives the identity and number of atoms of each element contained in the molecule. For example, C6H6 is the molecular formula for benzene. When the substance is not molecular or when the exact number of atoms in the molecule is indefinite, such as the infinite crystal sodium chloride, give the relative numbers of each element's atoms. You are required to enter the CAS registry number, if one has been assigned to the substance.

d. For a Class 1 substance, provide a structural diagram. The diagram should clearly indicate the identity of the atoms and the nature of bonds joining the atoms. Any ionic charges or stereochemistry should be shown clearly. In the description of the nature of the reaction or process, as much specific detail as possible should be provided on the reaction conditions, (i.e., temperature, time, etc.) and on the relative amounts of reactants. All known stereochemical details should be provided. Carbon atoms in ring systems and their attached hydrogen atoms need not be explicitly shown. Where applicable, specify the proportions of isomers or tautomeric forms, degree of neutralization, etc.

For a Class 2 substance:

(1) List the immediate precursor substances by chemical name and CAS Registry Number (if known).

(2) For substances prepared by chemical reaction, describe the nature of the reaction or process. A description should appear in the form of a reaction scheme:

The nature of the reaction must be described as specifically as possible (e.g., acetylation, alkaline hydrolysis, diazotization). For substances that have been produced without chemical reaction (e.g., by chemical extraction from a natural source), specify the source, the production process, and the nature of the product.

(3) If you intend to manufacture a Class 2 chemical substance within a limited range of possible compositions, report the range. For example, a manufacturer filing a notice for chlorinated naphthalene might specify a compositional range of 20–40 percent chlorine by weight. In determining the range, you may have to consider the reaction conditions, the catalyst, or the purification process that may be used to produce the substance, and other factors. You should provide the range of composition in weight percent for each specific component or class of components that you intend to manufacture for commercial purposes. Indicate the typical composition or any actual values for samples.

(4) Provide as complete a structural diagram as possible. The diagram should indicate the characteristic structure or variable compositional elements of the substance.

2. Polymers (Page 5)

Submitters should be aware of the PMN filing exemption, applicable to some polymers and useful to some manufacturers. The regulations for this exemption are at 40 CFR §723.250. Persons intending to manufacture polymers, and who have determined that their polymers do not meet the requirements of the exemption should file a PMN or LVE/LoREX. Persons intending to manufacture polymers, and who have determined that, for business reasons, an Inventory listing for their substance is desirable, should file a PMN.

a. Indicate the lowest number-average molecular weight of any composition of the polymer you intend to manufacture. Identify the method you used to make this determination (e.g., vapor pressure osmometry or other colligative property determinations, gel-permeation chromatography, light scattering, or various correlative techniques). If you have not determined number-average molecular weight by analytical methods, briefly explain the basis for your estimate. Indicate the **maximum** weight percent of low molecular weight species below 500 and below 1,000 absolute molecular weight. Include the weight of oligomeric reaction products (including molecules formed that are not polymer molecules) in your determination but do not include the weight of residual monomers or other reactants. Attach test data with two copies supporting your estimate. If you do $A + B \rightarrow C$ not have actual test data, provide an estimate and describe the basis for the estimate. NOTE: The lowest number-average molecular weight is NOT the lowest MW of any component of the polymer mixture, but the lowest number average of several samples of the same polymer, run over time.

b. Column (1)—Monomer or other reactant and CAS Registry Number: You are required to provide the chemical name and CAS Registry Number of each reactant used in the manufacture of the polymer or incorporated into the polymer, including those used or incorporated at 2 weight percent or less. Reactants include monomers, free radical initiators, and cross-linking, chain transfer, and other reactive agents that are used intentionally to become chemically a part of the polymer composition. If a prepolymer is used in the manufacture of the polymer, list the prepolymer and its CASRN as charged into the reaction vessel. If prepolymer compositional information is available to the submitter, identify by bracketing or another method the monomers which are components of the prepolymer. If compositional information of the prepolymer is not available to the submitter or if the prepolymer is represented for purposes of TSCA by a structural repeating unit (SRU) name (examples include silicones and polyethoxylated and -propoxylated substances), you should identify the prepolymer as it is listed in the TSCA Inventory. Solvents, emulsifiers, and non-reacting components should not be listed.

Column (2)—CBI claim.

Column (3)–% of reactant, typical composition: For each reactant (including monomers), indicate its typical weight percent in the polymer. The weight percent can be determined in one of two ways: according to the weight of the reactant charged to the reaction vessel or the weight of the chemically combined (incorporated) reactant in the polymer. For the first method, the weight percent of a reactant is the weight of the reactant charged to the reactor divided by the weight of the polymeric chemical manufactured (times 100). Thus, the weight percent of reactant A of a polymer manufactured from reactants A, B, and C is the weight of A charged to the reactor divided by the dry weight of the polymer A–B–C (times 100). For the second method, the weight percent of the reactant using the "incorporated method" is determined using theoretical calculations of the minimum weight of monomer or other reactant necessary to account for the polymer's actual weight. Manufacturers must maintain analytical data or theoretical calculations to demonstrate their determination.

Please note that zero percent is NEVER an acceptable level of a feedstock, if that feedstock is part of the chemical identity. To describe a substance as including "0–30%" of a given feedstock will only result in getting the PMN returned for a better identity. If you are notifying for a substance for which a feedstock will vary from very low levels you can enter ">0–30%" or "trace–30%" without, as you will if you enter "0–30%", forcing a new identity. "0–2%", however, would be acceptable for a chemical that is not part of the ID and is not marked in the ID field).

If you use a prepolymer in the manufacture of the polymer, you must determine the weight percent of its component reactants. For example, the weight percent of E used in the manufacture of a polymer (using the "amount charged" method) from reactants A, B, and C and prepolymer D–E is the total weight of monomer E in the prepolymer D–E used divided by the weight of the polymer A–B–C–D–E manufactured (times 100). You must provide the identity and typical weight percent of each monomer and other reactant used in the manufacture of the polymer regardless of the weight percent at which it is used. If you will typically manufacture the polymer using a reactant in a range of weight composition, you may indicate the range of weight percent instead of the typical weight percent.

Column (4)—Identity mark: Reactants used or incorporated at greater than 2 weight percent in the manufacture of the polymer are included as part of the description of the polymer listed on the TSCA Chemical Substance Inventory. However, you can choose to include a reactant used or incorporated at 2 weight percent or less in the Inventory description of the polymer by marking this column. Mark the identity column if you want a reactant present or incorporated at 2 weight percent or less to be included in the description of the polymer which is added to the Inventory.

Column (5)—CBI claim.

Column (6)—Maximum weight percent present: Indicate the **maximum** weight percent of each reactant that may be present as a residual (unreacted material) in the polymer as manufactured for commercial purposes.

Column (7)—CBI claim.

Note that you must make separate confidentiality claims for reactant identity, composition information, and residual reactant information.

c. Identify which method you used to develop or obtain the specified chemical identity—Method 1, CAS Inventory Expert Service (a copy of the identification report obtained from CAS Inventory Expert Service must be submitted as an attachment to the notice), or Method 2, other than CAS Inventory Expert Service.

d. Correct Chemical Abstracts (CA) name for the polymer that is consistent with TSCA Inventory listings for similar polymers.

e. Provide a simple, representative structural diagram that illustrates what you know or can reasonably ascertain concerning the key structural features of the polymer molecules. For example, you could identify the linkages formed during polymerization, the functional groups present, the range and typical values for the number of repeating structural units, and the relative molar ratios of the precursors. Indicate if the repeating substructures are arranged in a nonrandom order such as in graft or block arrangements. For example:

$$\text{HO-R-O-(C-R'-C-O-R-O)}_n\text{-H}$$

with the two O double-bonds ($\overset{O}{\underset{\|}{C}}$) above the two C's.

$3 \le n \le 10$, where R may be either

$$-CH_2CH_2- \text{ or } -CH_2CH-$$
$$\phantom{-CH_2CH_2- \text{ or } -CH_2CH}CH_3$$

and R′ may be either a 1,4-substituted benzene ring or $-(CH_2)_4-$

Provide approximate relative mole ratios of precursors, e.g.:

diethyl terephthalate	2.0
adipic acid	1.5
ethylene glycol	1.0
propylene glycol	3.0

3. Impurities (Page 6)

a. Identify each impurity you reasonably anticipate will be present in the substance as manufactured for commercial purposes. An impurity is any chemical substance that is unintentionally present in the new chemical substance. List all impurities (regardless of weight percent). If the substance contains some unidentified impurities, also enter "unidentified" in column (a). Do not include any substances that are mixed with the new substance after manufacture.

In addition to impurities, list in this section other chemical substances, such as solvents, inhibitors, etc., that also may be reasonably anticipated to be present in the chemical substance as manufactured for commercial purposes.

You should consider the following in identifying impurities:

(1) Chemical and instrumental analyses—Often performed on the chemical substance during research and development to characterize the substance before it undergoes health effects or environmental effects testing, to optimize product performance, or to understand process chemistry and optimize output.

(2) Manufacturing process chemistry—Including feedstocks, feedstock impurities, by-products, and intermediates both from the major reaction pathway and from significant side reactions.

(3) Quality control operations—Operations which determine the nature and level of impurities that may be present in the chemical substance.

Identify **impurities** as specifically as possible. You should provide the following information:

(1) The specific chemical name; or

(2) A class or range of structures (e.g., C_6–C_{18} fatty acid salts or poly-chlorinated cyclic and acyclic hydrocarbons in the range C_5–C_{12}); or

(3) The source (e.g., pyrolysis products of cellulose or coal tar residues).

Include the CAS Registry Number if available.

b. Enter the **maximum** weight percent of each impurity in the new chemical substance in column (b). If the substance contains unidentified impurities, enter the total weight percent of unidentified impurities in column (b).

Confidentiality claim: Must be made separately for each impurity.

4. Synonyms—Enter common chemical names by which the new chemical substance may be identified in the scientific or technical literature, including the names or codes used to identify the new chemical in test data or other data which are attached to the notice.

5. Trade identification—Enter any trade name under which the new substance has been or will be marketed. Report all trade names or brand names, even if they are not registered.

6. Generic chemical name—If the new substance's identity is claimed as confidential, enter a generic chemical name. This name should be only as generic as is necessary to protect the confidential chemical identity, and should reveal the chemical identity of the substance to the maximum extent possible. The generic name will be published in the FEDERAL REGISTER notice on the new substance. EPA will review the adequacy of the generic chemical name when the NOC for the substance is submitted. If the name seems more generic than necessary, EPA will contact you to develop an adequate name.

Generic chemical names are created for Class 1 chemical substances by masking structurally descriptive parts of their specific chemical names. Masking can be accomplished by substituting non-descriptive terms (e.g., "substituted") for descriptive parts of the name. Here is an example with the oil source masked: hydrogenated palm–oil fatty acids, esters with D-mannitol, ethoxylated, could become: hydrogenated fatty acids, esters with hexahydroxyalkane, ethoxylated. Guidelines for developing a generic name are provided in Appendix B of the TSCA Chemical Substance Inventory, Volume 1, 1985. This document is available in any Federal Depositary Library or through the TSCA hotline (see Contact List, Appendix C).

7. By-products—List the by-products that you reasonably anticipate will result from the manufacture, processing, use, and disposal of the new chemical substance at sites you control. Identify the byproducts as specifically as possible by name and CAS Registry Number (if available). You should give the following information:

(1) The specific chemical name; or

(2) A class or range of structures (e.g., C_6-C_{18} fatty acid salts or polychlorinated cyclic and acyclic hydrocarbons in the range C_5-C_{12}); or

(3) The source (e.g., pyrolysis products of cellulose or coal tar residues). If there are no byproducts, enter "None."

Section C—Production, Import, & Use Information (Page 7)

1. Production volume

Estimate the production volume for the first 12 months of production. Also estimate the maximum production volume for any consecutive 12-month period during the first 3 years of manufacture. Provide your estimates in kilograms. Maximum production volume is the maximum amount of the new chemical substance on a 100% basis. Report the amount of pure new chemical substance, not including solvents or other components if the new chemical substance is in a mixture that you expect to produce during any 12-month period (e.g., June 1990 through May 1991). Include in this total amounts produced by persons under contract to you. If part of the amount manufactured is for export, include this amount in your estimates. (You are not allowed to exclude exports from LVE quantity limits after any of the substance is manufactured for domestic use) If you submit a consolidated PMN, make your production volume estimates on a per chemical basis.

For an LVE, the Agency will generally perform the risk assessment under the exemption as if the total amount permissible under the exemption (10,000 kg) were being produced. However, submitters wishing their exemption to be based upon annual production volumes lower than 10,000 kg may so indicate in their exemption notice by marking the binding box adjacent to the production volume space on the form. Submitters who so elect are bound by their election, and if they subsequently wish to increase their maximum production volume under the exemption must submit a new exemption notice cross-referencing the original exemption number on the cover of the notice. If the new exemption is granted, it will supersede the previous exemption.

2. Use information

a. Column (1) category of use: Identify the intended category of use of the new chemical substance by describing its function and application. "Function" is related to the inherent physical and chemical properties of the substance (e.g., degreaser, catalyst, plasticizer, ultraviolet absorber). "Application" refers to the use of the substance in particular processes or products (e.g., a degreaser may be used for cleaning of fabricated metal parts).

Following are some examples of appropriate categories of use:

! A disperse dye carrier for finishing polyester fibers
! A cross-linking agent for epoxy-type coatings for metal surfaces
! A flame retardant for surface application on cotton apparel, textile home furnishings, and exterior canvas products
! A surfactant in automobile spray wax
! A colorant for paper and other cellulosics
! Fiber-reactive dye for nylon carpeting and upholstery
! An antioxidant in fuels oils and lubricants

Column (2) CBI claim.

Column (3) Binding option for category of use (binding option is described at I–F).

Column (4) Estimate the percent of the total production volume that you anticipate will be manufactured for each category of use.

Column (5) CBI claim.

Column (6) Estimate the weight percent of the new chemical substance that will be contained in any formulated mixture, suspension, emulsion, solution, or gel associated with each category of use as manufactured for commercial purposes at sites under your control. Where the substance is distributed from your site in a pure state, enter 100%.

In the example below, a PMN substance will be used for several different uses, including a cross-linking agent where the substance is distributed in a pure state and as a surfactant where the substance is manufactured, then formulated at a weight percent of 4%. Eighty percent of the production volume goes to the first use and 20 percent of the production volume will be used for the second use.

Category of Use	CBI	Binding Option	% of Production	CBI	% in Formulation	CBI
Cross-linking Agent	×	×	80	×	100	×
Surfactant in Automobile Spray Wax	×	×	20	×	4	×

Column (7) CBI claim.

Column (8) Mark to indicate if the category of use is site-limited or, if the substance is intended for industrial, commercial, and/or consumer use, as defined below, estimate the percent of the production volume expected for each category.

Site-limited—The substance will be used only on the contiguous property unit where it is manufactured and not intentionally distributed outside that site except for waste disposal. This includes all factories, storage places, and warehouses at the site. In most cases, this would be an intermediate which is further reacted on-site. If the substance is transported across a public road which bisects a site, it can still be site-limited.

Consumer—The new chemical substance **or products containing the substance** will be used by private individuals in or around a residence, or during recreation, or for any other personal use or enjoyment, e.g., automotive polish, dyed wearing apparel, household cleaners, hunting lure, etc. If a consumer use is identified, then the EPA requests the following information to be provided on an attached continuation sheet: a detailed description of the use of the new chemical substance expected in consumer products and any reactions that occur causing the substance to lose its identity in the consumer product. This section does not apply to situations where the PMN substance has clearly reacted away and lost its identity before the final article is used by consumers (e.g., coatings).

Industrial—The new chemical substance **or products containing the substance** will be used at the site of other manufacturers or processors, e.g., textile dyeing, paint formulation, use of a curable resin to manufacture an article.

Commercial—The new chemical substance **or products containing the substance** will be used by a commercial enterprise providing a consumer service, e.g., use by commercial dry cleaning establishments, use by painting contractors, or use by roofers in commercial building construction.

Mark all boxes, as appropriate. For example, a surfactant in an automobile wax may have a consumer use in liquid wax, a commercial use in auto washes, and an industrial use by automobile manufacturers. Mark the binding option statement where applicable.

Note: You must make separate confidentiality claims for the description of the category of use, the percent of production devoted to each category, the percent in formulation, and category of use information.

The information in this section is used to evaluate potential exposure and release of the new substance. If you wish to provide any additional information which would assist in this analysis, it may be submitted as optional information. You should be aware that the Agency uses conservative default assumptions to estimate exposure in the absence of specific information.

c. Generic use description—Includes degree of containment: For each category of use description which is claimed as confidential, provide a generic description of the category. If such a generic description does not provide a sufficient indication of potential exposure, the description can also describe the degree of containment of the new chemical substance, as shown in the list below; however, a generic use description that solely describes the degree of containment such as "open, non-dispersive use" is not acceptable.

Identify the category of use to which the generic description applies. The generic use should reveal the intended category of use to the maximum extent possible. For example, the specific use of a new substance as an antioxidant in a lubricant could be described generically as a lubricant additive; a fiber-reactive dye for nylon carpeting could be described generically as a dye for fibers.

Degree of Containment

(a) Destructive use	(e.g., fuels, fuel additives, chemical intermediates)
(b) Contained use	(e.g., catalysts used in closed processes, certain photographic chemicals, capacitor fluids)
(c) Open, non-dispersive use	(e.g., printing inks, textiles, dyes, plasticizers, adhesives, liquid paints, resins)
(d) Dispersive use	(e.g., cutting fluids, fabric softeners, automobile tire rubber)
(e) Highly dispersive use	(e.g., fertilizers, salt for de-icing, paint solvents, spray paints)
(f) Other	(describe)

3. Hazard information—Include in the notice a copy or reasonable facsimile of any hazard warning statement, label, material safety data sheet (MSDS), or other information which will be provided to any person regarding protective equipment, engineering controls, or practices for the safe transport, use, or disposal of the new chemical substance. If hazard warning information is not yet prepared, describe the statement you intend to provide, if any. You are not required to develop hazard warning statements for this notice if you do not otherwise intend to do so. Identify copies of hazard warning statements or other hazard information that you attach in Part III, List of Attachments.

Part II—HUMAN EXPOSURE AND ENVIRONMENTAL RELEASE (Pages 8–10)

In sections A and B, you must provide information on manufacturing, processing, and use operations involving the new chemical substance or products containing the new substance. Preparing a chemical for an end use application typically involves several production steps, or operations, where potential human exposure and environmental release could occur. Use section A to provide information on operations that occur at industrial sites you control. Typically, this will involve manufacturing and processing operations. Use section B to describe operations that occur at industrial sites controlled by others. Typically, this will involve processing and end use applications, e.g., dyeing of nylon carpeting, paint spraying of automobiles.

As an example, for a solvent used in automotive paint for automobile manufacture, there are solvent manufacture, paint manufacture, and industrial paint spraying operations. Paint manufacture may occur at a site you control or at a site you do not control. If paint manufacture occurs at a site you control, describe that operation in section A. If paint manufacture occurs at a site you do not control, describe that operation in section B. If processing or end use operations occur at industrial sites you control, they should be described in section A.

In most cases, you will have more specific information on sites you control than sites you do not control. If you do not have specific information on sites controlled by others, describe a typical operation involving the particular processing or end use application based on information available to you and on your experience with similar chemicals. Provide all information requested to the extent to which it is known to or reasonably ascertainable by you. Where EPA has available only limited information on worker exposure and environmental release, its evaluation will be based on reasonable worst case assumptions.

Note that if you are an importer, although you do not have to complete section A or B for operations outside the United States, you may still have to report information in these sections. If there are further industrial processing or use operations after import of the substance, you must describe these operations in section A or B as appropriate.

Because EPA assesses the potential domestic manufacture and processing of imported chemicals, you may choose to complete section A or B for operations outside the United States to assist EPA with this assessment. Providing this information is optional and is not required.

Section A—Industrial Sites Controlled by the Submitter (Pages 8–9)

Complete a separate section A for each type of manufacturing, processing, or use operation involving the new chemical substance at sites you control. If the same operation is performed at more than one site, you are not required to complete a separate section A for each operation, but simply describe the typical operation common to these sites. However, if operations or production rates vary substantially among the different sites, you must provide a separate section A for each different operation.

1. Operation Description

a. Identity—Identify the site which the section describes. If this section describes more than one site, provide additional site identities on a continuation sheet. Indicate the total number of sites at which the operation this section describes will occur. If you mark the confidential box next to this item, all information in it will be treated as confidential.

b. Type—Mark the appropriate box.

c. Amount and duration   Estimate the maximum amount of the new substance (on a 100% new chemical substance (i.e., pure) basis) manufactured, processed, or used in the operation and the duration of the operation. Provide information per batch for batch operations and per day for continuous operations. Base the estimates on the maximum 12 month production provided on page 7 of the PMN form.

d. Process description—Provide a process flow diagram which describes the manufacturing, processing, or use operation involving the new chemical substance.

(1) Identify the major unit operation steps and chemical conversions, including secondary operations involving the new chemical substance, such as interim storage and shipping containers. "Unit operation" means a functional step in a manufacturing, processing, or use operation where substances undergo chemical changes and/or changes in location, temperature, pressure, physical state, or similar characteristics. Include steps in which the new substance is formulated into gels, mixtures, suspensions, solutions, etc. Specify the shipping containers, including expected capacities (e.g., 5 gallon pails, 55 gallon drums, 5,000 gallon tank trucks, 20,000 gallon rail cars).

(2) Indicate in your diagram the entry and exit point of all feedstocks (e.g., reactants, solvents, catalysts) used in the operation, products, recycle streams, and wastes. Identify each feedstock and specify its approximate weight (by kg/day for continuous operations or kg/batch for batch operations). Include cleaning chemicals, and state how often they are used (e.g., every day, every batch, monthly).

(3) Number all points from which the new chemical substance and substances containing the new chemical substance will be released to the environment or to control equipment, including small or intermittent releases (e.g., some cleaning releases, drum residues, etc.) and trace amounts of the new chemical substance. Do not include accidental releases. Including fugitive emissions is optional.

(4) Mark the box if you wish to indicate your willingness to have your process description binding.

Sample Process Description for a Manufacturing Operation:
Sample Process Operation
All amounts in kg/day
NCS = New Chemical Substance

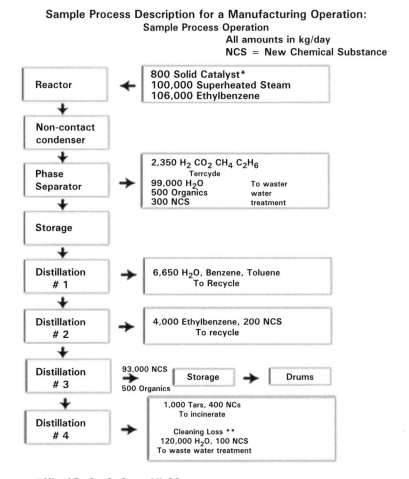

* Mix of Fe_2O_3, Cr_2O_3, and K_2CO_3

**Note: Entire process cleaned once annually with steam
and water (120,000 kg total). NCS loss from cleaning is 100 kg.

Sample Processing Operation

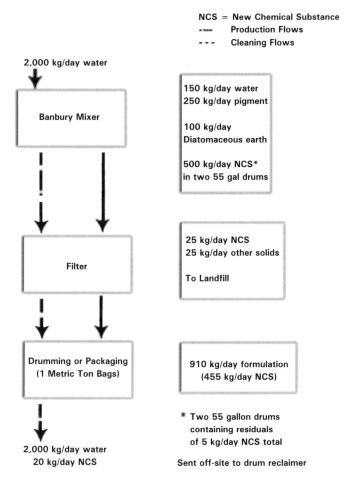

NCS = New Chemical Substance
-—   Production Flows
- - -   Cleaning Flows

2,000 kg/day water

Banbury Mixer

150 kg/day water
250 kg/day pigment

100 kg/day
Diatomaceous earth

500 kg/day NCS*
in two 55 gal drums

Filter

25 kg/day NCS
25 kg/day other solids

To Landfill

Drumming or Packaging
(1 Metric Ton Bags)

910 kg/day formulation
(455 kg/day NCS)

2,000 kg/day water
20 kg/day NCS

* Two 55 gallon drums
  containing residuals
  of 5 kg/day NCS total

Sent off-site to drum reclaimer

To on site waste water treatment

2. Occupational Exposure

Column (1)—Worker activity: Describe each specific activity in the operation during which workers may be exposed to the new chemical substance. Such activities may include charging reactor vessels, sampling for quality control, transferring materials from one work area to another, drumming, bulk loading, changing filters, and cleaning equipment, etc.. Activities must be described even if workers wear protective equipment. (Material Safety Data Sheets indicating recommended protective equipment should be submitted as part of Hazard Information in Part I, Section C, subsection 3 of the notice form.)

Column (2)—CBI claim for worker activity.

Column (3)—Protective equipment and engineering controls: Provide information on the specific types of protective equipment and engineering controls that will be employed to protect the worker from potential exposure to the new chemical substance, i.e., gloves, goggles, 21c respirator, 19c respirator, closed containment system, nitrogen blanket, etc.

Column (4)—Binding option for protective equipment and engineering controls.

Column (5)—Physical form: Indicate the physical form [e.g., solid (crystals, granules, powder, dust), liquid (solution, paste, slurry, emulsion, mist, spray), gas (vapor, fume), wet press cake] of the new substance and its weight percentage (if in a mixture) at the time of exposure, even if workers wear protective equipment.

Column (6)—Binding option for physical forms.

Column (7)—CBI claim for physical form.

Column (8)—Maximum number of workers exposed: Estimate the maximum number of workers involved in each specific activity, based on the estimated maximum 12-month production volume.

Column (9)—CBI claim.

Column (10)—Maximum duration in hours/day: Enter the maximum duration that any one worker will engage in the activity in hours/day, e.g., 8 hours/day.

Column (11)—Maximum duration in days/year: Enter the maximum duration that any one worker will engage in the activity in days/year, e.g., 200 days/ year, based on the estimated maximum production volume.

Column (12)—CBI claim.

Note that you must make separate confidentiality claims for the description of worker activity, physical form of the new substance, number of workers exposed, and duration of exposure. (See 2, 7, 9, 12.)

3. Environmental Release and Disposal

Column (1)—Release point number: For each release point indicated in the process description (Part II, Section A, subsection 1d(3) of the notice form, enter the corresponding number. If you indicated more than 5 release points, make a continuation sheet to cover them.

Column (2)—Amount of new chemical released at release point: Estimate the amount of new chemical (in kg/day for continuous operations or kg/batch for batch operations) that will be released from the release point directly into either (a) the environment or (b) into control technology (in kg/day or kg/batch). Base your estimate on the expected maximum twelve-month production volume. EPA is particularly interested in the amounts of chemicals used and frequencies of cleaning of equipment and releases from transport containers, including the location (if different from that of manufacture or processing) of drum recyclers, tank truck cleaning facility, etc.)

Column (3)—CBI claim.

Column (4)—Medium of release: Enter the medium [stack air, fugitive air (optional), surface water, on-site or off-site land or incineration, POTW, or other (specify)] into which the release stream discharges (whether or not control technology is used). You do not have to identify fugitive air releases, but if EPA reviewers consider that such air releases are likely and estimates are not provided in the submission the reviewers will estimate reasonable worst-case fugitive air releases. In estimating reasonable worst-case releases we generally consider vapor pressure and type of use, including the location if different from that of site manufacture or processing.

Column (5)(a) Release/control technology/efficiency: For a release to air or water, describe the type of technology used to control the release of the new chemical and the efficiency of the control technique. Examples of control technologies include carbon filter, scrubber, and biological treatment (primary, secondary, etc.). Use the optional binding box to indicate your willingness to be bound to control technology described. Attach optional information such as data and methods of waste treatment or purification efficiency studies.

(5)(b) Release after control technology: Enter the estimated amount released to the environment after control technology (in kg/day). Enter "none" if no control technology is used and the substance is released directly into the environment.

For disposal on land, describe the landfill site construction (including liners) and handling procedures. Describe landfill containers.

You may wish to optionally attach efficiency data for control technologies used. For example, carbon adsorption removal efficiency of the new chemical substance from an aqueous stream could help EPA estimate release of the new chemical substance from a facility using carbon adsorption treatment.

Column (6)—CBI claim.

Column (7)—Destination of water releases: Mark the appropriate box and/or specify other destinations of water releases. [i.e., POTW (Publicly Owned Treatment Works), navigable waterway or specify other]. Provide the name of the POTW receiving water releases and the NPDES (National Pollutant Discharge Elimination System) number for the POTW, navigable waterway or other direct discharger. This 9-digit number is assigned by EPA or the State under the authority of the Clean Water Act. When appropriate, contact your POTW to obtain its NPDES number.

Note that you must make separate confidentiality claims for the release number, amount of new chemical substance released, control technology disposal information, and the destination of the releases to water.

Section B—Industrial Sites Controlled by Others (Page 10)

Complete a separate Section B (continuation pages for second and subsequent) for each type of processing or use operation associated with each category of use specified in Part 1, Section C, subsection 2(a) at industrial sites you do not control. If the same operation is performed at more than one site, then describe the typical operation and enter the number of sites in the space provided.

Describe each typical processing or use operation to the maximum extent possible from information known to or reasonably ascertainable by you. Information may be provided as ranges or estimates.

1. Operation description—Generally self-explanatory—the information requested here differs from that requested for Section A Operation Description only by the additional worker activity information requested. EPA is particularly interested in the chemicals used and frequencies of cleaning of equipment and transport containers.

2. Worker Exposure/Environmental Release

Column (1)—Activity letter: Provide the letter for each activity from the diagram above and complete columns 2–8 for each worker activity associated with the letter.

Column (2)—Maximum number of workers exposed: Estimate the maximum number of workers exposed to the new substance during the activity.

Column (3)—CBI claim.

Column (4)(a)—Duration of exposure—hours/day: Estimate the duration of exposure of the new chemical substance per worker in hours per day.

Column (4)(b)—Duration of exposure—days/year: Estimate the duration of exposure of the new chemical substance per worker in days per year.

Column (5)—CBI claim.

Column (6)—Physical form, protective eqpt, eng'g controls: Provide information on the physical form [e.g., solid (crystals, granules, powder, dust), liquid (solution, paste, slurry, emulsion, mist, spray), gas (vapor, fume), wet press cake] of exposure and percent new chemical substance (if in a mixture), and protective equipment and engineering controls employed to safeguard the worker from potential exposure associated with the new chemical substance, i.e., gloves, goggles, respirators, etc.

Column (7)—Percent in the product formulation: Estimate the percent in the product formulation of the new chemical substance to which the worker is potentially exposed during the activity.

Column (8)—CBI claim.

Column (9)—Release points: Enter the number given to each release point in reference to the process diagram above and complete 9–13 for each release point identified (use continuation pages if necessary).

Column (10)—Maximum release: Provide an estimate for the maximum amount of new chemical substance in kg per year that may be released to the media specified in Column (12) (see below) under typical operating conditions. Provide this information for releases (a) directly to the environment or (b) into control technology to the environment in kg per day for continuous operations or kg per batch for batch operations. EPA is particularly interested in the amounts of chemicals used and frequencies of cleaning of equipment and releases from transport containers.

Column (11)—CBI claim.

Column (12)—Medium of release: Describe medium of release [stack air, fugitive air (optional), surface water, on-site or off-site land or incineration, POTW, or other (specify)] and the control technology(ies) used to limit the release of the new substance to the environment to the extent that the information is known to you.

Column (13)—CBI claim.

(14)—By-products: Identify all by-products resulting from the reaction to the extent that the information is available to you.

Note that you must make separate confidentiality claims for the numbers of workers exposed, the duration of exposure, the percentage of the product in formulation, the amount of new substance released, and the control technology.

(OPTIONAL) POLLUTION PREVENTION AND RECYCLING INFORMATION (Page 11)

On the form itself and in Section I of this Manual we describe the Agency's goals in providing this mechanism to inform us of environmental benefits which may be available from your new substance. Below are suggestions for organizing such information to be most useful to improve the review of your substance: Part of the material to provide is about characteristics of the substance and part relates to your process description and calls for a discussion of how a shift to the new chemical substance may enable the use of the more desirable process. If you wish your PMN substance to be considered for the Agency's EPA New Chemicals Pollution Prevention Recognition Project, you should tell us so in this information.

I. Product Information

Will this substance be produced to address an environmental problem or in response to environmental regulation? (One example would be CFC substitutes.) Do you wish your substance to be considered for the Agency's EPA New Chemicals Pollution Prevention Recognition Project?

II. Toxicity Information

EPA is interested in information describing possible reductions in toxicity of a new chemical substance as compared to that of already commercialized chemical substances for which the new substance may substitute. Such information may demonstrate that the new chemical substance may be a viable safer substitute.

III. Technical Information

a. Process Chemistry

Chemical approaches which may minimize waste production include the following:

1. Selection of synthetic method: high yield, minimal side reactions, minimization of coproducts unless they are of commercial value.

2. Choice and purity of feedstocks, reagents, catalysts, solvents, etc. Reduction/elimination of undesirable feedstocks, solvents (e.g., VOC's), catalysts, etc.

3. Optimization of reaction conditions to minimize side reactions.

4. Optimization of stoichiometry to minimize presence of excess reactants.

5. Product purification.

6. Use of advanced chemical technology such as shape-selective zeolite catalysts, stereoselective synthesis, and permselective membranes for molecular separations (liquid phase).

b. Product Substitution/Source Reduction

1. Perceived advantages related to the use of the new chemical substance compared to existing substances. Comparison of the relative risk of the new chemical to existing chemicals (for example, volumes, releases, exposure, human health effects and ecotoxicity) of the substances. A new chemical that causes less pollution may nonetheless be more toxic than an existing chemical and therefore the risk to health and the environment may be higher.

2. Changes in product composition or physical state.

Consider potential changes in product compositions or physical states which may reduce releases and exposures. For example, using a more concentrated product may decrease the amount of release or waste generated; or reformulating a powder into a paste to mitigate inhalation concerns.

3. Enhanced product life or durability.

Improvements which may be expected from enhancements to product life or durability of new chemical substances over existing substances. For example, a more durable product could result in less solid waste going to landfills.

4. Changes in product effectiveness/effect of product performance at lower concentrations (significant orders of magnitude).

New chemical substances which may be more effective than existing substances, or that may be as effective at significantly lower concentrations than an existing chemical may result in reduced releases and exposures.

5. Packaging (transportation, concentration, disposal, life-cycle analysis, weight loss, etc.).

Examples of packaging with pollution prevention in mind may include (A) reusable packaging materials, or (B) making the new chemical substance more concentrated to minimize size or volume of packaging.

c. Process Modifications

1. Engineering technology changes.

In some cases, existing equipment which may be used to produce new chemicals may not be the best technology from the standpoint of pollution prevention.

Consider alternative technologies for the major unit operations (i.e., separation, filtration, etc.) in the process and their potential impacts on releases and exposures. Some examples of alternative technologies include: (A) stream stripping instead of air stripping, (B) pan filter versus rotary drum filter or centrifuge or, (C) flash dryer versus spray dryer.

2. Equipment/piping/layout changes

Equipment/piping/layout alternatives can have an effect on releases and exposures. Some examples of these changes include: (A) redesigning equipment and piping to reduce the volume of material drained for batch changes and for cleaning operations, (B) installing bellow-sealed valves versus conventional valves to reduce VOC emissions and worker exposure, or, (C) use of new resources such as by-product steam from another process.

3. Operating conditions

Optimize operating conditions at major unit operations (including reactors and separation equipment). Available supporting laboratory information (i.e., experimental design data) may be collected and submitted describing reduction of chemical exposure using the new PMN substance, reduced emissions to the workplace or environment, etc. An example might be use of a new surfactant to reduce the amount of solvent needed in cleanup operations.

4. Automation

Potential automation/control steps may have pollution prevention or exposure implications. For example, a batch process can be successfully automated to reduce off-spec. product and spillage, which reduces waste. A manual packaging system can be automated to reduce worker exposures.

5. Closed systems

Reduced exposures, emissions, or elimination of isolated intermediates.

d. Operating Practices

Examples of operational and administrative changes which could reduce pollution and exposure are:

1. Procedural measures

Increasing drain time from 15 to 20 minutes could reduce leftover material in a tank.

2. Loss prevention

Installing overflow alarms in tanks could reduce overflows or releases.

3. Management practices

Reducing inventories of toxics to minimize the consequences of emergencies.

4. Waste stream segregation

Segregating waste streams to avoid cross-contaminating hazardous and nonhazardous materials could allow recycling of some waste. For example, changing waste segregation to separate organics from an aqueous waste stream.

5. Material handling improvements

Changing from small-volume containers to bulk or reusable containers could reduce releases of residue.

6. Production scheduling

Larger batch sizes can be made, which could reduce number of production runs, cleanup waste, and worker exposure.

7. Safety

Examples would be use of substances which are less explosive, less flammable, or less corrosive. Use of less corrosive substances may also serve to decrease cost and maintenance, as well as increase life, of equipment.

e. Reuse/Recycling/Reclamation

Identify potential reuse/recycling/reclamation (both on-site and off-site opportunities) of the waste streams. Some examples of the reuse/recycling/ reclamation opportunities are:

1. Use of recycled materials in original process

Returning recovered solids from a filtrate (through the use of a settling tank) to the reactor. Wastes from unrelated processes may be used in feedstock to make new substances.

2. Recovery and recycling of resources utilized in the process

Installing vapor recovery systems to capture and return vaporous emissions. Feedstock or off-spec. substance previously discarded may be recycled back into the process.

3. Reuse or processing of by-products/coproducts of the original process for other processes.

The off-acid gas from a reactor could be absorbed with water to convert it into dilute acid and sold as a by-product.

f. Energy Savings

Almost any change in an industrial process will entail some change in the amount and form of energy consumed in that process. How might the new chemical substance result in savings of energy resources in any phases of manufacture, processing, or use?

Energy conservation is a form of pollution prevention. For example, reduced energy use decreases the quantity of fossil fuels burned and the amount of air pollutants generated. Reduced boiler operation also reduces the discharge of waste cooling water blowdown and boiler blowdown. Purification of raw water to produce boiler feedwater by ion exchange or other processes produces wastes such as regeneration chemicals. Reduced boiler operation could also reduce this waste stream. Another example might be shifting from thermal polymerization/ curing to radiation polymerization/curing.

g. Alternative Treatment Methods

Consider efficiency of alternative treatment methods and the compositions and quantities of the input PMN waste streams and output release streams. For example:

1. Use of ammonia instead of sodium hydroxide to neutralize a process, allowing the excess to be vacuum stripped and possibly recycled, rather than removal by water washes which create a caustic waste requiring treatment and/or disposal as a hazardous waste.

2. Replacement of organic solvents with water, eliminating VOC emissions.

h. Alternative Waste Disposal Methods

Compare release media (land, air, water) in light of the PMN waste streams being disposed, including quantity of release, composition of the waste, and applicable regulatory limitations. For example, an aqueous ammonia waste stream may be converted into fertilizer for agricultural use.

Part III—LIST OF ATTACHMENTS (Page 12)

Attach any continuation sheets for sections of the form, test data and other data (including structure-activity information), and optional information after the last page of the form. Clearly identify the attachment and the section to which it relates, if appropriate. Number consecutively the pages of the attachments. Enter the inclusive page numbers of each attachment. Enter the total number of pages in the notice on page 1 of the form.

Mark the "Confidential" box next to any attachment **name** you claim as confidential. Read "E" and "H" in Section I of this manual for guidance on how to claim any information in an attachment as confidential. You must include with the sanitized copy of the notice a sanitized version of any attachment in which you claim information as confidential.

Physical and Chemical Properties Worksheet (Optional)

A worksheet which assists EPA's review of the physical/chemical property information you submit is provided on the last page of the form. Providing physical/chemical property information in this format is optional. However, all physical/chemical properties data in your possession or control must be submitted with your notice. If you submit this worksheet, identify it on the List of Attachments.

APPENDIX A

EXAMPLES OF TEST DATA

Following is a list of the types of test data which you must attach to the notice form if it is in your possession or control. This list is illustrative, not exhaustive.

Physical and Chemical Properties and Environmental Fate Data

- Chromatograms
- Spectra (ultraviolet, visible, infrared)
- Density/relative density
- Solubility in water
- Melting temperature
- Boiling/sublimation temperature
- Softening point
- Vapor pressure
- Dissociation constant
- Particle size distribution
- Octanol/water partition coefficient
- Henry's law constant
- Volatilization from soil
- pH
- Flammability
- Explodability
- Adsorption/desorption characteristics
- Photochemical degradation
- Viscosity
- Odor
- Hydrolysis
- Thermal analysis
- Chemical analysis
- Chemical oxidation
- Chemical reduction
- Biodegradation
- Transformation to persistent or toxic products

Health Effects Data

- Mutagenicity
- Carcinogenicity
- Teratogenicity
- Neurotoxicity/behavioral effects
- Pharmocological effects
- Mammalian absorption
- Distribution
- Metabolism and excretion
- Cumulative, additive, and synergistic effects
- Acute, subchronic, and chronic effects
- Structure/activity relationships
- Epidemiology
- Reproductive effects
- Clinical studies
- Dermatoxicity
- Phototoxicity
- Irritation
- Sensitization
- Allergy
- Skin staining

Environmental Effects Data

- Microbial bioassay
- Algal bioassay
- Aquatic macrophyte bioassay
- Seed germination and root elongation
- Seedling growth
- Plant uptake
- Acute toxicity to invertebrates
- Life cycle test on invertebrates
- Acute toxicity to fish
- Early life stage (fish)
- Avian dietary/reproduction
- Bioaccumulation/ bioconcentration
- Model ecosystem studies
- Physical environment impairment effects
- Flesh staining of aquatic organisms

APPENDIX B

This Appendix is an annotated version of the form distributed internally by the Minnesota Mining and Manufacturing Co., for its personnel, which many submitters may find useful as an adjunct to the instructions provided by the Agency:

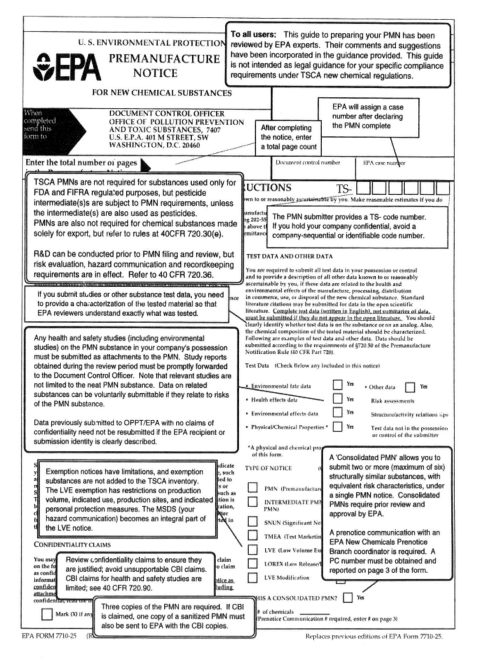

Public reporting burden for this collection of information is estimated to average 110 hours per response, including time for searching existing data sources, gathering and maintaining the data needed, and completing and reviewing the collection of information. Send comments regarding the burden estimate or any other aspect of this collection of information, including suggestions for reducing this burden to Chief, Information Policy Branch, PM-223, U.S. Environmental Protection Agency, 401 M. St., S.W., Washington, D.C. 20460 and to the Office of Management and Budget, Paperwork Reduction Act (2070-0012), Washington, D.C. 20503.

CERTIFICATION

I certify that to the best of my knowledge and belief:

1. The company named in Part I, section A, subsection 1a of this notice form intends to manufacture or import for a commercial purpose, other than in small quantities solely for research and development, the substance identified in Part I, Section B.

2. All information provided in this notice is complete and truthful as of the date of submission.

3. I am submitting with this notice all test data in my possession or control and a description of all other data known to or reasonably ascertainable by me as required by §720.50 of the Premanufacture Notification Rule.

Additional Certification Statements:

If you are submitting a PMN, Intermediate PMN, statement that applies:

> PMN notices are to be based on actual commercial intent, not speculation. The person who certifies with their signature below attests that the information provided is correct, to the best of his/her knowledge, and that test data are submitted in compliance with 40 CFR 720.50.

☐ The Company named in Part I, Section A has remitted the fee of $2500 specified in 40 CFR 700.45 (b), or

☐ The Company named in Part I, Section A has remitted the fee of $1000 for an Intermediate PMN (defined @ 40 CFR 700.43) in accordance with 40 CFR 700.45(b), or

☐ The Company named in Part I, Section A is a small a fee of $100 in accordance with 40 CFR 700.45 (b).

> Notice(s) for chemical intermediates submitted together in sequence with a final PMN are eligible for a reduced review fee of $1,000 for each intermediate notice.

If you are submitting a **low volume exemption (LVE)** application in accordance with 40 CFR 723.50 (c) (1) or a **Low release and low exposure exemption (LoREX)** application in accordance with 40 CFR 723.50 (c) (2), check the following certification statements:

☐ The manufacturer submitting this notice intends to manufacture or import the new chemical substance for commercial purposes, other than in small quantities solely for research and development, under the terms of 40 CFR 723.50.

☐ The manufacturer is familiar with the terms of this section and will comply with those terms; and

☐ The new chemical substance for which the notice is submitted meets all applicable exemption conditions.

☐ If this application is for an LVE in accordance with 40 CFR 723.50 (c)(1), the manufacturer intends to commence manufacture of the exempted substance for commercial purposes within 1 year of the date of the expiration of the 30 day review period.

> Amended LVE rules took effect May 30, 1995, and are listed at 40 CFR 723.50. EPA must be able to conclude that manufacture, processing and use of the LVE substance will not result in an unreasonable risk to health or the environment. Review applicable exemption rules for LVE or LoREX applications.

The accuracy of the statements you make in this notice should reflect your best prediction of the anticipated facts regarding the chemical substance described herein. Any knowing and willful misinterpretation is subject to criminal penalty pursuant to 18 USC 1001.

	Confidential
Signature and title of Authorized Official (Original Signature Required)	**Date**
Signature of agent - (if applicable)	

> If your company's identity is CBI, the sanitized copy cannot be signed. Otherwise, both the CBI and sanitized original copies require an original certifcation signature.

Part I -- GENERAL INFORMATION

Section A -- SUBMITTER IDENTIFICATION

Mark (X) the Confidential box next to any subsection you claim as confidential.

			Confidential

1a. Person Submitting Notice (in U.S.)

Name of authorized official — Position

If the company identity is CBI, this section is also CBI and will be so marked.

Mailing address (number and street)

City, State, ZIP Code

b. Agent (if applicable)

Name of authorized official — Position

Complete the 'Agent' section only if an agent outside your company assists with preparing the notice.

The agent must also sign the general certification on page two of the form.

Company

Mailing address (number)

City, State, ZIP Code

Telephone — Number

c. If you are submitting this notice as part of a joint submission, mark (X) this box.

Joint Submitter (if applicable)

A Joint Submitter files a PMN form containing information known to them and not available to the original submitter. They must certify their information and sign their PMN notice. The PMN review period will not commence until EPA has received all required notice information from you and the joint submitter. Note: It is also possible for joint submitters to be equal partners, with all information shared between them.

A joint submitter is not the same as a third party that files a 'letter of support' which provides confidential information to EPA in support of your premanufacturing notice. (e.g., chemical identity of their proprietary reactant used in production of your polymer PMN; or, documentation of their polymer exemption substance you use to manufacture a polymer, or chemical identity of a PMN substance, whose identity is held confidential from you, that wish to import.)

2. Technical Contact (in U.S.)

The technical contact should be a person available by telephone during normal business hours who can provide EPA with additional information on the new chemical substance during the notice review period. EPA also encourages listing a back-up contact person.

Company

Mailing address

If the company is held as CBI, the contact is usually, but not necessarily, claimed as CBI.

City, State, ZIP Code

3. If you — and E

List any prenotice communication numbers, exemption notice number, or Bona fide notice identifier assigned by EPA. If there are none that apply, mark the negative certification boxes to the right.

Mark (X) if none ☐

4. If you substance by EP PMN

Mark (X) if none ☐

5. If you have submitted a notice of Bona fide intent to manufacture or import for the chemical substance covered by this notice, enter the notice number assigned by EPA.

Mark (X) if none ☐

6. Check the boxes to indicate your intent to manufacture in the US, to import the PMN chemical, or both.

1. ☐ Manufacture Only   ☐ Binding Option Mark (x)
2. ☐ Import Only   ☐ Binding Option Mark (x)
3. ☐ Both

Part I -- GENERAL INFORMATION -- Continued

▶ Section B -- CHEMICAL IDENTITY INFORMATION

> If a third party will file a letter of support or act as a joint submitter and provide proprietary chemical substance information, check this box.

Mark (X) the "Confidential" box next to ... substance inventions.

Complete either item 1 (Class 1 or 2 substances) or 2 (Polymers) as appropriate. Complete all other items.

If another person will submit chemical identity information for you (for either item 1 or 2), mark (X) the box at the right. Identify the name, company, and address of that person in a continuation sheet. ☐ Confi-

> Class 1: a single molecular entity that can be represented by a single, definite structural diagram

... itions of class 1 and class 2 substances, see the ...

☐ 1 ☐ Class 1 or 2 ☐ Class 2

... mical Abstracts (CA) Name that is consistent ...

For Class 1 substances ... index name must be provided. For Class 2 substances ... be provided, whichever is appropriate based on CA 9CI nomenclature rules and cont ...

> Class 2: a substance whose composition cannot be represented by a single, definite chemical structural diagram. Class 2 substances include substances of unknown, variable, or uncertain composition, complex mixtures or reaction products, or well-defined substances without specific structures.

> CAS is a contractor to EPA. The CAS Inventory Expert Service can provide a TSCA-compliant Ninth Collective Index (9CI) Chemical Abstracts (CA) name for your PMN substance; only a name developed by this procedure meets 'Method 1' criteria. An error by CAS will not hold up review of your PMN if your chemical name was developed by Method 1. Include the nomenclature assignment documentation as an attachment to the PMN.
>
> An alternative method of providing a name, such as a STN database report, qualifies as 'Method 2'. If you provide a chemical name using Method 2, and EPA determines the name to be in error, review of your PMN will be halted until the name is corrected and the entire notice may need to be resubmitted along with the correct name.

d. Molecular formula and CAS Registry Number (if a number already exists for the substance)

> The molecular formula must give the correct identity and number of atoms of each element contained in the PMN molecule. This is required for a Class 1 substance and for Class 2 substances that have a definite molecular formula.

CAS #

> Provide a CAS Registry number if one has already been assigned by CAS. Seeking the assignment of a new CAS registry number may jeopardize confidentiality, and is not required for your notice.

e. For a **class 1** substance, provide a complete and correct chemical structure diagram. For a c ... precursor substances with their respective CAS Registry Numbers. (2) Describe the nature of the reaction or process. (3) indicate the range of composition and the typical composition (where appropriate). (4) Provide a correct representative or partial chemical structure diagram, as complete as can be known, if one can be reasonably ascertained.

> A Class 1 structural diagram should clearly and completely indicate the identity of the atoms and the nature and location of the bonds joining the atoms. Ionic charges and known stereochemical details should be provided. Carbon atoms in ring systems do not need to be explicitly shown.

> Class 2 substances: follow the requirements in (e.) above. Provide partial, or complete, or representative structural diagrams to the extent possible. The diagram should indicate the characteristic or variable compositional structural components of the substance.
>
> If you are unsure of the exact structure, provide the best information you have and indicate that it is your best estimate of the chemical structure.

> Polymer information is reported on Page 5.

☐ Mark (X) this box if you attach a continuation sheet.

Part I -- GENERAL INFORMATION -- Continued

▶ **Section B -- CHEMICAL IDENTITY INFORMATION -- Continued**

2. **Polymers** (For a definition of polymer, see the **Instructions Manual**.)

Confi-
dential

a. Indicate the number-average weight of the lowest molecular weight composition of the polymer you intend to manufacture.
Indicate **maximum** weight percent of low molecular weight species (not including water)
500 and below 1,000 absolute molecular weight of that composition.

Describe the methods of measurement or the basis for your estimates: GPC ☐

> GPC is a well-established method for establishing Mn and weight fractions below 500 and 1,000 Daltons. It may not be useful for all polymer evaluations, however. Solubility problems and misleading calibration standards can lead to questionable GPC data.

i) **lowest** number average molecular weight: _____

ii) maximum weight % below 500 molecular weight: _____

iii) maximum weight % below 1000 molecular weight: _____

☐ Mark (X) this box if you attach a continuation sheet.

> Correction estimates can be documented and presented in cases where only a soluble fraction of the polymer can be analyzed.

b. You must make separate confidentiality claims for monomer or other reactant identity.
  Mark (X) the "Confidential" box next to any item you claim as confidential.
  (1) -- Provide the specific chemical name and CAS Registry Number (if a number exists) for each monomer used in the manufacture of the polymer.
  (2) -- Mark (X) this column if entry in column (1) is confidential.
  (3) -- Indicate the **typical** weight percent of each monomer or other reactant in the polymer.
  (4) -- **Mark (X) the identity column if you want a monomer or other reactant used at two weight percent or less to be listed as part of the polymer description on the TSCA Chemical Substance Inventory.**
  (5) -- Mark (X) this column if entries in columns (3) and (4) are confidential.
  (6) -- **Indicate** the **maximum** weight percent of each monomer or other reactant that may be present in the polymer manufactured for commercial purposes.
  (7) -- Mark (X) this column if entry in column (6) is confidential.

> Indicate the maximum weight percent of each unreacted monomer or other reactant.

Monomer or other reactant and CAS Registry Number (1)	Confi-dential (2)	Typical composition (3)	Identity Mark (X) (4)	Confi-dential (5)	Maximum residual (6)	Confi-dential (7)

> List all reactants that are used in the manufacture of the polymer including those used at 2 percent by weight or less. This includes monomers, free radical initiators, chain transfer agents, chain terminating agents, and cross-linking agents.
>
> Each prepolymer to be charged should be identified by a correct chemical name and applicable CAS Registry number (for the prepolymer as a whole)
>
> Review which components & data require confidential check-off.

> Composition: list the typical percentage you intend to make. You will be able to modify the % level as long as all components used to name the polymer are still present. (This is true for reactants used at >2% or those at 2% or less with column 4 checked. If listed reactants are at 2% or less, without column 4 checked, they can be eliminated, but they cannot exceed 2%. See note on Identity Mark.
>
> Reactants at > 2% by weight are automatically included in the TSCA Inventory description of the polymer.
>
> Identity Mark, column 4: check this only if you need to include a minor (2% or less) reactant in the polymer name. This will allow a minor component to be increased above 2% by weight without the need for a new PMN. However, an Identity Marked reactant must always be intentionally present.

☐ Mark (X) this box if you attach a continuation sheet.

c. Please identify which method you used to develop or obtain the specified chemical name.
  ☐ Method 1

> Indicate the method of nomenclature determination, and provide the assigned chemical name in (d.) Include the nomenclature documentation as an attachment to the PMN.
>
> Method 1 is reserved for nomenclature determinations made by CAS Inventory Expert Service. (Contact CAS at 1-800-631-1884 and ask for the Inventory Expert Service.)

d. The currently correct chemical name.

Inventory listings for similar polymers.

e. Provide a correct structural diagram, if one can be reasonably ascertained.

> Represent the polymer with a structural diagram, to the extent it can be shown.
>
> Random polymers cannot be definitively represented, as different hypothetical monomer combinations can occur, but a representative structural model must still be provided, as complete as can be reasonably ascertained by a chemist. Merely indicating "Random polymer" is not acceptable to EPA.
>
> Provide a basic representation of the monomer structural units and linkages formed during polymerization, the functional groups present, the range and typical values for the number of repeating structural units, and the relative molar ratios of precursors.
>
> Polymers that are siloxanes, silicones, certain polyglycols, and vinylacetals are named on the TSCA inventory by the repeat groups rather than their starting materials.

☐ Mark (X)

Part I -- GENERAL INFORMATION -- Continued

Section B -- CHEMICAL IDENTITY INFORMATION -- Continued

3. Impurities
(a) – Identify each impurity that may be reasonably anticipated to be present in the chemical substance as manufactured for commercial purposes. Provide the CAS Registry Number if available. If there are unidentified impurities, enter "unidentified."
(b) --Estimate the maximum weight % of each impurity. If there are unidentified impurities, estimate the total weight %.

Impurity and CAS Registry Number (a)	Maximum Percent (b)	Confi-dential
	%	
	%	
	%	
	%	
	%	
	%	
	%	

An impurity is a chemical substance that you do not intend to be present with your PMN substance. Your intent is thus the primary determinant as to impurity status.

List all known impurities that are unintentionally present, regardless of their weight percent. List the expected maximum percentage of each impurity and indicate its confidentiality status. Unreacted feedstocks to the reaction should be listed.

An impurity may be suggestive of chemical precursors or chemical processes. Review whether the identity of certain impurities justifies CBI protection.

☐ Mark (X) this box if you attach a continuation sheet.

4. Synonyms – Enter any chemical synonyms for the new chemical substance identified in subsection 1 or 2. [Confidential]

Synonyms include common chemical names used in scientific or technical literature, and code numbers or code names referenced in the PMN and its attachments. Review which synonyms may be proprietary and require CBI protection. Chemical name synonyms must be consistent with the chemical structure and thus not misleading.

☐ Mark (X) this box if you attach a continuation sheet.

5. Trade identification -- List trade names for the new chemical substance identified in subsection 1 or 2.

List any trade name that is used or will be used for the neat chemical substance, whether or not they are registered brands / trade marks. Review CBI status.

☐ Mark (X) this box if you attach a continuation sheet.

6. Generic chemical name – If you claim chemical identity as confidential, you **must** provide a generic chemical name for your substance that reveals the specific chemical identity of the new chemical substance to the maximum extent possible. Refer to the TSCA Chemical Substance Inventory, 1985 Edition, Appendix B for guidance on developing generic names.

If the chemical identity is confidential, a generic chemical identity that is as specific as possible is required. The generic name should reveal the chemical identity to the maximum extent possible, must be consistent with the chemical structure, and must not be misleading.

☐ Mark (X) this box if you attach a continuation sheet.

7. Byproducts -- Describe any byproducts resulting from the manufacture, processing, use, or disposal of the new chemical substance. Provide the CAS Registry Number if available.

Byproduct (1)	CAS Registry Number (2)	Confi-dential

List any byproducts that you reasonably anticipate will result from the manufacture, processing, use and disposal of the new chemical substance at sites under your control. Provide the specific chemical name, CAS numbers if they can be determined, and the confidential status of the byproducts formed. If no byproducts are formed, enter "None".

☐ Mark (X) this box if you attach a continuation sheet.

Part I -- GENERAL INFORMATION -- Continued

Section C -- PRODUCTION, IMPORT, AND USE INFORMATION:

Your production volumes should be estimated as accurately as possible, but you are not bound to your estimates unless you check the binding option box. Include amounts you intend to produce for export. Provide an estimate of the first 12 months of production and the maximum production volume for any consecutive 12 month period during the first 3 years of commercial manufacture.

LVE notices are limited to a maximum of 10,000 kg for any 12 month period.

Are there some potential concerns for your PMN/LVE substance? Setting a binding production limit may be a useful risk management option.

Maximum first 12-month production (kg/yr) (100% new chemical substance basis)	Maximum 12-month production (kg/yr) (100% new chemical substance basis)	Confidential	Binding Option Mark (x)

2. Use Information -- You must make separate confidentiality claims for the description of the category of use, the percent of production volume devoted to each category, the formulation of the new substance, and other use information. Mark (X) the "Confidential" Box next to any item you claim as confidential.

a. (1) -- Describe each intended category of use of the new chemical substance by function and application.
   (2) -- Mark (X) this column if entry in column (1) is confidential business information (CBI).
   (3) -- Indicate your willingness to have the information provided in column (1) binding.
   (4) -- Estimate the percent of total production for the first three years devoted to each category of use.
   (5) -- Mark (X) this column if entry in column (4) is confidential business information (CBI).
   (6) -- Estimate the percent of the new substance as formulated in mixtures, suspensions, emulsions, solutions, or gels as manufactured for commercial purposes at sites under your control associated with each category of use.
   (7) -- Mark (X) this column if entry in column (6) is confidential business information (CBI).
   (8) -- Indicate % of product volume expected for the listed "use" sectors. Mark more than one box if appropriate. Mark (X) to indicate your willingness to have the use type provided in (8) binding.
   (9) -- Mark (X) this column if entry(ies) in column (8) is (are) confidential business information (CBI).

Production volume estimates relate to market intent and are normally claimed confidential by submitters.

Category of use (1) (by function and application i.e. a dispersive dye for fin	CBI (2)	Binding Option Mark (x) (3)	Production % (4)	CBI (5)	% in Formulation (6)	CBI (7)	% of substance expected per use (8)					CBI (9)
							Site-limited	Con-sumer	Indus-trial	Com-mercial	Binding Option	
			%		%							
			%		%							
			%		%							
			%		%							
			%		%							
			%		%							

At least one TSCA regulated use must be provided.

Provide the requested information for each separate use of the PMN substance. The category of use information should be specific enough to allow EPA reviewers to anticipate potential exposure patterns for the new substance. (e.g., merely indicating "Dye" is not sufficient - is it a photographic dye, a reactive dye for nylon fibers, or a dye to be incorporated in a polymer marking film?)

LVE notices limit you to your stated category of use(s), so make sure your description fully encompasses all intended uses for any LVE.

Break down the use(s) by percentage for the indicated end user categories.

Site-limited means the substance will only be used on the contiguous property unit where manufactured and will not intentionally be removed from the site except for waste disposal.

Note that consumer uses require expanded detail of how the PMN substance will be used, what its concentration will be, and how it may be further reacted when used as a consumer product. A continuation sheet will be needed to provide this level of detail.

*If you have ide substance in c consumer products and describe the chemical reactions by which this substance loses its identity in the consumer product.

☐ Mark (X) this box if you attach a continuation sheet.

b. Generic use description

If you claim any category of use description in subsection 2a as confidential, enter a generic description of that category. Read the **Instructions Manual** for examples of generic use descriptions.

If necessary to protect confidential use information, the category of use data provided above can be claimed CBI, and a generic (non-CBI) use will be provided here.

e.g., "Surfactant for mineral ore aqueous extraction" (CBI) vs. "surfactant" (generic)

☐ Mark (X) this box if you attach a continuation sheet.

3. Hazard I or other equipme

Hazard Information: provide the MSDS for the PMN substance or intermediate or product mixture (that contains it) as an attachment to the PMN. Also include hazard warning statements or labels, if relevant. The personal protection requirements of the MSDS become integral requirements of a Low Volume Exemption notice.

☐ Mark (X

Binding Option Mark (x)

Part II -- HUMAN EXPOSURE AND ENVIRONMENTAL RELEASE

Section A -- INDUSTRIAL SITES CONTROLLED BY THE SUBMITTER — Mark (X) the "Confidential" box next to any item you claim as confidential.

Complete section A for each type of manufacture, processing, or use operation involving the new chemical substance at industrial sites you control. Importers do not have to complete this section for operations outside the U.S. ; however, you may still have reporting requirements if there are further industrial processing or use operations after import. You must describe these operations. See instructions manual.

1. Operation description	Confidential
a. Identity -- Enter the identity of the site at which the operation will occur.	

Name

Site address

City, County

> Where will the PMN substance be manufactured (or processed or used)? List the plant locations. If the submitting company is confidential, this data will also be claimed as CBI.
>
> This page is duplicated to provide separate information pages for manufacturing, processing, and use activities. Each stage of PMN substance handling results in potential human exposure and environmental release.

# of sites

If the same operation will [...] additional sites on a continuation sheet, and if any of the sites have significantly different production rates or operations, include all the information requested in this section for those sites as attachments.

☐ Mark (X) this box if you attach a continuation sheet.

b. Type --
Mark (X) ☐ Manufacturing   ☐ Processing   ☐ Use

c. Amount and Duration -- Complete 1 or 2 as appropriate

	Maximum kg/batch (100 % new chemical substance)	Hours/batch	Batches/year
1. Batch			
2. Continuous	Maximum kg/day (100 % new chemical substance)		

> Provide batch or continuous process information as required above, based on your maximum production (generally for the third year of estimated production). These data are normally kept confidential by submitters.
>
> The maximum daily or batch production multiplied by the days per year or batches should agree with the maximum production estimate provided on page 7.

d. Process description   ☐ Mark (X) to indicate your willingness t[...]

(1) Diagram the major unit operation steps and chemical conversions. I[...] gallon drum, rail car, tank truck, etc.).

(2) Provide the identity, the approximate weight (by kg/day or kg/bat[...] materials and feedstocks (including reactants, solvents, and catalyst[...] chemicals (note frequency if not used daily or per batch).

(3) Identify by number the points of release, including small or intermit[...]

> The required manufacturing (and processing) flow diagram is frequently provided as a full page attachment on a following page, to allow sufficient room for detail.
>
> The diagram should show all reactants and other starting materials as inputs to typical process unit operations. Input and output amounts should be listed, and should allow the EPA reviewer to follow all process steps, the disposition of the PMN substance, and all waste streams.
>
> Number all waste streams for reference on page 9, 3.(1) of the PMN form.

Simplified Example Only:

_ kg (chem. name), CAS# _
_ kg (chem. name), CAS# _ } MIX

_ kg catalyst, CAS# _

Releases of non-PMN substances from the chemical mfg and processing steps should also be indicated.

Your diagram should show all critical intermediate steps, separations and isolations of the PMN substance.

REACT (1)

Clean-up with methanol wash, x kg of PMN substance and y kg methanol to off-site incinerator

(Also, indicate the frequency of clean-up releases; are they intermittent, after every run, every 30 days, etc.)

If you are importing the substance, but considering domestic mfg, provide a prospective mfg process diagram.

DRUM — x kg of PMN substance drained to 15 gallon and 55 gallon drums

☐ Mark (X) this box if you attach a [...]

Manufacturing / Processing Operations Diagram
Manufacturing Reference Example

(Mark This Page if it Contains TSCA CBI)

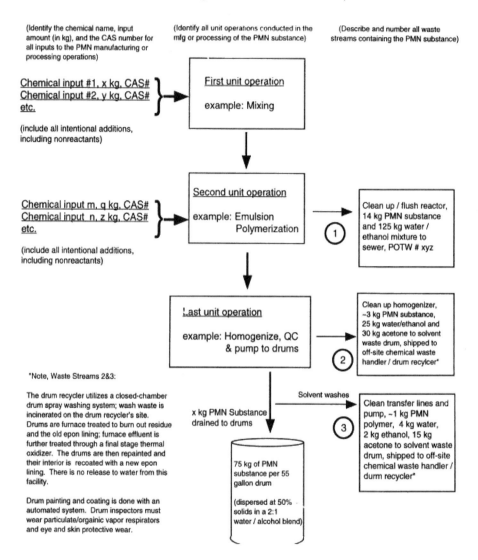

(Identify the chemical name, input amount (in kg), and the CAS number for all inputs to the PMN manufacturing or processing operations)

(Identify all unit operations conducted in the mfg or processing of the PMN substance)

(Describe and number all waste streams containing the PMN substance)

Chemical input #1, x kg, CAS#
Chemical input #2, y kg, CAS#
etc.

(include all intentional additions, including nonreactants)

First unit operation

example: Mixing

Chemical input m, q kg, CAS#
Chemical input n, z kg, CAS#
etc.

(include all intentional additions, including nonreactants)

Second unit operation

example: Emulsion
            Polymerization

①

Clean up / flush reactor, 14 kg PMN substance and 125 kg water / ethanol mixture to sewer, POTW # xyz

Last unit operation

example: Homogenize, QC
            & pump to drums

②

Clean up homogenizer, ~3 kg PMN substance, 25 kg water/ethanol and 30 kg acetone to solvent waste drum, shipped to off-site chemical waste handler / drum recylcer*

*Note, Waste Streams 2&3:

The drum recycler utilizes a closed-chamber drum spray washing system; wash waste is incinerated on the drum recycler's site. Drums are furnace treated to burn out residue and the old epon lining; furnace effluent is further treated through a final stage thermal oxidizer. The drums are then repainted and their interior is recoated with a new epon lining. There is no release to water from this facility.

Drum painting and coating is done with an automated system. Drum inspectors must wear particulate/orgainic vapor respirators and eye and skin protective wear.

x kg PMN Substance drained to drums

Solvent washes

③

Clean transfer lines and pump, ~1 kg PMN polymer, 4 kg water, 2 kg ethanol, 15 kg acetone to solvent waste drum, shipped to off-site chemical waste handler / durm recycler*

75 kg of PMN substance per 55 gallon drum

(dispersed at 50% solids in a 2:1 water / alcohol blend)

Manufacturing / Processing Operations Diagram
Processing Reference Example

(Mark This Page if it Contains TSCA CBI)

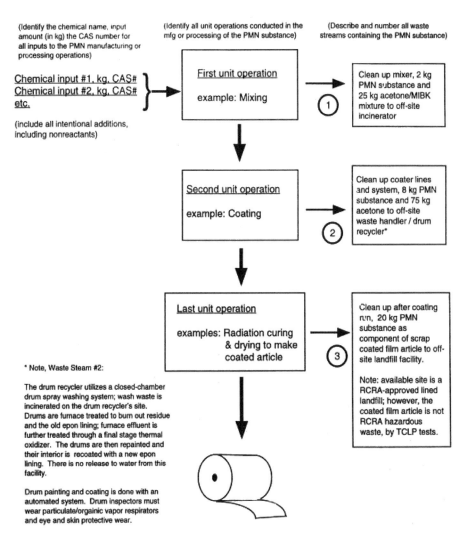

(Identify the chemical name, input amount (in kg) the CAS number for all inputs to the PMN manufacturing or processing operations)

<u>Chemical input #1, kg, CAS#</u>
<u>Chemical input #2, kg, CAS#</u>
<u>etc.</u>

(include all intentional additions, including nonreactants)

(Identify all unit operations conducted in the mfg or processing of the PMN substance)

(Describe and number all waste streams containing the PMN substance)

<u>First unit operation</u>

example: Mixing

(1)

Clean up mixer, 2 kg PMN substance and 25 kg acetone/MIBK mixture to off-site incinerator

<u>Second unit operation</u>

example: Coating

(2)

Clean up coater lines and system, 8 kg PMN substance and 75 kg acetone to off-site waste handler / drum recycler*

<u>Last unit operation</u>

examples: Radiation curing & drying to make coated article

(3)

Clean up after coating run, 20 kg PMN substance as component of scrap coated film article to off-site landfill facility.

Note: available site is a RCRA-approved lined landfill; however, the coated film article is not RCRA hazardous waste, by TCLP tests.

* Note, Waste Steam #2:

The drum recycler utilizes a closed-chamber drum spray washing system; wash waste is incinerated on the drum recycler's site. Drums are furnace treated to burn out residue and the old epon lining; furnace effluent is further treated through a final stage thermal oxidizer. The drums are then repainted and their interior is recoated with a new epon lining. There is no release to water from this facility.

Drum painting and coating is done with an automated system. Drum inspectors must wear particulate/orgainic vapor respirators and eye and skin protective wear.

Part II -- HUMAN EXPOSURE AND ENVIRONMENTAL RELEASE -- Continued

▶ Section A -- INDUSTRIAL SITES CONTROLLED BY THE SUBMITTER -- Continued

EPA is attempting to estimate exposure levels, the number of workers potentially exposed, and possible routes and duration of exposure to the PMN substance.

Work with your plant engineers to develop accurate data. Review the CBI status of data you provide.

Enter the maximum days/year any one worker will engage in each activity

Enter maximum hours/day any one worker will engage in specific activity

Estimate the number of all workers (all sites, all shifts) involved in each specific activity listed.

Worker activity (e.g. bag dumping, filling drums)	CBI	Protective / Engineering Controls	Option Mark (x)	(e.g. solid, powder) and % new substance	Binding Option Mark (x)	CBI	# of Workers Exposed	CBI	Maximum duration Hrs/day Days/yr		CBI
(1)	(2)	(3)	(4)	(5)	(6)	(7)	(8)	(9)	(10)	(11)	(12)

List all worker activities with potential exposure to the PMN substance; e.g., QC Sampling, Draining, Packaging, Clean-up.

List one activity per line.

Be as specific as possible on physical form - see item (5) above on the PMN form. (e.g., is the PMN substance isolated as a crystalline or amorphous powder, wet cake, flake solid, granular solid, or a fused solid?)

If data on particle size of solids are available, it should be included as an attachment. For wet cakes, a % moisture estimate may prevent EPA from making an estimate that is too conservative.

☐ Mark (X) this box if you attach a continuation sheet.

The information on exposure and environmental releases is related to the flow diagram from the previous page. This page is duplicated and filled out for each flow diagram supplied in the PMN (manufacturing, each processing description, and any use diagrams). The information provided should focus on the PMN substance.

EPA reviewers pay particular attention to human exposure and environmental release data under their TSCA mandate to protect against unreasonable risks from new chemical substances. In the absence of this information, EPA will make its own estimates of worker exposure and and environmental release, which are usually conservative. Present data that is accurate and as complete as possible.

Release estimates should be consistent with the release levels shown on the preceding flow diagram.

Include intermittent release estimates from cleaning of equipment and transportation containers, even if cleaned off-site.

Environmental release and control technology data: Refer to the instructions listed in 3. (4) & (5) on page 9 of the PMN form.

(7) Mark (X) the destination(s) of releases to water. ☐ POTW provide name(s) below: CBI ☐ Navigable waterway ☐ Other - Specify provide **NPDES #** CBI

If there are releases to water, provide information in this bottom section regarding POTW(s), releases to navigable waterways, and NPDES numbers. Review confidentiality, since these data may identify your company by location, and may require CBI protection.

☐ Mark (X) this box if you

Part II -- HUMAN EXPOSURE AND ENVIRONMENTAL RELEASE -- Continued

Section B -- INDUSTRIAL SITES CONTROLLED BY OTHERS

Complete section B for typical processing or use operations involving the new chemical substance at sites you do not control. Importers do not have to complete this section for operations outside the U.S.; however, you must report any processing or use activities after import. See the Instructions Manual. *Complete a separate section B for each type of processing, or use operation involving the new chemical substance.* If the same operation is performed at more than one site describe the **typical** operation common to these sites. Identify additional sites on a continuation sheet.

1. **Operation Description** - To claim information in this section as confidential, circle or bracket the specific information that you claim as confidential.
   (1) -- Diagram the major unit operation steps and chemical conversions, including interim storage and transport containers (specify e.g. tank cars, 55gallon drums, rail cars, tank trucks, etc). On the diagram, identify by letter and briefly describe each worker activity. (2) -- Provide the identity, the approximate weight (by kg/day or kg/batch, on an 100% new chemical substance basis), and entry point of all feedstocks (including the new substance, solvents and catalysts, etc) and of all products, recycle streams, and wastes. Include cleaning chemicals (note frequency if not used daily or per batch). (3) -- Identify by number the points of release, including small or intermittent releases, to the environment of the new chemical substance.
   (4) Please enter the # of sites (remember to identify the locations of these sites on a continuation sheet) :

   # of sites _____

   > Provide to the maximum extent possible process information and diagrams for your 'downstream' industrial processors and users.
   >
   > This will enable EPA to better understand an important part of the life cycle of your PMN substance. In the absence of this information, EPA will make its own estimates and assumptions about worker exposure and environmental releases, which are usually conservative.
   >
   > This information can be claimed confidential. Confidential information provided in this section should be bracketed or circled and marked to indicate its CBI status.

   ☐ Mark (X) this box if you attach a continuation sheet.

2. **Worker Exposure/Environmental Release**
   (1) -- From the diagram above, provide the letter for each worker activity. Complete 2-8 for each worker activity described.
   (2) -- Estimate the number of workers exposed for all sites combined.
   (4) -- Estimate the typical duration of exposure per worker in (a) hours per day and (b) days per year.
   (6) -- Describe physical form of exposure and % new chemical substance (if in mixture), and any protective equipment and engineering controls used to protect workers.
   (7) -- Estimate the percent of the new substance as formulated when packaged or used as a final product.
   (9) -- From the process diagram above, enter the number of each release point. Complete 9-13 for each release point identified.
   (10) -- Estimate the amount of the new substance released (a) directly to the environment or (b) into control technology to the environment (in cg/day or kg/batch).
   (12) -- Describe media of release i.e. stack air, fugitive air (optional-see Instructions Manual), surface water, on-site or off-site land or incineration, POTW, or other (specify) and control technology, if any, that will be used to limit the release of the new substance to the environment.
   (14) -- Identify byproducts which may result from the operation.
   (3), (5), (8), (11), (13) and (15) - Mark (X) this column if any of the proceeding entries are confidential business information (CBI).

Letter of Activity (1)	# of Workers Exposed (2)	CBI (3)	Duration of Exposure (4a) | (4b)	CBI (5)	Protective Equip. / Engineering Controls/ Physical Form and % new substance (6)	% in Form-ulation (7)	CBI (8)	Release Number (9)	Amount of New Substance Released (10a) | (10b)	CBI (11)	Control Technology (12)	CBI (13)

> Provide as much information as you can gather from industrial processors or users of your PMN substance. Providing this information will eliminate the need for EPA to make default assumptions about exposure and releases at sites beyond your control, which will likely be conservative assumptions that could alter their risk assessments of your premanufacturing notification.
>
> Providing sound information enables EPA to develop a more accurate risk assessment of your PMN substance, and improves your own risk assessment of your PMN substance prior to submitting the notice to the Agency.

(14) -- Byproducts: _____ (15)

☐ Mark (X) this box if you attach a continuation sheet.

OPTIONAL POLLUTION PREVENTION INFORMATION

To claim information in this section as confidential circle or bracket the specific information that you claim as confidential.

In thi[...] our
effort[...]
manu[...]
infor[...]s
and s[...]on
inclu[...]
modi[...]
subst[...]l
comp[...]er
disch[...]
recyc[...]
regul[...]in
this i[...]
expos[...]
(e.g.,[...]
perfo[...]
All i[...]w
of thi[...]h
manu[...]

Review the guidance of the PMN form for this page and the comments provided here. EPA strongly encourages submitters to provide pollution prevention data in their notices, as this information assists the Agency in balancing the benefits of a PMN substance against its unique risks.

While Agency staffers frequently receive information on the benefits of the PMN substance with respect to its use(s), data on pollution prevention effects are rarely included. This section should be utilized to indicate to EPA benefits in terms of the raw materials incorporated, the manufacturing process, emission reductions, less toxic waste byproducts, or lower risks upon end use and disposal. Consider the entire life cycle of the PMN substance in your evaluation.

Also consider opportunities for product stewardship or other risk management controls that could significantly reduce risk characteristics of the new substance. (e.g., disposal techniques that can be easily implemented and substantially reduce waste toxicity.)

Initiatives in the New Chemicals Program such as the Environmental Technology Initiative (ETI) for Chemicals, and the Pollution Prevention Recognition Project, are geared towards incorporating this type of information into the Agency's risk management decision making and recognizing innovative chemical design or technological innovation that shows potential for environmental and human health benefits.

Confidential information provided in this section should be bracketed or circled and marked to indicate its CBI status.

Describ[...]on in
the volume manufactured; (3) a reduction in the generation of waste materials through recycling, source reduction or other means; (4) a reduction in potential toxicity or human exposure and/or environmental release; (5) an increase in product performance, a decrease in the cost of production and/or improved operation efficiency of the new chemical substance in comparison to existing chemical substances used in similar applications; or (6) the extent to which the new chemical substance may be a substitute for an existing substance that poses a greater overall risk to human health or the environment.

Pollution Prevention Information

Are there Pollution Prevention benefits associated with the new PMN chemical? Does it allow reduction or elimination of (more) toxic solvents? Does it reduce disposal to land, or in safer forms, compared to the chemistry it replaces? Can it be recycled more easily than existing materials? Does your manufacturing process eliminate or reduce the use / release of hazardous substances?

Describe any pollution prevention benefits here, and indicate if this information is confidential business information. Attempt to quantify the expected benefits to the extent possible and avoid exaggeration and highly speculative statements.

☐ Mark (X) this box if you attach a continuation sheet.

Part III -- LIST OF ATTACHMENTS

Attach continuation sheets for sections of the form and test data and other data (including physical/chemical properties and structure, activity information, and optional information after this page. Clearly identify the attachment and the section of the form to which it relates, if appropriate. Number consecutively the pages of the attachments. In the column below, enter the inclusive page numbers of each attachment.

Mark (X) the Confidential box next to any attachment **name** you claim as confidential. Read the **Instructions Manual** for guidance on how to claim any information in an attachment as confidential. You must include with the sanitized copy of the notice form a sanitized version of any attachment in which you claim information as confidential.

Attachment name	Attachment page number(s)	Confi-dential
Material Safety Data Sheet (MSDS)		

The MSDS and all attachments to the complete PMN are listed on this page.

The List of Attachments includes individual toxicity studies, GPC data, spectra, exposure monitoring evaluations, physical and chemical properties data, SAR modeling, test sample characterization data, explanatory comments, etc.

Analytical tests should be provided with their complete data set. For example, provide a GPC test report with time interval data and all graphs.

As a PMN submitter, you are required to submit all test data in your possession or control and provide a description of all other data known to or reasonably ascertainable by you, if those data are related to the health and environmental effects of the manufacture, processing, distribution, use or disposal of the new chemical substance.

For ease of review, group attachments by type of data, e.g., physical/chemical properties, fate information, environmental toxicity, and human health toxicity.

Number the pages of the attachments in consecutive order. Enter the inclusive page numbers of each attachment. Enter the total number of pages for the notice on page 1 of the form.

Confidential check box: check this box if the attachment name or part of the attachment name reveals CBI information.

☐ Mark (X) this box if you attach a continuation sheet. Enter the attachment name and number.

APPENDIX C

CONTACT LIST

The New Chemicals Program is organized in two Branches within the Chemical Control Division in EPA. They are the New Chemicals Pre-Notice Branch and the New Chemicals Notice Management Branch. The Notice Management Branch coordinates review of submissions and negotiations with submitters on Section 5 Consent Orders, etc.

Staff in the Notice Management Branch include:

New Chemicals Notice Management Staff Member	
Flora Chow, Branch Chief	(202) 260–3406
Rose Allison, Senior Notice Specialist	(202) 260–3391
Jim Alwood, SNUR Specialist	(202) 260–1857
Jeff Bauer, Notice Manager	(202) 260–4219
Mary Begley, Notice Manager	(202) 260–1769
Audrey Binder, Notice Manager	(202) 260–3380
Geraldine Hilton, Notice Manager	(202) 260–3992
Mark Howard, Notice Manager	(202) 260–4143
Shirley Howard, Notice Manager	(202) 260–3780
Darlene Jones, Notice Manager	(202) 260–2279
Loraine Passe, Notice Manager	(202) 260–0467
Edna Pleasants, Notice Manager	(202) 260–4142
Laura Stalter, Notice Manager	(202) 260–0028

The New Chemicals Prenotice Branch gives guidance for persons considering a submission through Prenotice Coordinators. Prenotice Coordinators are staff in the New Chemicals Program who specialize in assisting with status questions and questions on how to properly complete the notifications. At the time of this writing (May 1999) staff in the New Chemicals Pre-Notice Branch include:

New Chemicals Pre-Notice Branch Staff Member	Telephone Number
Rebecca Cool, Branch Chief	(202) 260–8539
Dave Schutz, Prenotice Coordinator (Schutz.David@epa.gov)	(202) 260–8994
Nancy Vogel, Prenotice Coordinator (Vogel.Nancy@epa.gov)	(202) 260–4183
Adella Watson, Prenotice Coordinator (Watson.Adella@epa.gov)	(202) 260–3752
Miriam Wiggins-Lewis, Prenotice Coordinator (Wigginslewis.Miriam@epa.gov)	(202) 260–3937
Roy Seidenstein, Lawyer (Seidenstein.Roy@epa.gov)	(202) 260–2252

Written inquiries may be sent by mail to:

Prenotice Coordinator, New Chemicals Program
Chemical Control Division
Mail Stop 7405
USEPA, 401 M Street, S.W.
Washington, D.C. 20460

or by facsimile at (202) 260–0118. Persons considering using facsimile to send confidential information to the Agency have to be aware that the 260–0118 telephone line is not a secured line, however the facsimile machine itself is in a secured area. EPA does not regard email as an appropriate way to transmit confidential information to or from the Agency.

EPA uses a contractor, the TSCA Assistance Information Service (TAIS, TSCA Hotline) to distribute forms and to answer some questions. TAIS can be reached on (voice) (202) 554–1404, (facsimile) (202) 554–5603, (email) tsca-hotline@epa.gov.

Mail inquiries may be sent to:

TSCA Assistance Information Service
Mail Stop 7408
USEPA 401 M Street, S.W.
Washington, D.C. 20460

Appendix 7.9

Polymer Exemption Guidance Manual

(Reproduced from:
www.epa.gov/opptintr/newchems/polyguid.pdf)

United States Environmental Protection Agency	Office of Pollution Prevention and Toxics (7406)	EPA 744-B-97-001 June 1997

 Polymer Exemption Guidance Manual

Polymer Exemption Guidance Manual
5/22/97

A technical manual to accompany, but not supersede the "Premanufacture Notification Exemptions; Revisions of Exemptions for Polymers; Final Rule"
found at 40 CFR Part 723, (60) FR 16316–16336, published Wednesday,
March 29, 1995.

Environmental Protection Agency
Office of Pollution Prevention and Toxics
401 M St., S. W.,
Washington, DC 20460-0001

Copies of this document are available through the TSCA Assistance Information Service at (202)
554–1404 or by faxing requests to (202) 554–5603.

TABLE OF CONTENTS

LIST OF EQUATIONS

LIST OF FIGURES

12. NOMOGRAPH FOR DETERMINING FUNCTIONAL GROUP EQUIVA-
LENT WEIGHT

LIST OF TABLES

1. DISTRIBUTION CRITERIA EXAMPLES 6, 7, AND 8: ETHOXYLATED
   ALCOHOLS
2. COMBINED FUNCTIONAL GROUP EQUIVALENT WEIGHT SUMMARY
3. (e)(3) MONOMER AND REACTANT LIST
4. ALLOWABLE THRESHOLDS FOR REACTIVE FUNCTIONAL GROUPS

1. INTRODUCTION

The Environmental Protection Agency (EPA) published a series of proposed
rules (USEPA 1993a–1993d) in the **Federal Register** on February 8, 1993 to
announce the Agency's plan to amend premanufacture notification (PMN) regulations for new chemical substances under §5 of the Toxic Substances Control
Act (TSCA). Included were proposed amendments to the polymer exemption
rule originally published on November 21, 1984 (USEPA 1984) under the aus-
pices of §5(h)(4) of TSCA and entered into the Code of Federal Regulations
(CFR), the administrative rules under which the U.S. Government operates, at
40 CFR Chapter I, Subchapter R, part 723.250.

After the proposed polymer exemption rule was published, the Agency considered public comments, consulted with European counterparts, and utilized
the experience gained in the review of over 12,000 polymers in publishing its
new final rule for polymer exemptions on March 29, 1995, amending 40 CFR
§723.250 (USEPA 1995). The new polymer exemption rule is notably different
from that originally published in 1984 and it is the purpose of this technical
manual to provide the regulated community with additional insight, so that
manufacturers and importers will be able to determine if their new chemical
substances are eligible for the polymer exemption under the new rule. Substances submitted before May 30, 1995 are subject to the original rule (USEPA
1984) and its requirements. On or after that date, all polymer exemptions are
subject to the new rule and its requirements.

A few notable features of the 1995 Polymer Exemption are as follows:

- Manufacturers and importers are no longer required to submit notice prior
  to manufacture or import. However, manufacturers and importers must
  submit an annual report for those exempt polymers whose manufacture or
  import has commenced for the first time during the preceding calendar
  year, as stipulated in §723.250(f), and the manufacturer or importer of
  an exempt polymer must comply with all recordkeeping requirements at
  §723.250(j).

- A new method can be used for determining which monomers and reactants are considered part of the polymer's chemical identity (modification of the so-called "Two Percent Rule").
- More polymers are now eligible for exemption because previous exclusions have been modified or eliminated. Some of the changes are in regard to halogens, cyano groups, biopolymers and reactive group limitations.
- Certain high molecular weight polymers once considered eligible for submission under the 1984 exemption are not eligible for this exemption.

The EPA hopes this technical manual will: (1) assist the chemical manufacturer or importer in determining whether the PMN substance is a polymer as defined by the polymer exemption rule, (2) guide the manufacturer or importer in determining whether the polymer meets the exemption criteria of the rule and (3) assist the manufacturer or importer in determining whether the polymer is excluded from exemption by certain factors. In addition, this manual provides technical guidance and numerous pertinent examples of decision-making rationales.

The Agency hopes that after reviewing this document prospective manufactures and importers will be able to decide easily whether the polymer exemption is applicable to any of their new substances. This technical guidance manual is not intended to substitute for or supersede the regulations as found at 40 CFR §723.250 and the **Federal Register** (USEPA 1995). Manufacturers and importers must read those provisions to assure compliance with all the procedural and recordkeeping requirements of the polymer exemption.

2. HISTORY

Section 5 of TSCA contains provisions that allow the Agency to review new chemical substances before they are manufactured or imported. Section 5(a)1 of TSCA requires that persons notify EPA at least 90 days prior to the manufacture or import of a new chemical substance for commercial purposes. A "new" chemical substance is one that is subject to TSCA but is not already included on the TSCA Chemical Substance Inventory. If the Agency determines that a new chemical substance may present an unreasonable risk of injury to human health or the environment or if there is insufficient information to establish that no such risk exists, the Agency may limit the manufacture, processing, distribution in commerce, use, or disposal of the new chemical substance under the authority provided in TSCA §5(e).

From the beginning of the PMN program in 1979 until 1984 all new chemical substances, including polymers, were subject to the full reporting requirements of the premanufacture notification process. Under §5(h)(4) the Agency has authority to promulgate rules granting exemptions from some or all of the premanufacture requirements for new chemicals if the Agency determines that the manufacturing, processing, distribution in commerce, use, or disposal of a new chemical substance will not present an unreasonable risk of injury to human health or the environment.

Through its experience in reviewing new chemical substances, the Agency identified certain criteria to determine which polymers were most unlikely to present an unreasonable risk of injury to human health or the environment. This experience led to the original polymer exemption rule under §5(h)(4) allowing polymers that met certain criteria under these conservative guidelines to be exempt from some of the reporting requirements for new chemicals (USEPA 1984).

Since the EPA published the 1984 TSCA polymer exemption rule, the Agency has reviewed over 10,000 polymer submissions under the standard 90 day PMN review process and an additional 2,000 polymer exemption notices. With the experience gained by the review of this large number of submissions, the Agency reevaluated the criteria used to identify those polymers which were unlikely to present unreasonable risks. This led to the proposal of a revised polymer exemption rule that would increase the number of polymers qualifying for exemption and enable the Agency to concentrate its limited resources on those polymers that do not meet the polymer exemption criteria and on non-polymeric new chemical substances that may present greater risks. The amendments are expected to result in resource savings for industry as well as the EPA.

The new polymer exemption rule amends appropriate sections of 40 CFR 723.250 to allow certain polymers to be exempt from the reporting requirements for new chemicals and imposes new restrictions on a limited set of polymers that were previously eligible for the exemption (USEPA 1993d). To be eligible for the exemption, a new chemical substance must: 1) meet the polymer definition, 2) meet one of three exemption criteria and 3) not be excluded. The definition of polymer, for purposes of the new exemption, is found at 40 CFR §723.250(b). There are now three exemption types, located at 40 CFR §723.250(e)(1), (e)(2), and (e)(3), subsequently referred to as the (e)(1), (e)(2), and (e)(3) criteria. Excluded categories are listed at 40 CFR §723.250(d) of the new rule.

The definition of polymer, the key components of each of the three exemption types, and the categories excluded from the exemption are discussed below. The remainder of this technical manual provides prospective submitters with information helpful for establishing whether or not their new chemical substances meet the exemption criteria.

The (e)(1) **exemption** concerns polymers with a number-average molecular weight (NAVG MW) in a range that is greater than or equal to 1,000 ($\geq 1,000$) daltons and less than 10,000 ($< 10,000$) daltons.

> **Dalton**—precisely 1.0000 atomic mass unit or 1/12 the mass of a carbon atom of mass 12. Hence, a polymer with a molecular weight of 10,000 atomic mass units has a mass of 10,000 daltons.

For the (e)(1) **exemption**, oligomer content must be less than 10 percent by weight below 500 daltons and less than 25% by weight below 1,000 daltons. The polymer must also meet functional group criteria to be described in a later section of this manual.

> **Oligomer** (in the context of the rule and this manual)—a low molecular weight species derived from the polymerization reaction. The Organization for Economic Cooperation and Development (OECD) has a draft guidelines document[1] for determining the low molecular weight polymer content.

For the (e)(2) **exemption**, the NAVG MW for eligible polymers must be greater than or equal to 10,000 daltons and these polymers must have oligomer content less than two percent below 500 daltons and less than 5 percent below 1,000 daltons.

The (e)(3) **exemption** concerns certain polyester polymers (as defined at §723.250(b)) composed solely of monomers and reactants from the list as found at §723.250(e)(3).

In addition to meeting the specific criteria of one of the three exemption types described above, the new polymer must not fall into any of the prohibited categories listed at §723.250(d) of the new rule. This section of the amended rule specifically excludes certain polymers from the reduced reporting requirements of the polymer exemption: certain cationic polymers; polymers that do not meet elemental restrictions; polymers that degrade, decompose, or depolymerize; and polymers that are produced from monomers and/or other reactants that are not on the TSCA inventory or otherwise exempted from full PMN reporting under a §5 exemption. Some highly water-absorbing, high molecular weight polymers are also specifically prohibited. Any new chemical polymer substance that does not meet the polymer definition, does not meet any of the (e)(1), (e)(2), or (e)(3) exemptions, or is specifically excluded from the polymer exemption is subject to the full PMN reporting requirements.

3. DEFINITIONS

For a new polymer to be eligible for the exemption it must meet distinct criteria set forth in the 1995 polymer exemption rule. Much of the terminology used in these criteria is explained in this and subsequent sections of the guidance manual. Note that the definitions provided herein are those used in the new polymer exemption rule, and that these terms may not necessarily have the same meaning as commonly used in an academic or industrial setting. Careful attention must be paid to the definitions contained in the new polymer exemption rule when determining eligibility.

The polymer definition has been revised to conform with the international definition recently adopted by the OECD as a result of the *Experts on Polymers Meetings* held in Toronto, Canada (January 1990), Paris, France (October 1991), and Tokyo, Japan (April 1993), in which the Agency participated. The definition was agreed upon in May 1993 by the OECD member countries, including the United States, Canada, Japan, and member nations of the European Union. The definitions of polymer and other important terms as used in the new polymer exemption rule are:

Polymer—a chemical substance consisting of molecules characterized by the sequence of one or more types of monomer units and comprising a simple weight majority of molecules containing at least 3 monomer units which are covalently bound to at least one other monomer unit or other reactant and which consists of less than a simple weight majority of molecules of the same molecular weight. Such molecules must be distributed over a range of molecular weights wherein differences in the molecular weight are primarily attributable to differences in the number of monomer units.

Monomer—a chemical substance that is capable of forming covalent bonds with two or more like or unlike molecules under the conditions of the relevant polymer-forming reaction used for the particular process.

Monomer unit—the reacted form of the monomer in a polymer.

Sequence—a continuous string of monomer units within the molecule that are covalently bonded to one another and are uninterrupted by units other than monomer units.

Reactant—a chemical substance that is used intentionally in the manufacture of a polymer to become chemically a part of the polymer composition. (Reactants include monomers, chain transfer and crosslinking agents, monofunctional groups that act as modifiers, other end groups or pendant groups incorporated into the polymer. For example, sodium hydroxide is considered a reactant when the sodium ion becomes part of the polymer molecule as a counter ion.)

Other reactant—a molecule linked to one or more sequences of monomer units but which under the relevant reaction conditions used for the particular process cannot become a repeating unit in the polymer structure. (This term is used primarily in applying the concept of sequence in the definition of a polymer.)

Polymer molecule—a molecule that contains a sequence of at least 3 monomer units, which are covalently bound to at least one other monomer unit or other reactant.

Internal monomer unit—a monomer unit of a polymer molecule that is covalently bonded to at least two other molecules. Internal monomer units of polymer molecules are chemically derived from monomer molecules that have formed covalent bonds between two or more other monomer molecules or other reactants.

Number-average molecular weight—the arithmetic average (mean) of the molecular weights of all molecules in a polymer. (This value should not take into account unreacted monomers and other reactants, but must include oligomers.)

4. ELIGIBILITY REQUIREMENTS

In order for a new chemical substance to be eligible for exemption under the amended rule, it must meet the following requirements:

- The substance must meet the definition of a polymer as defined in §723.250(b);
- The substance must not be specifically excluded from the polymer exemption by §723.250(d); and
- The substance must meet one of the (e)(1), (e)(2), or (e)(3) criteria.

4.1. MEETING THE DEFINITION OF A POLYMER AT 40 CFR §723.250(b)

For deciding if a substance meets the definition of a polymer these sequence and distribution criteria must be met:

- >50 percent of molecules must be composed of a sequence of at least 3 monomer units plus at least one additional monomer unit or other reactant. (In other words, >50 percent of the substance must be polymer molecules.)
- The amount of polymer molecules of any one molecular weight cannot exceed 50 weight percent.

The following examples illustrate the analysis of substances with regard to the polymer definition criteria mentioned above. Some of these have been taken from the Chairman's Report[2] of the Chemicals Group and Management Committee at the Third Meeting of OECD Experts on Polymers (Tokyo, April 14–16, 1993) in which the Agency participated. In the figures, "o.r." refers to "other reactant," and "m.u." refers to "monomer unit." Examples 1–5 illustrate the sequence criteria for defining a polymer molecule. In Examples 1–3 the relevant polymer-forming reaction is ethoxylation with ethylene oxide.

Example 1:

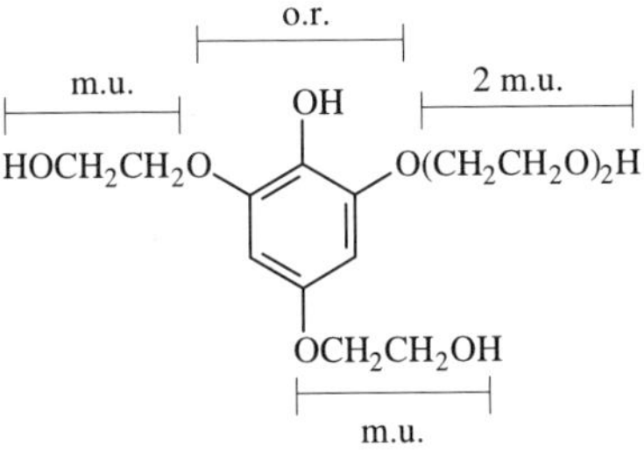

Figure 1   Ethoxylated Benzenetetrol

Example 1 does not meet the sequence criterion and is therefore not a polymer molecule. Under the reaction conditions, the phenol hydroxy group can neither react with another phenol hydroxy nor an opened epoxide. Therefore, phenolic precursor is an "other reactant," (o.r.). In the molecule shown, there is no sequence of three monomer units (m.u.) from ethylene oxide.

Example 2:

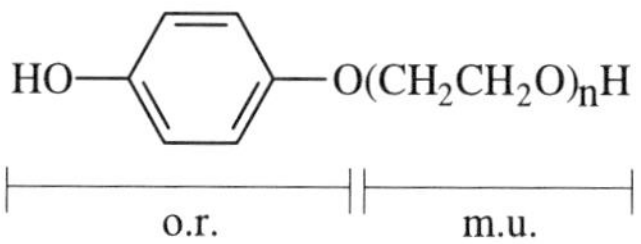

$$HO-\langle\text{benzene ring}\rangle-O(CH_2CH_2O)_nH$$

o.r.   m.u.

Figure 2   *Ethoxylated Hydroquinone*

The Example 2 molecule, produced from the ethoxylation of hydroquinone, would meet the sequence criterion if $n \geq 3$ and therefore would be a polymer molecule. Hydroquinone would be an "other reactant" because the phenol hydroxyl can react with neither another phenol hydroxyl nor an opened epoxide, under the reaction conditions.

Example 3:

o.r.   2 m.u.   m.u.

$$CH_2O-(CH_2CH_2O)_2H$$
$$CHO-(CH_2CH_2O)_2H$$
$$CH_2O-(CH_2CH_2O)_2-CH_2CH_2OH$$

Figure 3   *Ethoxylated Glycerol*

Example 3 meets the sequence criterion and would be considered a polymer molecule. If polymer formation is desired, at least 7 equivalents of EO should be charged to the reactor. With less EO charged, each hydroxyl may only be ethoxylated twice or less, which would not satisfy the sequence criterion.

Example 4:

o.r.   o.r.

$$CH_2O-CO(CH_2)_{16-18}CH_3$$
$$CHO-CO(CH_2)_{16-18}CH_3$$
$$CH_2O-CO(CH_2)_{16-18}CH_3$$

Figure 4   *Glycerol Triester*

Example 4 does not meet the sequence criterion. There are no repeating units. Neither the glycerol other reactant nor the fatty acid other reactant can repeat under the reaction conditions. Methylene (CH2) is not a monomer unit, because it is not the reacted form of a monomer present in the polymer.

Example 5:

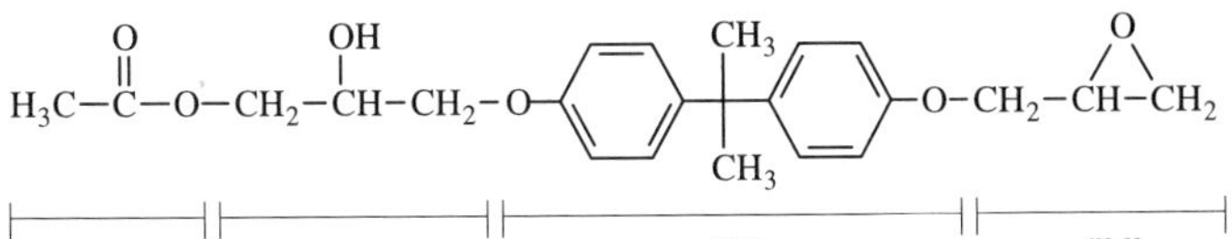

Figure 5 *Epoxy Resin*

Example 5 meets the sequence criterion and therefore would be a polymer molecule. It has an unbroken chain of three monomer units and one other reactant.

Examples 6, 7, and 8:

Examples 6–8 illustrate the sequence and distribution criterion of the new polymer exemption rule.

TABLE 1   Distribution Criteria Examples 6, 7, and 8: Ethoxylated Alcohols

Species	o.r. + m.u.	Example 6	Example 7	Example 8
RO.EO.H	1 + 1	5%	25%	8%
RO.EO.EO.H	1 + 2	20%	35%	20%
RO.EO.EO.EO.H	1 + 3	30%	20%	52%
RO.EO.EO.EO.EO.H	1 + 4	40%	10%	10%
RO.EO.EO.EO.EO.EO.H	1 + 5	5%	10%	10%

For these examples, "EO" is a monomer unit derived from ethylene oxide, and "RO" is an other reactant derived from an alcohol. Example 6 meets the definition of a polymer because >50 percent of the substance has molecules of at least 3 monomer units in sequence and <50 percent of each species (same molecular weight components) is present. Example 7 does not meet the definition of polymer because <50 percent of substance has molecules of at least 3 monomer units plus at least one additional monomer unit or other reactant. Example 8 does not meet the definition of polymer because >50 percent of one molecular weight species is present.

Example 9:

Consider the enzyme pepsin and the sequence and distribution criteria of the new polymer exemption rule's definition of a polymer substance. Although pepsin meets the sequence requirements of the polymer definition, the molecules will always have the same distinct molecular weight, corresponding to the sum of the molecular weights of the amino acid monomer units which comprise the specific protein sequence of the enzyme. As such it has a majority of molecules having identical weight and will not meet that portion of the new rule's definition of a polymer.

On the other hand, a lipoprotein or mucoprotein with its attachments intact might satisfy the sequence and distribution criteria. The lipo- or muco- portions can be quite variable in quantity and this could cause enough variation in weight of the polymer molecules.

4.2. SUBSTANCES EXCLUDED FROM THE EXEMPTION AT 40 CFR §723.250(d)

Certain categories of polymers are ineligible for exemption under the new polymer exemption rule because the Agency cannot determine whether these substances can be reasonably anticipated to present an unreasonable risk of injury to human health or the environment. For a discussion of the history behind the selection of these categories, consult the preamble to the 1995 polymer exemption rule (USEPA 1995). The following sections discuss the excluded categories.

4.2.1. EXCLUSIONS FOR CATIONIC AND POTENTIALLY CATIONIC POLYMERS

Cationic polymers and those polymers which are reasonably anticipated to become cationic in the natural aquatic environment are excluded from the exemption and may not be manufactured under it. The principal concern is the toxicity toward aquatic organisms.

Cationic polymer—a polymer that contains a net positively charged atom(s) or associated group(s) of atoms covalently linked to the polymer molecule. This includes, but is not limited to phosphonium, sulfonium, and ammonium cations.

Potentially cationic polymer—a polymer containing groups that are reasonably anticipated to become cationic. This includes, but is not limited to, all amines (primary, secondary, tertiary, aromatic, etc.) and all isocyanates (which hydrolyze to form carbamic acids, then decarboxylate to form amines).

Reasonably anticipated—means that a knowledgeable person would expect a given physical or chemical composition or characteristic to occur, based on such factors as the nature of the precursors used to manufacture the polymer, the type of reaction, the type of manufacturing process, the products produced in the polymerization, the intended uses of the substance, or associated use conditions.

4.2.1.1. CATIONIC POLYMERS NOT EXCLUDED FROM EXEMPTION

Through its experience reviewing thousands of polymers, the Agency has determined that two categories of cationic and potentially cationic polymers would not pose an unreasonable risk of injury to human health or the environment. These two types are **not** excluded from consideration for the exemption and are as follows:

- Cationic or potentially cationic polymers that are solids, are neither water soluble nor dispersible in water, are only to be used in the solid phase, and are not excluded from exemption by other factors, and
- Cationic or potentially cationic polymers with low cationic density (the percent of cationic or potentially cationic species with respect to the overall weight of polymer) which would not be excluded from the exemption by other factors.

For a polymer to be considered to have low cationic density, the concentration of cationic functional groups is limited to a functional group equivalent weight of greater than or equal to 5,000 daltons.

Functional group equivalent weight (FGEW)—the weight of polymer that contains one equivalent of the functional group; or the ratio of number-average molecular weight (NAVG MW) to the number of functional groups in the polymer. The methods for calculating the FGEW are described in a later section.

Example 10:

As an example of the cationic density requirement, consider the reaction of precisely equal molar amounts of ethanediamine and phthalic acid, resulting in a polyamide (polymer) with an equal number of unreacted amine and unreacted carboxylic acid groups. This would be equivalent to a sample of polymer molecules that would have (on average) one end group that was an unreacted amine (potentially cationic) and the other end group an unreacted carboxylic acid. For this polymer to be eligible for the exemption it must have a minimum NAVG MW of 5,000 daltons which would give the amine FGEW as 5,000 daltons (1 amine termination per 5,000 MW of polymer).

4.2.2. EXCLUSIONS FOR ELEMENTAL CRITERIA

A polymer manufactured under the 1995 rule must contain as an integral part of its composition at least two of the atomic elements of carbon, hydrogen, nitrogen, oxygen, sulfur, or silicon (C, H, N, O, S, Si).

In addition to the six elements listed above, only certain other elements are permitted either as counterions or as an integral part of the polymer. These additional elements are as follows: fluorine, chlorine, bromine and iodine (F, Cl, Br, and I) when covalently bonded to carbon, and the monatomic counterions chloride, bromide, and iodide (Cl−, Br−, and I−). The fluoride anion, (F−) is not permitted. This decision was based on data obtained by the Agency. Other permitted monatomic cations are sodium, magnesium, aluminum, potassium, and calcium (Na+, Mg+2, Al+3, K+, and Ca+2). Allowed at less than 0.20 weight percent total (in any combination) are the atomic elements lithium, boron, phosphorus, titanium, manganese, iron, nickel, copper, zinc, tin and zirconium (Li, B, P, Ti, Mn, Fe, Ni, Cu, Zn, Sn, and Zr). No other elements are permitted, except as impurities.

4.2.3. EXCLUSIONS FOR DEGRADABLE OR UNSTABLE POLYMERS

A polymer is not eligible to be manufactured under the new exemption rule if the polymer is designed or reasonably anticipated to substantially degrade, decompose, or depolymerize, including those polymers that could substantially decompose after manufacture and use, even though they are not actually intended to do so. For purposes of this section the following definition applies:

> **Degradation, decomposition, or depolymerization**—a type of chemical change in which a polymeric substance breaks down into simpler, smaller weight substances as the result of (for example) oxidation, hydrolysis, heat, sunlight, attack by solvents or microbial action.

4.2.4. EXCLUSIONS BY REACTANTS

A polymer may contain at more than two percent by weight only those reactants and monomers that are either: on the TSCA Chemical Substance Inventory, granted a §5 exemption, (a low-volume exemption; a polymer exemption under the 1984 rule; etc.), excluded from reporting or a non-isolated intermediate. Monomers and reactants that do not fit one of these categories would render a polymer ineligible for the polymer exemption. This applies to both manufactured and imported polymers. (See section 5.2. of this manual for a discussion of the so-called "Two Percent Rule").

Monomers and reactants incorporated or charged at greater than two percent in a polymer are considered part of the chemical identity of the new polymer. (See Section 5.2. on the "Two Percent Rule.") Monomers and reactants which are not on the Inventory and do not have a §5 exemption may be used at less than or equal to two percent provided that those monomers and reactants will not introduce into the polymer any elements, properties, or functional groups that would render the polymer ineligible for the exemption. However, in practice, the use of non-Inventory monomers or reactants at two percent or less applies only to imported polymers since domestic manufacturers may not distribute or use any substance unless it is on the TSCA Inventory or exempt from TSCA reporting requirements. In other words, non-Inventory monomers and reactants may be handled domestically only if they are intermediates made *in situ* and are not isolated, or if they are already exempt.

4.2.5. EXCLUSIONS FOR WATER-ABSORBING POLYMERS

Water-absorbing polymers with number-average molecular weight (NAVG MW) of 10,000 daltons and greater are excluded from exemption. A history describing how the EPA came to select this NAVG MW and the level of water absorptivity to be excluded is given in the preamble to the new rule. The Agency's definition of water-absorbing is given below:

> **Water-absorbing polymer**—means a polymeric substance that is capable of absorbing its weight of water.

4.3. CATEGORIES WHICH ARE NO LONGER EXCLUDED FROM EXEMPTION

Three exclusions have been dropped from the new polymer exemption rule because the Agency now believes that other provisions of the new rule will exclude any polymers that would pose an unreasonable risk of injury to human health or the environment. The three types of polymers that are no longer automatically excluded from the exemption are: (1) polymers containing less than 32 percent carbon; (2) polymers manufactured from reactants containing halogen atoms (see section 4.2.2 of this manual) or cyano groups; and (3) biopolymers. To be manufactured under the exemption these polymers must meet all of the criteria of the new rule. For example, in the biopolymer category, most enzymes and polypeptides will not meet the polymer definition because of the requirement that the molecular weight of the polymer must be distributed over a range (no one molecular weight species can be present in a simple majority). Some DNA, RNA or polysaccharide substances may meet the molecular weight distribution criterion but fail because of reactivity (reactive group content, degradability, etc.), cationic potential, or water-absorbing properties.

4.4. MEETING EXEMPTION CRITERIA at 40 CFR §723.250(e)

Providing the new polymer meets the definition of a polymer at §723.250(b) and the polymer is not automatically excluded by section §723.250(d), the polymer must also meet one or more of the criteria listed in §723.250(e)(1), (e)(2), or (e)(3) to be manufactured or imported under a polymer exemption.

4.4.1. THE (e)(1) EXEMPTION CRITERIA

In order to be manufactured or imported under §723.250(e)(1), the polymer must have a NAVG MW equal to or greater than 1,000 daltons and less than 10,000 daltons. (See section 5.1., for determining NAVG MW.) The polymer also must contain less than 10 percent oligomer content of molecular weight below 500 daltons and less than 25 percent oligomer content of molecular weight below 1,000 daltons. In addition, (e)(1) polymers have reactivity constraints. The polymer must have either: no reactive functional groups; only low-concern functional groups; or it must have a functional group equivalent weight (FGEW) above threshold levels for moderate- and high-concern functional groups in order to remain eligible for the exemption. (See section 5.3., "Calculating Functional Group Equivalent Weight," in this manual.)

> **Reactive functional group**—an atom or associated group of atoms in a chemical substance that is intended or can be reasonably anticipated to undergo facile chemical reaction.

The following sections describe the reactive functional groups in the low-concern, moderate-concern and high-concern categories.

4.4.1.1. LOW-CONCERN FUNCTIONAL GROUPS AND THE (e)(1) EXEMPTION

Low-concern functional groups defined in §723.250(e)(1)(ii)(A) may be used without limit. These groups are so categorized because they generally lack reactivity in biological settings. The low-concern reactive functional groups are: carboxylic acid groups; aliphatic hydroxyl groups; unconjugated olefinic groups that are considered "ordinary;" butenedioic acid groups; those conjugated olefinic groups contained in naturally-occurring fats, oils, and carboxylic acids; blocked isocyanates (including ketoxime-blocked isocyanates); thiols; unconjugated nitrile groups; and halogens (not including reactive halogen-containing groups such as benzylic or allylic halides).

> **Ordinary olefinic groups**—unconjugated olefinic groups that are **not** specifically activated either by being part of a larger functional group, such as a vinyl ether, or by other activating influences, such as the strongly electron-withdrawing sulfone functionality (in a vinyl sulfone system).

In addition, carboxylic esters, ethers, amides, urethanes and sulfones are implicitly permitted because polyesters, polyethers, polyamides, polyurethanes, and polysulfones are among the types of polymers allowed under the exemption, as long as these functional groups have not been modified to enhance their reactivity. One such group that would **not** be allowed is the dinitrophenyl ester of a carboxylic acid, which is far more reactive due to the activating functionality.

In summary, if a substance (1) meets the definition of a polymer, (2) is not excluded by §723.250(d), (3) has a NAVG MW greater than or equal to 1,000 daltons and less than 10,000 daltons, (4) contains only the low-concern reactive functional groups, and (5) meets oligomer content criteria (<10 percent below 500 daltons and <25 percent below 1,000 daltons), the new substance may be manufactured under a polymer exemption.

4.4.1.2. MODERATE-CONCERN FUNCTIONAL GROUPS AND THE (e)(1) EXEMPTION

Moderate-concern groups defined in §723.250(e)(1)(ii)(B) may be used with functional group equivalent weight (FGEW) constraints. Each functional group present from category (B) must have a FGEW of greater than or equal to 1,000 daltons. For a polymer containing no type (C) groups (see section 4.4.1.3 for when type (C) groups are present), the $FGEW_{combined}$ must be greater than or equal to 1,000 daltons. (The method for calculating a $FGEW_{combined}$ is covered in section 5.3. of this manual). The moderate-concern reactive functional groups are: acid halides; acid anhydrides; aldehydes; hemiacetals; methylolamides; methylolamines; methylolureas; alkoxysilanes bearing alkoxy groups greater than C2; allyl ethers; conjugated olefins (except those in naturally-occurring fats, oils, and carboxylic acids); cyanates; epoxides; imines (ketimines and aldimines); and unsubstituted positions ortho- and para- to a phenolic hydroxyl group.

In summary, if a substance (1) meets the definition of a polymer, (2) is not excluded by any of the provisions of §723.250(d), (3) has a NAVG MW greater than or equal to 1,000 daltons and less than 10,000 daltons, (4) has individual FGEWs and a $FGEW_{combined}$ greater than or equal to 1,000 daltons for moderate-concern groups (when high-concern groups are not present, but low-concern groups may be present without limit), and (5) meets oligomer content criteria (<10 percent below 500 daltons and <25 percent below 1,000 daltons), the new substance may be manufactured under a polymer exemption.

4.4.1.3. HIGH-CONCERN FUNCTIONAL GROUPS AND THE (e)(1) EXEMPTION

Reactive groups not defined by (e)(1)(ii)(A) or (B) fall into category (e)(1)(ii)(C), the high-concern reactive functional groups. These may be used with more restriction than moderate-concern groups alone. If a polymer contains type (C) reactive functional groups, each type (C) functional group present must meet a 5,000 dalton minimum permissible limit, each type (B) group present must meet the 1,000 dalton limit and the polymer must have a $FGEW_{combined}$ of greater than or equal to 5,000 daltons. A $FGEW_{combined}$ calculation takes into account all type (C) and type (B) reactive groups. (This type of calculation is covered in section 5.3. of this manual.) Therefore, if a substance containing category (e)(1)(ii)(C) functional groups meets the definition of a polymer, is not excluded by any of the provisions of §723.250(d), has a NAVG MW greater than or equal to 1,000 daltons and less than 10,000 daltons, has a $FGEW_{combined}$ greater than 5,000 daltons, meets the individual type (B) and (C) FGEW limits of 1,000 and 5,000, respectively, and the polymer meets oligomer content criteria (<10 percent below 500 daltons and <25 percent below 1,000 daltons) the new substance may be manufactured under a polymer exemption.

Table 2 summarizes the $FGEW_{combined}$ minimum permissible levels as discussed in the preceding (e)(1) exemption criteria section of this manual. In the table, the 'X' marks which type of group (or groups) is present from the categories: low-concern, moderate-concern, and high-concern.

TABLE 2 $FGEW_{combined}$ Summary

Low-Concern	×			×	×		×
Moderate-Concern		×		×		×	×
High-Concern			×		×	×	×
Minimum $FGEW_{combined}$	None*	1,000	5,000	1,000**	5,000**	5,000	5,000**

*There are no FGEW limits for polymers containing *only* low-concern (type A, also known as (e)(1)(ii)(A)) functional groups.

**When calculating $FGEW_{combined}$ for substances with moderate (Type (B)) and/or high-concern (Type (C)) functional groups, low-concern groups (Type (A)) are *not* included in the calculation.

4.4.2. THE (e)(2) EXEMPTION CRITERIA

Those polymers having NAVG Mws exceeding the limits of §723.250(e)(1) are subject to §723.250(e)(2). Hence, this section covers polymers with NAVG Mws greater than or equal to 10,000 daltons. The oligomeric content of these higher molecular weight polymers must be less than two percent for species with molecular weight less than 500 daltons, and must be less than 5 percent for species of molecular weight less than 1,000 daltons. There are no functional group restrictions for the (e)(2) exemption, but the substance must not be excluded from exemption by any of the provisions found at §723.250(d). For example, water-absorbing polymers and cationic or potentially cationic polymers in this weight range are excluded from exemption by §723.250(d).

Therefore, if a substance meets the definition of a polymer, is not excluded by any of the provisions of §723.250(d), has a NAVG MW greater than or equal to 10,000, and meets the oligomer content criteria (less than two percent below 500 daltons and <5 percent below 1,000 daltons), the new substance may be manufactured under a polymer exemption.

4.4.3. THE (e)(3) EXEMPTION CRITERIA

Section 723.250(e)(3) provides for the exemption of manufactured or imported polyesters which have been prepared exclusively from a list of feedstocks identified in section (e)(3) of the new rule. To qualify for this exemption, each monomer or reactant in the chemical identity of the polymer (charged at any level) must be on the list. At this writing (5/22/97), six entries on the list are not on the TSCA Inventory. Therefore, these six monomers and reactants are not allowed for use in domestic manufacture.

Just as for all other exempted polymers, polyesters that are allowed an exemption under (e)(3) must meet the definition of a polymer and must not be excluded from exemption by §723.250(d). For example, excluded from an (e)(3) exemption are biodegradable polyesters and highly water-absorbing polyesters with number-average molecular weights (NAVG MW) greater than 10,000 daltons.

The following is the list from which all monomers and reactants in (e)(3)-exempted polymers must be taken. They are listed by Chemical Abstracts Index Names and Registry Numbers (where available). A "√" identifies the six substances not on the TSCA Inventory, as of this writing.

TABLE 3   The (e)(3) Monomer and Reactant List (in order by CAS Registry Number)

[56-81-5]	1,2,3-Propanetriol
[57-55-6]	1,2-Propanediol
[65-85-0]	Benzoic acid
[71-36-3]**	1-Butanol
[77-85-0]	1,3-Propanediol, 2-(hydroxymethyl)-2-methyl-
[77-99-6]	1,3-Propanediol, 2-ethyl-2-(hydroxymethyl)-
[80-04-6]	Cyclohexanol, 4,4'-(1-methylethylidene)bis-
[88-99-3]	1,2-Benzenedicarboxylic acid
[100-21-0]	1,4-Benzenedicarboxylic acid
[105-08-8]	1,4-Cyclohexanedimethanol
[106-65-0]	Butanedioic acid, dimethyl ester
[106-79-6]	Decanedioic acid, dimethyl ester
[107-21-1]	1,2-Ethanediol
[107-88-0]	1,3-Butanediol
[108-93-0]	Cyclohexanol
[110-15-6]	Butanedioic acid
[110-17-8]	2-Butenedioic acid (E)-
[110-40-7]	Decanedioic acid, diethyl ester
[110-63-4]	1,4-Butanediol
[110-94-1]	Pentanedioic acid,
[110-99-6]	Acetic acid, 2,2'-oxybis-
[111-14-8]	Heptanoic acid
[111-16-0]	Heptanedioic acid
[111-20-6]	Decanedioic acid
[111-27-3]	1-Hexanol
[111-46-6]	Ethanol, 2,2'-oxybis-
[112-05-0]	Nonanoic acid
[112-34-5]	Ethanol, 2-(2-butoxyethoxy)-

TABLE 3 *(Continued)*

[115-77-5]	1,3-Propanediol, 2,2-bis(hydroxymethyl)-
[120-61-6]	1,4-Benzenedicarboxylic acid, dimethyl ester
[121-91-5]	1,3-Benzenedicarboxylic acid
[123-25-1]	Butanedioic acid, diethyl ester
[123-99-9]	Nonanedioic acid
[124-04-9]	Hexanedioic acid
[126-30-7]	1,3-Propanediol, 2,2-dimethyl-
[141-28-6]	Hexanedioic acid, diethyl ester
[142-62-1]	Hexanoic acid
[143-07-7]	Dodecanoic acid
[144-19-4]	1,3-Pentanediol, 2,2,4-trimethyl-
[505-48-6]	Octanedioic acid
[528-44-9]	1,2,4-Benzenetricarboxylic acid
[624-17-9]	Nonanedioic acid, diethyl ester
[627-93-0]	Hexanedioic acid, dimethyl ester
[629-11-8]	1,6-Hexanediol
[636-09-9]	1,4-Benzenedicarboxylic acid, diethyl ester
[693-23-2]	Dodecanedioic acid
[818-38-2]	Pentanedioic acid, diethyl ester
[1119-40-0]	Pentanedioic acid, dimethyl ester
[1459-93-4]	1,3-Benzenedicarboxylic acid, dimethyl ester
[1732-08-7]	Heptanedioic acid, dimethyl ester
[1732-09-8]	Octanedioic acid, dimethyl ester
[1732-10-1]	Nonanedioic acid, dimethyl ester
[1852-04-6]	Undecanedioic acid
[2163-42-0]	1,3-Propanediol, 2-methyl
[3302-10-1]	Hexanoic acid, 3,3,5-trimethyl-
[8001-20-5]*	Tung oil
[8001-21-6]*	Sunflower oil
[8001-22-7]*	Soybean oil
[8001-23-8]*	Safflower oil
[8001-26-1]*	Linseed oil
[8001-29-4]*	Cottonseed oil
[8001-30-7]*	Corn oil
[8001-31-8]*	Coconut oil
[8002-50-4]*	Fats and glyceridic oils, menhaden
[8016-35-1]*	Fats and glyceridic oils, oiticica
[8023-79-8]*	Palm kernel oil
[8024-09-7]*	Oils, walnut
[13393-93-6]	1-Phenanthrenemethanol, tetradecahydro-1,4a-dimethyl-7-(1- methylethyl)-
[25036-25-3]	Phenol, 4,4'-(1-methylethylidene)bis-, polymer with 2,2'-[(1-methylethylidene)bis(4,1-phenyleneoxymethylene)]-bis[oxirane]
[25119-62-4]	2-Propen-1-ol, polymer with ethenylbenzene
[25618-55-7]	1,2,3-Propanetriol, homopolymer
[61788-47-4]*	Fatty acids, coco
[61788-66-7]*	Fatty acids, vegetable-oil
[61788-89-4]*	Fatty acids, C18-unsatd., dimers
[61789-44-4]*	Fatty acids, castor oil
[61789-45-5]*	Fatty acids, dehydrated castor oil

TABLE 3 *(Continued)*

[61790-12-3]*	Fatty acids, tall-oil
[67701-08-0]*	Fatty acids, C16-18 and C18-unsatd.
[67701-30-8]*	Glycerides, C16-18 and C18-unsatd.
[68037-90-1]*	Silsesquioxanes, Ph Pr
[68132-21-8]*	Oils, perilla
[68153-06-0]*	Fats and glyceridic oils, herring
[68308-53-2]*	Fatty acids, soya
[68424-45-3]*	Fatty acids, linseed oil
[68440-65-3]*	Siloxanes and silicones, di-Me, di-Ph, polymers with Ph silsesquioxanes, methoxy-terminated
[68957-04-0]*	Siloxanes and silicones, di-Me, methoxy Ph, polymers with Ph silsesquioxanes, methoxy-terminated
[68957-06-2]*	Siloxanes and silicones, Me Ph, methoxy Ph, polymers with Ph silsesquioxanes, methoxy- and Ph-terminated
[72318-84-4]*	Methanol, hydrolysis products with trichlorohexylsilane and trichlorophenylsilane
[84625-38-7]*	Fatty acids, sunflower-oil
[68649-95-6]*	Linseed oil, oxidized
[68953-27-5]*	Fatty acids, sunflower-oil, conjugated
[91078-92-1]* √	Fats and glyceridic oils, babassu
[93165-34-5]* √	Fatty acids, safflower-oil
[93334-41-9]* √	Fats and glyceridic oils, sardine
[120962-03-0]*	Canola oil
[128952-11-4]* √	Fats and glyceridic oils, anchovy
[No Registry #]* √	Fatty acids, tall-oil, conjugated
[No Registry #]* √	Oils, cannabis

*Designates chemical substances of unknown or variable composition, complex reaction products, or biological materials (UVCB substances). The CAS Registry Numbers for UVCB substances are not used in Chemical Abstracts and its indexes.

**1-Butanol may not be used in a substance manufactured from fumaric or maleic acid because of potential risks associated with esters which may be formed by reaction of these reactants.

5. NUMERICAL CONSIDERATIONS

There are several numerical criteria to consider when deciding if a polymer is eligible for an exemption:

1) **Number-average molecular weight** (NAVG MW) is one of the criteria defining whether an eligible polymer fits an (e)(1) or (e)(2) exemption; in the case of a water-absorbing polymer, NAVG MW defines whether or not the polymer will be excluded from exemption due to the 10,000 dalton restriction. Section 5.1., expounds on NAVG MW determination.

2) The **"Two Percent Rule"** governs whether a monomer or other reactant is part of the chemical identity. Substances that are considered by the Agency as automatically part of the chemical identity of the polymer

(those monomers or reactants used at greater than two percent composition) must be on the TSCA Inventory, excluding from reporting or otherwise exempt under section 5 of TSCA. For (e)(1) and (e)(2) exemptions, imported polymers may have monomers or reactants at less than or equal to 2 percent which are not on the Inventory; whereas in the case of domestic manufacture under an (e)(1) or (e)(2) exemption, all monomers and reactants must either be: on the Inventory; a non-isolated intermediate; otherwise exempt; or excluded from reporting. For (e)(3) polymers, all monomers and reactants, regardless of charge must be from the (e)(3) list, but only those charged at greater than 2 percent will be part of the identity. Section 5.2. explains the so-called "Two Percent Rule" and its determination.

3) The **functional group equivalent weight** (FGEW) is a measure of the concentration of functional groups of moderate- and high-concern in the polymer. This is an important factor in determining eligibility for polymers with NAVG MW greater than or equal to 1,000 daltons and less than 10,000 daltons. It is also important in determining whether a cationic polymer is excluded. Section 5.3., below, explains the determination of FGEW.

5.1. CALCULATING NUMBER-AVERAGE MOLECULAR WEIGHT

The rationale and theoretical basis for determining number-average molecular weight (NAVG MW) and brief summaries of the preferred analytical methods for determining the value follow.

The Agency uses number-average molecular weight (NAVG MW) instead of the weight-average molecular weight (WAVG MW) for defining polymer exemption categories and criteria. The NAVG MW takes into account the number of molecules of various molecular weights in the polymer sample and therefore is representative of the average weight of the typical (major) components of a polymer sample. The WAVG MW takes into account the total weight of all molecules, placing no emphasis on the number of molecules at each individual weight. When the WAVG MW is calculated, a small percentage of large molecules can bias the average and give a false representation of the majority of molecules in the sample.

The equations for determining NAVG MW (M_n) and WAVG MW (M_w) are taken from the OECD guidelines draft proposal entitled "Determination of the Number-Average Molecular Weight and the Molecular Weight Distribution of Polymers using Gel Permeation Chromatography"[3] and "Determination of the Low Molecular Weight Polymer Content"[1]. In the equations, N_i is the number of molecules at a given molecular weight (which in gel permeation chromatography (GPC) is proportional to the detector signal for the retention volume V_i). M_i is the molecular weight of the polymer fraction at the retention volume V_i.

Equation 1:

$$M_n = \frac{\sum_i N_i}{\sum_i \dfrac{N_i}{M_i}}$$

Equation 2:

$$M_w = \frac{\sum_i (M_i N_i)}{\sum_i N_i}$$

Example 11:

The reason for using the NAVG MW instead of the WAVG MW in the criteria is best demonstrated by an example. Suppose a polymer contains 200 molecules that weigh 1,000 daltons, 300 molecules that weigh 1,500 daltons, 400 molecules that weigh 2,000 daltons and 2 molecules that weigh 1,000,000 daltons. In this case 99.8 percent of the molecules in this sample weigh ≤ 2000 daltons. Clearly, one might say that typically, the polymer has a molecular weight from 1,000 to 2,000 daltons. The NAVG MW and the WAVG MW are calculated below:

$$M_n = \frac{902}{\dfrac{200}{1000} + \dfrac{300}{1500} + \dfrac{400}{2000} + \dfrac{2}{1,000,000}} = 1503$$

$$M_w = \frac{200,000 + 450,000 + 800,000 + 2,000,000}{902} = 3825$$

Of these two calculations, the M_n at 1503 daltons more accurately represents 99.8 percent of the molecules in the polymer batch. The M_w is biased by the two incidental 1,000,000 dalton molecules to the extent that the M_w average is a considerably greater weight than 99.8 percent of the sample.

The Agency requires that the manufacturer of an exempt polymer keep records of the "lowest" number-average molecular weight at which the polymer is to be made. This is not the value for the lowest MW species in a sample, but rather the lowest value of the NAVG MW obtained from polymer samples taken from a series of batches in the production of the polymer.

There are several analytical techniques for determining NAVG MW. Two literature references[4,5] as well as OECD's guidelines document for testing of chemicals[3] discuss methodologies in some detail and provide additional references. Brief summaries of the information provided in these references are given below. The techniques are based on molecular size (a function of the NAVG MW); colligative properties of polymer solutions (osmotic pressure, boiling point, freezing point, vapor pressure, etc.); or the number of chemically reactive groups present in the polymer. Any method that can be verified is acceptable for purposes of the polymer exemption. The following are most commonly used:

- Gel permeation chromatography (polymer size),
- Membrane osmometry (colligative property),
- Vapor-phase osmometry (colligative property),
- Vapor-pressure lowering (colligative property),
- Ebulliometry (colligative property),
- Cryoscopy (colligative property), and
- End-group analysis (chemical reactivity).

5.1.1. GEL PERMEATION CHROMATOGRAPHY

Gel permeation chromatography (GPC), the most frequently used and generally most reliable method for determining NAVG MW of polymers and oligomer content below 500 and 1,000 daltons, is suitable for substances ranging from very low to very high molecular weights. In an ideal situation, separation of the polymer sample is governed by hydrodynamic radius (size) of each molecular species as it passes through a column filled with porous material, typically an organic gel. Smaller molecules penetrate the pores and thereby travel a longer path and elute after larger molecules. The GPC column must be calibrated using polymers of known weight and, ideally, similar structure. Polystyrenes are used quite extensively as internal standards. Detection techniques used for GPC are refractive index and UV-absorption.

One potential problem with GPC is band broadening, especially when measuring low molecular weight polymers, or as the result of unevenly packed columns or dead volumes. Empirical calibrations of the instrument can be made to minimize broadening[6], but become unimportant when the ratio of the WAVG MW to the NAVG MW is greater than two. Another limitation with GPC is that many high molecular polymers are insoluble in usable solvents, and therefore can't be analyzed by GPC.

5.1.2. MEMBRANE OSMOMETRY

Membrane osmometry exploits the principle of osmosis for determining NAVG MW[7]. Polymer is placed in a membrane osmometer on one side of a semi-permeable membrane while a solvent is placed on the other. Solvent is drawn through the membrane as the system progresses toward equilibrium, creating a pressure differential that is dependent on the concentration difference and the molecular weight of the polymer.

The major disadvantage to this method is that accuracy and reliability may be compromised by diffusion of low weight oligomers through the membrane. Generally, diffusion is absent for unfractionated polymers with NAVG MWs greater than 50,000 daltons. The upper limit of the NAVG MW that may be measured with confidence is generally 200,000 daltons (OECD guidelines[3]).

5.1.3. VAPOR-PHASE OSMOMETRY

This method is based on the comparison of evaporation rates for a solvent aerosol and at least three other aerosols with varying polymer concentration in the same solvent. The technique is most accurate for polymers with NAVG MW less than 20,000 daltons (OECD guidelines[3]). This method is best applied to samples with molecular weight too low to be measured in a membrane osmometer.

5.1.4. VAPOR-PRESSURE LOWERING

For this technique the basic principle is similar to vapor phase osmometry, however, vapor pressure is measured instead of the rate of aerosol evaporation. The vapor pressure of a reference solvent is compared against the vapor pressure of at least three concentrations of the polymer mixed with the solvent. Theoretically this technique may be applicable for polymers of up to 20,000 dalton NAVG MWs. In practice, however, it is of limited value.

5.1.5. EBULLIOMETRY

This technique exploits the boiling point elevation of a solution of a polymer to determine NAVG MW[8]. This method makes accurate determinations for polymers with NAVG MW approaching 30,000 daltons; however, it is limited by the tendency of polymer solutions to foam upon boiling. The polymer may even concentrate in the foam due to the foam's greater surface area, making the observed concentration of the polymer in solution less than the actual. It is customary to calibrate the ebulliometer with a substance of known molecular weight. Octacosane, with a molecular weight of 396 daltons, is a common choice.

5.1.6. CRYOSCOPY

Freezing point depressions of polymer solutions can also be used to determine NAVG MW. Although the limitations associated with cryoscopy are fewer than those of ebulliometry, care must be taken to avoid supercooling. The use of a nucleating agent to provide controlled crystallization of the solvent is helpful. Reliable results may be obtained for molecular weights of up to 30,000 daltons. As with ebulliometry, calibration with a substance of known molecular weight is customary.

5.1.7. END-GROUP ANALYSIS

This method is generally the least useful since a fair amount of prior knowledge, such as overall structure and the nature of the chain-terminating end groups, is needed about the polymer. Basically, end-group analysis methods take into account the number of molecules in a given weight of a sample, which in turn, yields the NAVG MW. End-group analysis is best suited to linear condensation polymers. For branched condensation polymers or addition polymers no general procedures can be established because of the variety and origin of the end-groups. However, when the polymerization kinetics are well known, the degree of branching may be estimated based on the amount of feedstock charged. For addition polymerization, end-group analysis can be used to determine molecular weight by analyzing for specific initiator fragments containing identifiable functional groups, elements, or radioactive atoms; for chain terminating groups arising from transfer reactions with solvent; or for unsaturated end groups such as in polyethylene and poly-α-olefins.

The analytical method used must distinguish the end groups from the main polymer skeleton. The most widely used methods are NMR, titration, or derivatization. For example, carboxyl groups in polyesters are usually titrated directly with a base in an alcoholic or phenolic solvent. Infrared spectroscopy is used when the polymer cannot be titrated due to insolubility in certain solvents. This technique is useful for NAVG MWs up to 50,000 daltons (with decreasing reliability as the NAVG is increased).

5.2. THE TWO PERCENT RULE AND CHEMICAL IDENTITY

According to the polymer exemption rule at §723.250(d)(4), a polymer is not eligible for exemption if it contains at greater than two weight percent monomers and/or reactants that are not: included on the TSCA Inventory, manufactured under an applicable TSCA §5 exemption, excluded from exemption, or an non-isolated intermediate. Monomers and reactants at greater than two percent make up the "chemical identity" of the polymer. For an exempt polymer, monomers and reactants at less than or equal to two weight percent are not considered part of the "chemical identity" of the polymer; and the use of these monomers and reactants creates a different set of issues, which are discussed below.

A manufacturer or importer must carefully decide at what weight percent level each monomer or other reactant is to be used in the preparation of the exempt polymer. This choice (which must be obvious from the manufacturing data kept by the manufacturer or importer) limits the manufacturer or importer of an exempt polymer in two major ways. First, if a certain monomer or reactant is used in an exempt polymer at less than or equal to two weight percent, the manufacturer may **not** later use that reactant at greater than two weight percent (under the exemption for the same polymer). The new polymer substance that results when the reactant is increased to greater than two weight percent is different, by definition, from the polymer that contains the reactant at less than or equal to two weight percent. Second, if a reactant or monomer is used at greater than two weight percent in an exempt polymer, the reactant or monomer must **not** be eliminated completely from the polymer (if the manufacturer is trying to satisfy the exemption for the same polymer). If either of these "identity-changing" events occur, the manufacturer must do one of the following: 1) find the new polymer identity on the TSCA Inventory, 2) submit a PMN at least 90 days prior to manufacture if the new polymer is not on the Inventory, or 3) meet the conditions of a PMN exemption to cover the new polymer identity.

Non-Inventory monomers and reactants cannot be used in domestic manufacture (unless they are subject to another §5 exemption or are non-isolated intermediates). Therefore, a manufacturer cannot use such monomers and reactants for an exemption even at levels of two percent or less. A manufacturer will be able to exchange Inventory-listed monomers and reactants at less than or equal to two weight percent under one exemption, as long as such changes do not affect the eligibility of the polymer and records for such changes are maintained as stated in the rule. An exempt imported polymer under the new rule may contain non-Inventory monomers and reactants at two percent or less as long as they do not introduce into the polymer elements, properties, or groups that would render the polymer ineligible for the polymer exemption. The exception to this last statement is the (e)(3) type exempted polymer, for which monomers and reactants must only come from the (e)(3) list, even if at levels less than or equal to two percent. For all polymer types, restrictions unique to the polymer exemption must be applied in addition to the "Two Percent Rule."

> **Percent by weight**—has been defined as the weight of the monomer or other reactant used expressed as a percentage of the dry weight of polymer.

The Agency has long recognized that when calculating the percentage of each reactant, it is a matter of convenience rather than a matter of science to use the amount charged to the reactor, rather than the amount of a monomer incorporated into the polymer. EPA believes that the actual content of a polymer (what is actually incorporated into the polymer) is a better indicator of its physical, chemical, and toxicological properties, but has accepted calculations based upon the amount charged to the reaction vessel in order to facilitate PMN reporting for industry. Under the 1995 PMN rule revisions, the Agency now accepts two methods for determining the "percent by weight" of each reactant for the purpose of establishing the chemical identity of a polymer:

1) **The percent charged method:** The percent composition of each monomer or reactant is established by the amounts charged to the reaction vessel.
2) **The percent incorporated method:** The percent composition is based on the minimum theoretical amount of monomer or reactant needed to be charged to the reactor in order to account for the amount analytically determined to be incorporated in the polymer. The percent composition of each whole monomer or reactant whose fragment is present in the polymer should be established by analytical determination of the incorporated fragment, or may be established by theoretical calculations if

it can be documented that an analytical determination cannot or need not be made to demonstrate compliance with the new polymer exemption rule.

At 40 CFR §723.250(g) the Agency specifies what identity information is required to be kept by the manufacturer or importer. By paragraph (1) of section (g), the Agency requires that a manufacturer or importer must identify, to the extent known or reasonably ascertainable, the specific chemical identity and CAS Registry Number (or EPA Accession Number) for each "reactant" used at *any* weight in the manufacture of an exempt polymer. This criterion is considered reasonable by the Agency based on the requirement that any reactant used at greater than two percent must already be listed on the TSCA Inventory or otherwise exempt under an appropriate §5 rule. There may be cases where a monomer or reactant was the subject of a previous PMN, exempted or excluded from exemption, hence the requirement to have a CAS registry number for such a monomer or reactant may not be necessary. However, manufacturers and importers should maintain in their records the CAS registry number for the monomer or reactant, if one exists for that substance).

At paragraph (2) of section (g), the Agency requests that a structural diagram be provided if possible, to further clarify the identity of an exempt polymer. The Agency believes it is possible to provide a representative chemical structure diagram for nearly all polymers. It is often the structure that best illustrates the intended identity of a substance. For instance, if 2,2-bis(hydroxymethyl)propionic acid and an amine are among the feedstocks, would these two feedstocks react in such a way as to form amides or carboxylic acid salts? A structure makes clear the intent of the manufacturer or importer. All monomers and reactants at greater than two percent by weight in the polymer should be represented by the polymer structural diagram kept in the records.

5.2.1. PERCENT CHARGED METHOD

The calculations required to determine the percent by weight of a reactant charged to the reaction vessel are straightforward. The weight percent of the reactant is the weight of the material charged to the reactor (weighed before addition into the reaction), expressed as a percentage of the dry weight of the manufactured polymer (weighed after isolation from the reaction). The following equation applies, where "GFC" is grams of feedstock charged, and "GPF" is grams of dry polymer formed:

Equation 3:

$$\text{Percent by weight charged} = \frac{(\text{GFC})}{(\text{GPF})} \times 100$$

Calculations by percent charged to the reaction vessel can cause confusion if monomers or reactants lose a substantial portion of their molecular structure when incorporated into a polymer. Under these circumstances, the sum of the weights of reactants charged significantly exceeds, 100 percent. This type of calculation is demonstrated by Example 12, the formation of polyvinyl alcohol (PVA) produced from the polymerization of vinyl acetate followed by hydrolysis.

Example 12:

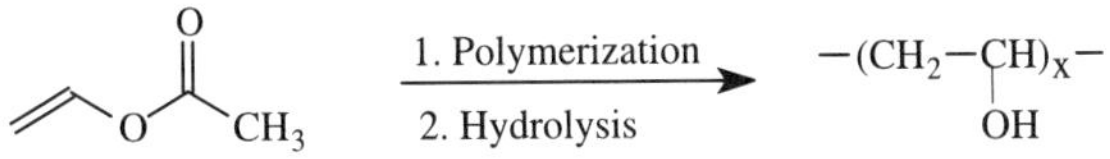

Figure 6 Polyvinyl Alcohol and Weight Percent

The molecular weight of vinyl acetate is 86 daltons, and the molecular weight of the repeating unit for PVA $[-CH_2-CH(OH)-]$ is 44 daltons. In this example, because only one monomer is used to form the polymer, and one monomer fragment is present in the polymer, the ratio of (GFC/GPF) is the same as (Feedstock MW/Fragment MW), so the equation can be simplified as follows:

$$\text{Wt. \% of vinyl acetate} = \frac{(86)}{(44)} \times 100 = 195$$

The weight percent of the vinyl acetate charged to the reactor is 195 percent!

5.2.2. PERCENT INCORPORATED METHOD

In the percent incorporated method, as stated in the 1995 PMN rule amendments, "the weight percent is based on ... the minimum weight of monomer or other reactant required in theory to account for the actual weight of monomer or other reactant molecule or fragments chemically incorporated (chemically combined) in the polymeric substance manufactured." Therefore, if a percent incorporated is to be calculated for a monomer or reactant, the degree of incorporation of the fragment resulting from the monomer or reactant must be measured.

It is not always possible or feasible to determine analytically the degree of incorporation for every type of reactant, especially for random polymerizations where no repeating subunits exist and for polymerizations using chemical reactants where the structures are not completely specified (such a reactant as conjugated sunflower-oil fatty acids, for example). Complete or efficient incorporation cannot be assumed, even if the reaction equilibrium and kinetics predict a certain result. It is also necessary to identify a structural unit within the polymer that corresponds to the specific monomer from which it came. Often the same monomer unit may originate from more than one monomer. For example, empirically determining the exact chemical incorporation of oxirane, methyloxirane, ethylene diamine, and epichlorohydrin in a polymer would require a complicated study, perhaps using radioisotope-labeled reactants. If the percent incorporated cannot be deduced by measurement or reliably estimated, the manufacturer must use the percent charged method.

In order to calculate a weight percent incorporated for a reactant, certain data must be known: the molecular weight of the reactant charged; the molecular weight of the fragment that is incorporated into the polymer (if the feedstock is not entirely incorporated); and the analytically determined amount of the incorporated reactant that is present in the polymer (the weight percent of the polymer that consists of the fragment). From these data the number of moles of fragment present in the polymer can be calculated, which is proportional to the amount of feedstock that reacted to form the polymer. The following ratio is useful:

Equation 4:

$$\frac{\text{Wt. \% Frag.}}{\text{MW of Frag.}} = \frac{(\text{g Frag.})}{(100\text{g Polym.})} \times \frac{(\text{mol Frag.})}{(\text{g Frag.})} = \frac{\text{Moles of Frag.}}{100\text{g of Polym.}} = \text{Ratio A}$$

The weight percentage of reactant incorporated is calculated by converting moles of incorporated fragment per 100 g of polymer (Ratio A), to moles of reactant and then multiplying by the reactant molecular weight. (The specific units used are irrelevant; gram-moles per 100 grams or ton-moles per 100 tons are equally valid for the calculation.) This is accomplished by the following equation:

Equation 5:

$$\text{Wt. \% React. Incorp.} = (\text{Ratio A}) \times \frac{(\text{Moles React.})}{(\text{Moles Frag.})}$$

$$\times (\text{MW of Reactant}) \times 100$$

Example 13:

For an example of the calculation for the percent incorporated method consider the polymerization of ethylene glycol with a dialkyl terephthalate. It is known that both oxygen atoms of the glycol are incorporated into the resulting polyester while two alkoxy groups of the terephthalate ester are lost in the process. The following calculation determines the weight percentage incorporated for dialkyl terephthalate: The polymer was empirically shown to contain 13.2 percent by weight of the terephthaloyl unit $[-C(=O)-C_6H_4-C(=O)-]$, which has a MW of 132 daltons. Ratio A for terephthaloyl is calculated as follows:

$$\text{Ratio A} = \frac{(13.2 \text{ g})}{(100 \text{ g polymer})} \times \frac{(1 \text{ mole})}{(132 \text{ g})} = \frac{(0.10 \text{ moles terephthaloyl})}{(100 \text{ g polymer})}$$

The weight percent of reactant incorporated is calculated as shown below. Each mole of parent dialkyl terephthalate ester would result in one mole of fragment, so the molar conversion factor is 1. If the dialkyl terephthalate charged is dimethyl terephthalate, the MW used for the calculation is 194 g/mole.

$$\text{Wt. \% React. Incorp.} = \frac{(0.10 \text{ moles})}{(100 \text{ g Polymer})} \times \frac{(1 \text{ mole of Feedstock})}{(1 \text{ mole of Fragment})}$$
$$\times \ 194 \times 100 = 19.4$$

Example 14:

For a comparison to Example 13, consider if the dialkyl terephthalate charged were diethyl terephthalate. The MW for the diethyl terephthalate is 222 g/mole. The calculation would show a weight percent of reactant incorporated as 22.2 percent. This would mean that diethyl terephthalate would have to be charged to the reaction vessel at 22.2 percent for the terephthaloyl fragment to be incorporated into the polymer at 13.2 percent; whereas dimethyl terephthalate would have to be charged at only 19.4 percent to have the terephthaloyl fragment incorporated at 13.2 percent. These percentage values make sense because a larger alkoxy group is lost when the diethyl terephthalate is the source of the terephthaloyl group than when dimethyl terephthaloyl is the source of the terephthaloyl groups and the methoxy group is lost. Therefore, to provide the same fragment incorporated in the polymer, more weight of diethyl terephthalate would have to be charged in comparison to dimethyl terephthalate.

Example 15:

Neutralizers are often used in considerable excess over the amount actually incorporated into the polymer. If the amount of incorporation is two percent or less, neutralizer may be omitted from the identity of the polymer. A sample calculation of the "weight percent incorporated" for a neutralizing base is given below:

A polymer containing free carboxylic acid functional groups was neutralized using a large excess of sodium hydroxide (NaOH; formula weight = 40); the total amount of base charged to the reactor was 10 percent. Analysis of the resulting polymer salt revealed that the polymer contained 0.92 weight percent of sodium (atomic weight = 23), coming only from the base. This amount of sodium corresponds to 0.04 moles of sodium per hundred grams of polymer, or 1.6 grams of NaOH per hundred grams of polymer—that is, 1.6 weight percent

NaOH incorporated, despite the large excess charged. Because the weight percent of NaOH is not greater than two percent, the polymer substance would not have to be described as the sodium salt.

$$\text{Ratio A for Na} = \frac{(0.92 \text{ g Na})}{(100 \text{ g Polym.})} \times \frac{(1 \text{ mol Na})}{(23 \text{ g Na})} = \frac{(0.04 \text{ moles Na})}{(100 \text{ g Polym.})}$$

$$\text{Wt. \% Inc. NaOH} = \frac{(0.04 \text{ mol Na})}{(100 \text{ g Polym.})} \times \frac{(40 \text{ g NaOH})}{(1 \text{ mol NaOH})} \times 100 = 1.6$$

If sodium bicarbonate ($NaHCO_3$; formula weight = 86) had been the neutralizing agent, the same number of moles of sodium per hundred grams of polymer would have corresponded to 3.36 weight percent of $NaHCO_3$. Because the weight percent of $NaHCO_3$ is greater than two percent, the polymer substance must be described as the sodium salt.

$$\text{Wt. \% Inc. NaHCO}_3 = \frac{(0.04 \text{ mol Na})}{(100 \text{ g Polym.})} \times \frac{(84 \text{ g NaHCO}_3)}{(1 \text{ mol NaHCO}_3)} \times 100 = 3.36$$

If a combination of bases is used for neutralization, the amounts incorporated should be prorated according to the mole ratios of the neutralizing agents charged if the reactivities are similar. Otherwise, assume the most reactive neutralizing agents is consumed first, etc.

Example 16:

For calculating the weight percent incorporated of an initiator, the computation will be similar to that for an excess neutralizing base. Initiator may be charged to the reaction vessel at a higher percentage than what is actually incorporated into the polymer. If the amount of incorporation is consistently below two percent, the initiator will not be in the chemical identity of an exempted polymer. (For polymers with PMNs and NOCs, the submitter has the option of leaving the initiator out of the identity, or including it.) In the case where initiator is not in the identity of the either an exempted polymer or in the identity of a polymer covered by a PMN and NOC, a change in initiator could be made without having to establish another polymer exemption or PMN for the change in the polymer manufacture, as long as the alternate initiator remained at or under two percent and in the case of the exemption, the initiator did not exclude the polymer in other ways. A sample calculation of the "weight percent incorporated" for an initiator is given below:

A polyolefin with a NAVG MW of 9,000 daltons was produced using azobis[isobutyronitrile] (AIBN, MW = 164) charged at 3 percent. This class of initiator is known to produce radicals that contain the nitrile moiety (CN, FW = 26), which can be analytically determined. The polymer sample was found to contain 0.29 weight percent nitrile, which was assumed to originate only from AIBN. This 0.29 g of fragment in 100 g of polymer corresponds to

0.011 moles of fragment [(0.29 g/26 g/mol) = 0.011 moles] in 100 g of polymer. Since every 1 mole of AIBN reactant produces 2 moles of fragment, a molar conversion factor of 1/2 is used to relate the amount of fragment present to the amount of reactant incorporated. The weight percent of reactant incorporated is calculated as follows:

$$\text{Wt. \% React. Inc.} = (\text{Ratio A}) \times \frac{(\text{Moles of React.})}{(\text{Moles of Frag.})}$$

$$\times (\text{MW of React.}) \times (100)$$

or

$$\text{Wt. \% AIBN Incorp.} = \frac{(0.011 \text{ moles Frag.})}{(100 \text{ g Polym.})} \times \frac{(1 \text{ Mole React.})}{(2 \text{ Moles Frag.})}$$

$$\times (164)(100) = 0.9$$

As stated at the beginning of this example, AIBN was charged to the reaction vessel at 3 weight percent, but only 0.9 percent was actually incorporated into the polymer. After establishing that the weight percent of AIBN incorporated is less than or equal to two percent, the submitter need not include it in the polymer identity.

5.2.3. METHODS FOR DETECTION OF POLYMER COMPOSITION

There are many methods available for chemical analysis of polymers and for detecting weight percent of fragments incorporated. Although any analytical method that can be verified is acceptable, this section explores some of the more common approaches. The following list of options is not meant to be exhaustive:

- Classical chemical analysis (elemental analysis, titration, etc.),
- Mass spectrometry,
- Gas chromatography,
- Infrared spectroscopy,
- Nuclear magnetic resonance spectroscopy, and
- X-ray diffraction analysis.

A brief description for each of the non-chemical methods of analysis follows.

5.2.3.1. MASS SPECTROMETRY

In mass spectrometry, an electron beam bombards a sample and creates from it positive ion fragments that are separated by mass to charge ratio in an electromagnetic field and measured quantitatively. From the abundance of the various ionic species found, the structure and composition of the original substance can be inferred. When mass spectrometry is used for analyzing polymers, the polymer is usually thermally degraded first to form fragments of low molecular weight. These fragments are volatized, ionized, and then separated as per the standard technique.

5.2.3.2. GAS CHROMATOGRAPHY

In gas chromatography (GC) gaseous or vaporized components of the sample are distributed between a moving gas phase and a fixed liquid phase or solid absorbent. By a continuous succession of elution steps, occurring at different rates for each species, separation is achieved. In the resulting gas chromatogram the peaks are proportional to the instantaneous concentration of the components. Therefore, information about the number, nature, and weight percentages of the components can be derived.

5.2.3.3. INFRARED SPECTROSCOPY

Infrared frequencies are in the wavelength range of 1 to 50 microns and are associated with molecular vibration and vibration-rotation spectra. Often for polymers, the infrared absorption spectra are surprisingly simple. This is because many of the normal vibrations have the same frequency and the strict selection rules for absorption prevent many of the vibrations from causing absorption peaks in the spectrum. Rarely can infrared be used for quantitative analysis of polymer composition.

5.2.3.4. NUCLEAR MAGNETIC RESONANCE SPECTROSCOPY

Nuclear magnetic resonance (NMR) spectroscopy is a powerful tool in the study of chain configuration, sequence distribution, and microstructures in polymers. NMR spectroscopy utilizes the property of spin angular momentum possessed by nuclei whose atomic number and mass number are not both even. Irradiation of the sample by a strong magnetic field splits the energy level into two: one corresponding to alignment of electrons with the field and the other with an antiparallel alignment. Transitions between these states lead to spectra. Peak intensity is proportional to concentration for proton NMR.

5.2.3.5. X-RAY DIFFRACTION ANALYSIS

X-ray diffraction is a useful method for detecting the presence of structures that are arranged in an orderly array or lattice. The interferences that result from the lattice interaction with electromagnetic radiation provides information regarding the geometry of the structures. Since single crystals of polymer as now prepared are too small for X-ray diffraction experiments, the crystal structure is generally derived from a fiber drawn from the polymer.

5.3. CALCULATING FUNCTIONAL GROUP EQUIVALENT WEIGHT

Reactive functional groups that come from monomers and reactants at greater than or equal to two percent in a polymer must meet the minimum FGEW requirements for the exemption category under which the polymer is manufactured or imported. Polymers that are exempt under the (e)(1) criteria must meet or exceed the minimum permissible equivalent weights for reactive functional groups (FGEW). There are no functional group restrictions for polymers meeting the (e)(2) exemption except for cationic and potentially cationic group concerns, as specified in §723.250(d). For (e)(3) polymers, reactive functional groups of moderate and high concern would not be present in any polymer derived from monomers on the allowed list. In addition, the monomers on the list are not expected to bear reactive functional groups of moderate or high concern once they are incorporated into the polymer. Hence, the (e)(3) section of the new polymer exemption rule does not have any FGEW requirements.

For (e)(1) polymers the allowable thresholds for certain reactive functional groups are listed in the Table 4. Note: this is not an exhaustive list. Consult the 1995 polymer exemption rule for groups not mentioned. Note that if a functional group is not mentioned in the rule among the low-(e)(1)(ii)(A) or moderate-concern groups (e)(1)(ii)(B), it is considered to be a high-concern functional group. Low-concern reactive groups may be used without limit, and no thresholds have been set for them.

Unless a functional group equivalent weight can be determined empirically by recognized, scientific methodology (typically titration), a worst-case estimate must be made for the FGEW, in which all moderate- and high-concern functional moieties must be factored. A generalized approach for performing equivalent weight estimations with specific methods and examples is provided below. The following is limited guidance on how to calculate functional group equivalent weights. The methods discussed are end-group analysis (Section 5.3.1), calculation based on percent charged (Section 5.3.2.), and nomograph use (Section 5.3.3.).

5.3.1. END-GROUP ANALYSIS

Most condensation polymers (polyesters, polyamides, etc.) contain reactive functional groups only at the chain ends, because all other reactive functionality in the monomers is consumed to produce the condensation polymer backbone in the final product. For this type of polymer, FGEW determination may be as simple as theoretical end group analysis and can be performed regardless of the reactive group type.

TABLE 4   Allowable Thresholds for Reactive Functional Groups

Moderate-concern: The minimum permissible FGEW is 1,000 daltons

Acid halides
Acid anhydrides
Aldehydes
Alkoxysilanes where alkyl is greater than C2
Allyl ethers
Conjugated olefins
Cyanates
Epoxides
Hemiacetals
Hydroxymethylamides
Imines
Methylolamides
Methylolamines
Methylolureas
Unsubstituted position ortho- or para- to phenolic hydroxyl

High-concern: The minimum permissible FGEW is 5,000 daltons*

Acrylates
Alkoxysilanes where alkyl = methyl or ethyl
Amines
Aziridines
Carbodiimides
Halosilanes
Hydrazines
Isocyanates
Isothiocyanates
.alpha.-Lactones; .beta.-Lactones
Methacrylates
Vinyl sulfones

---

* For polymers containing high-concern functional groups, the $FGEW_{combined}$ must be greater than or equal to 5,000 daltons taking into account high-concern (e)(1)(ii)(c) and, if present, moderate-concern (e)(1)(ii)(b) functional groups.

For a linear polymer (two reactive groups per monomer) with either the nucleophilic or electrophilic reagents in excess, the FGEW is half the NAVG MW, as described.

Example 17:

A polyamide with a NAVG MW of 1,000 daltons, made from excess ethylene diamine (two nucleophiles) and adipic acid (two electrophiles), would be anticipated to be amine-terminated at both ends, assuming a worst-case scenario (the greatest content of reactive functional groups present). The amine equivalent weight would be $\frac{1}{2}$ the NAVG molecular weight, or 500 daltons.

For simple, branched condensation polymers (having only one monomer possessing more than 2 reactive sites), the FGEW must be calculated from the total number of end groups present in the polymer. This is calculated from an estimated degree of branching, which is derived by knowing the number of reactive groups in the polyfunctional monomer. If reasonable, it should be assumed that the monomer responsible for the branching will be incorporated in its entirety to form the polymer.

The mathematics for estimating the FGEW for simple branched condensation polymers follows. The equivalent weight of the monomer is the molecular weight of the monomer divided by the weight percent charged to the reaction vessel. The monomer equivalent weight of 1,000 daltons means that there is one mole of monomer for every 1,000 daltons of polymer.

Equation 6:

$$\text{Monomer Equivalent Wt.} = \frac{(\text{Monomer molecular weight})}{(\text{Weight Percent Charged})}$$

The degree of branching is calculated by dividing the NAVG MW value by the monomer equivalent weight, multiplied by the number of reactive groups that are not used to make the polymer backbone, which is $(\text{NRG} - 2)$. (The NRG value is the number of reactive groups originally in the monomer.)

Equation 7:

$$\text{Degree of Branching} = \frac{(\text{NAVG MW})}{(\text{Monomer Eq. Wt.})} \times (\text{NRG} - 2)$$

The total number of end-groups in the polymer is the degree of branching value plus two, where the two in this equation is the number of end-groups of the polymer backbone.

Equation 8:

$$\text{Total number of Polymer End Groups} = \text{Degree of Branching} + 2$$

The FGEW is then derived by simply dividing the NAVG MW by the number of end-groups in the polymer.

Example 18:

Consider the polymerization of pentaerythritol (PE, 4 reactive groups) with polypropylene glycol (PPG, 2 reactive groups) and an excess of isophorone diisocyanate (2 reactive groups). The polyfunctional feedstock (PE) is added to the reaction at 10 percent by weight to produce an isocyanate-terminated polymer having a NAVG molecular weight equal to 2,720 daltons. The monomer equivalent weight of pentaerythritol is 1360, obtained by dividing the monomer molecular weight by the weight percent charged $(136 \div 0.10)$. PE has four reactive alcohol moieties, two are used to form the polymer backbone and the other two form branches. Following the equations given above, the degree of branching for this polymer example is $[(2720 \div 1360) \times (4 - 2)] = 4$. The total number of end-groups is $[4 + 2] = 6$. Due to the excess of isophorone diisocyanate, we assume that each end-group is an isocyanate group. Finally, the FGEW can be calculated by simply dividing the NAVG MW by the total number of end groups theoretically present. Therefore, FGEW = $(2720 \div 6) = 453$ daltons.

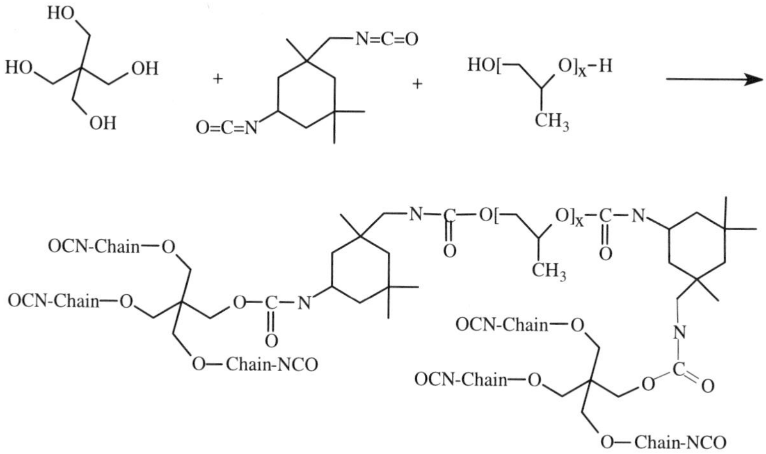

Figure 7 *Isocyanate-Teriminated Urethane and Functional Group Equivalent Weight*

For condensation polymers derived from a more complex mixture of feed-stocks, computer programs that simplify the complicated FGEW calculations may be used. (There are a few commercial programs which perform a "Monte Carlo" simulation of a random condensation polymerization that directly estimates the NAVG MW and FGEW from the types of data described earlier.) Analytical data should be used periodically to confirm computer estimates and verify eligibility.

5.3.2. MORE COMPLEX FGEW CALCULATIONS

Some condensation and addition reactions create polymers where not all reactive functional groups along the backbone of the polymer are consumed during the reaction, so a simple end-group analysis will not suffice for determining an accurate FGEW. In many of these cases the equations in this section may be used to estimate FGEWs. These equations aid in calculating FGEWs for elements (for example, basic nitrogen), for reactive groups that are unchanged under the reaction conditions and for multiple types of functional groups that remain in the polymer molecule.

Equation 9 can be used for any reactive functional group in a polymer. This may even be an atom, such as basic nitrogen, as in an example that follows. In the equation, "FWG" is the formula weight of the group; and "W%G" is the weight percent of the group:

Equation 9:

$$FGEW = \frac{(FWG) \times 100}{(W\%G)}$$

Example 19:

To calculate the amine FGEW for a polymer containing 2.8 weight percent basic nitrogen (using 14.0, the atomic weight of nitrogen, as the formula weight of the group), the equation becomes:

$$FGEW = \frac{14.0 \times 100}{2.8} = 500$$

Functional groups are typically introduced into polymers from the precursor monomers. Using Equation 10 one may calculate the weight percent of the functional group in the polymer, as long as the monomer is included in its entirety and the functional groups are introduced unchanged. In Equation 10, "FWG" is formula weight of the group, "NGM" is the number of groups in the monomer, "W%M" is the weight percent of the monomer, and "FWM" is the formula weight of the monomer:

Equation 10:

$$Weight\ \%\ of\ Group = \frac{(FWG) \times (NGM) \times (W\%M)}{(FWM)}$$

Substituting Equation 10 into Equation 9, FGEW Equation 11 is obtained, where "FWM" is the formula weight of the monomer, "W%M" is the weight percent of the monomer, and "NGM" is the number of groups in the monomer:

Equation 11:

$$FGEW = \frac{(FWM) \times 100}{(W\%M) \times (NGM)}$$

Example 20:

For an acrylic polymer containing 5.4 weight percent of acryloyl chloride (formula weight 90.5) as a monomer, the FGEW of acid chloride groups in the polymer is:

$$FGEW = \frac{(90.5)(100)}{(5.4)(1)} = 1676$$

If the various moderate- and high-concern functional groups in the polymer arise from more than a single monomer, the $FGEW_{combined}$ may be calculated using Equation 12. Also, if several different monomers contain the same groups, for example, if three monomers contribute epoxides which remain intact in the polymer, Equation 12 may be used to calculate the epoxide FGEW. This combined epoxide FGEW should be compared to the minimum permissible FGEW for epoxides when determining eligibility of the polymer.

In Equation 12, $FGEW_n$ is the FGEW for each particular functional group in the polymer:

Equation 12:

$$FGEW_{combined} = \frac{1}{\dfrac{1}{FGEW_1} + \dfrac{1}{FGEW_2} + \cdots + \dfrac{1}{FGEW_n}}$$

Example 21:

This calculation of FGEW demonstrates the use of end-group analysis and equation estimations.

Some condensation polymers contain unreacted reactive functional groups in addition to the end groups of interest; for example, an epoxide-capped phenol-formaldehyde novolak resin. The FGEW for each type of reactive group present in the molecule (end groups and unreacted groups) should be calculated separately and then summed using Equation 12. Assume a para-cresol and formaldehyde copolymerization produced a condensation polymer that was reacted with one percent epichlorohydrin. The NAVG MW of this product was determined

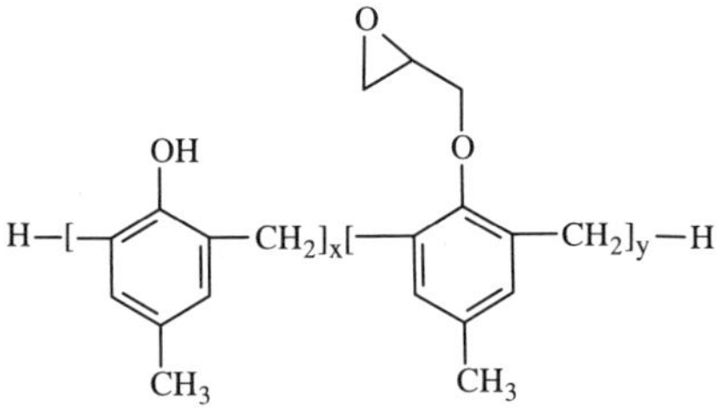

Figure 8   *Epoxide-capped Novolak and Functional Group Equivalent Weight*

by GPC to be 8,000 daltons. It would be difficult to show empirically that the polymer would not be phenol-terminated. Therefore, the polymer is assumed to be phenol-terminated as a worst case scenario. This would mean phenol groups with reactive ortho positions reside at the polymer backbone termini. The FGEW for the terminal phenolic ortho positions is (NAVG MW/2), or 4,000 daltons. This is above the minimum permissible functional group equivalent weight for the phenol reactive group which is of moderate concern (1,000 daltons minimum permissible weight). If the terminals are the only reactive groups in the polymer, this polymer would be eligible for exemption. However, epoxy rings from the epichlorohydrin are also present, so the FGEW for epoxide must also be considered. Even though epichlorohydrin would not be included in the chemical identity of the polymer being considered for exemption, (it is charged at less than two percent by weight), the FGEW for the epoxide must be included for the FGEW$_{combined}$ calculation. Following Equation 11, the epoxide FGEW is calculated to be 9,250. (The molecular weight for epichlorohydrin, 92.5 was used; along with 1 percent for the amount charged, and 1 as the number of reactive epoxides.) The FGEW of 9,250 means that there is one epoxide moiety present for every 9,250 daltons of polymer. If epoxide were the only reactive group in the polymer the minimum equivalent weight requirement for moderate concern groups would be exceeded and the polymer would meet the FGEW criteria for exemption. However, for a polymer with more than one type of reactive group of concern, a FGEW$_{combined}$ must be calculated to determine exemption eligibility.

For the polymer, the phenolic FGEW is 4,000 and for epoxides the FGEW is 9,250. The FGEW$_{combined}$ would be calculated following Equation 12, as follows:

$$FGEW_{combined} = \cfrac{1}{\cfrac{1}{4000} + \cfrac{1}{9250}} = 2792$$

With a FGEW$_{combined}$ of 2,792 daltons, this polymer would be eligible for exemption because the FGEW$_{combined}$ is greater than the required 1,000 minimum permissible equivalent weight (threshold level). Although there are two reactive functional groups from the moderate-concern list, there are no high-concern groups present.

However, note that if instead of epichlorohydrin, 1 percent of acryloyl chloride (high-concern reactant with a molecular weight 90.5) had been used, the same type of calculation would produce a polymer that is excluded from the exemption. In this further example, groups from (e)(1)(ii)(B) and (e)(1)(ii)(C) are both present and give a FGEW$_{combined}$ of 2,774 daltons. The threshold of 5,000 is daltons is not satisfied.

$$FGEW_{combined} = \cfrac{1}{\cfrac{1}{4000} + \cfrac{1}{9050}} = 2774$$

Example 22:

Similar calculations may be done for addition reaction polymers. Consider a radical polymerization of acrylates, which react via the alkene leaving reactive functionality in the molecule. In this case it would be reasonable to assume that each monomer charged to the reaction vessel will be incorporated in its entirety to form polymer.

Assume that polyacrylate was produced from 10 percent glycidyl methacrylate (MW = 142), two percent hydroxymethyl acrylamide (MW = 101) and 88 percent acrylic acid. (See Figure 9.) The reactive functional groups of concern are the epoxide (1,000 dalton threshold) from glycidyl methacrylate and the hydroxymethyl amide from the acrylamide (1,000 dalton threshold). The carboxylic acid moiety from acrylic acid may be used without limit. (See the rule, section (e)(1)(A); and also the tables in this manual.)

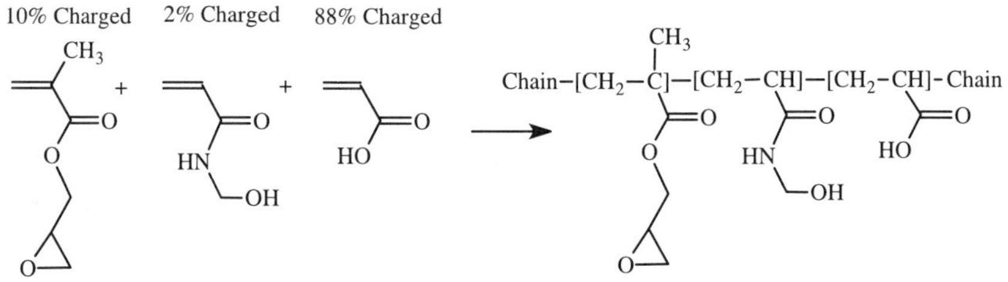

Figure 9 *Acrylate with Multiple Functional Groups*

Using Equation 11, one can calculate the FGEW for the epoxide to be 1,420 daltons (142/0.10), and the FGEW for hydroxymethyl amide to be 5,050 daltons (101/0.02). (If either of these monomers had been used separately in the stated proportions, the polymer FGEW eligibility restrictions would have been met.) The FGEW$_{combined}$ for the polymer calculated using Equation 12 is 1,108 daltons $(1/[(1/1420) + (1/5050)])$. This polymer would be eligible for the exemption because the 1,000 dalton threshold for two or more moderate-concern reactants was met. Because 1,108 daltons is fairly close to the 1,000 dalton threshold, the manufacturer will not have a lot of flexibility to increase the epoxide or amide in future batches. Also, each batch must meet the exemption. If it is anticipated that some batches will not qualify for the exemption, the manufacturer or importer must file a regular PMN 90 days prior to the manufacture of the commercial product, to cover those particular production runs.

In some addition reactions the reactive groups that effect the desired polymerization reaction are consumed and in others they are not. Examples 23 and 24 contrast these two types.

Example 23:

An example of an addition reaction that consumes the reactive functional groups is the addition of an amine to an isocyanate molecule. The reactive amine adds to the isocyanate to produce a "urea" polymeric backbone which is unreactive. Typically, an end-group analysis would be used to determine if the FGEW falls within the allowable limits for the exemption.

Example 24:

An addition reaction where the reactive group involved in the polymerization is not consumed (is still considered reactive) involves a more complicated calculation of FGEW.

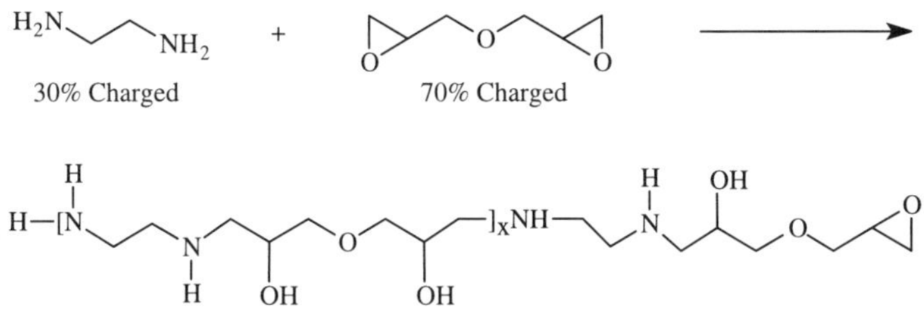

Figure 10 *Unconsumed Amines and Combined Functional Group Equivalent Weight*

Consider the reaction between ethanediamine ($MW = 60$) charged at 30 percent, and diglycidyl ether ($MW = 130$) charged at 70 percent. In the reaction, amine nitrogens react with the epoxides. This results in consumption of the epoxide to form an aliphatic alcohol, which is on the low-concern list and may be present in any quantity. The amine functionality remains intact and the FGEW for the amine is proportional to the amount of feedstock containing the amine charged to the reaction vessel. The FGEW for the amines in this type of reaction is estimated using Equation 11, the molecular weight of the feedstock (60), the percent of the monomer charged to the reaction vessel (30), and the number of reactive functional groups in the feedstock (2):

$$FGEW = \frac{60 \times 100}{30 \times 2} = 100$$

The minimum permissible equivalent weight for amines is 5,000 daltons. Because adding more groups to the $FGEW_{combined}$ calculation can only lower the value, no further calculation would be necessary since the polymer would not be eligible by amine content alone. This is demonstrated by factoring in the epoxide contribution. The polymer would likely be epoxide-terminated because of the excess molar amount of glycidyl ether charged. If this polymer had a NAVG $MW = 5,000$ daltons, the epoxide FGEW would be 2,500 daltons by end group analysis, assuming linear polymerization. The epoxide-terminated polymer containing reactive amines would have a $FGEW_{combined}$ equal to 96 daltons $[1 \div [(1/100) + (1/2500)]]$.

In some addition polymer processes one reactant (or group of reactants) is used in large excess compared to the other reactants. The reporting of residual amounts of monomers or other reactants is not required under the new rule. (The amount of reactant that does not form polymer is not regulated by the new polymer exemption rule, since these residual, unreacted materials must be on the TSCA inventory and are covered by different Agency authority, as existing chemicals.) For polymers made under these conditions, a simple repeating unit of known molecular weight can be assumed. The FGEW can be calculated by dividing the unit molecular weight by the number of groups in the unit.

Example 25:

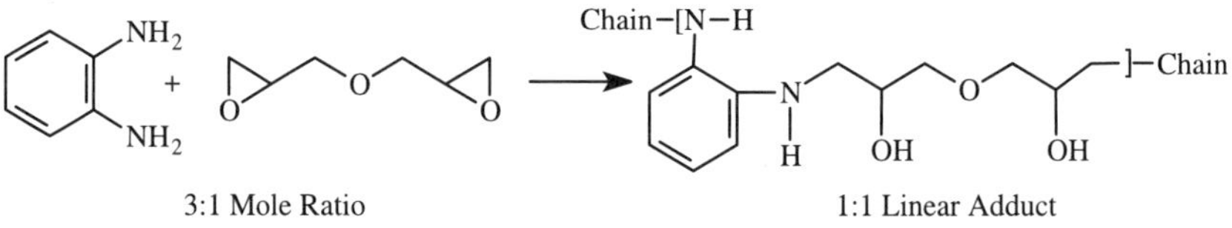

Figure 11 Repeating Units, a Polyamine and Functional Group Equivalent Weight

A polyamine was made from the addition of 70 weight percent 1,2-benzenediamine (MW = 108) to 30 weight percent of diglycidyl ether (MW = 130). The diamine:diepoxide ratio equals about 3:1, as charged to the reaction vessel. A linear polymer of a 1:1 adduct (MW = 238) is the most likely representative repeating unit. The amine FGEW would be 119 daltons (the repeating unit MW of 238 daltons divided by two, the number of reactive amines in the repeating unit). The FGEW will not change regardless of the number of repeating units in the polymer or the amount of excess diamine monomer.

5.3.3. DETERMINING FGEW BY NOMOGRAPH

The nomograph in Figure 12 has been developed to aid in the estimation of FGEW. The logarithmic axes on the nomograph are "Formula weight of group or monomer," "FGEW," and "Weight percent of group or monomer in the polymer." Choosing the axis points for the first and last of these data and drawing a line between the two points will intersect the FGEW axis at the point representing the FGEW for the monomer or group being estimated. For monomers containing several identical groups, the FGEW should be divided by the number of identical groups in the monomer. For the case of several different monomers containing the same groups use FGEW equation 4 instead of the nomograph.

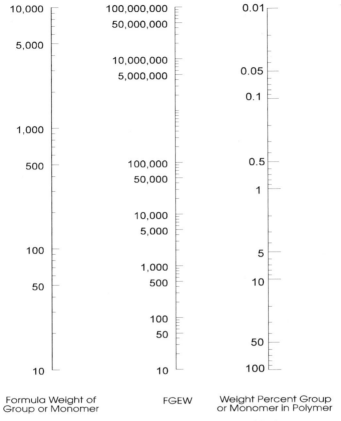

Figure 12 Nomograph for Determining FGEW

6. OTHER REGULATIONS AND REQUIREMENTS

Please consult the new rule at 60 FR 16316–16336 (USEPA 1995) for any of the following topics:

- Exemption Report and Requirements
- Chemical Identity Information
- Certification
- Exemptions Granted under Superseded Regulations
- Recordkeeping
- Inspections
- Submission Information
- Compliance

- Inspections
- Confidentiality

7. COMMON QUESTIONS AND ANSWERS

POLYMER DEFINITION:

1. In determining whether a polymer is on the Inventory, does the "new" polymer definition under the polymer exemption apply? For example, if I manufacture a substance of the type $R(OCH_2CH_2)_nOSO_3Na$ where n = an average of 7, will I have to submit a PMN even though >3 units of consecutive monomer are present? The Inventory currently considers all the ethoxylates with >3 units as polymeric, and therefore as the same substance. What if n = exactly 7? Exactly 15?

The alkyl ethoxylate sulfates with definite numbers of repeating units that you describe would not meet the polymer definition, because they would consist of molecules of a single molecular weight. Chemical Abstracts nomenclature rules and the TSCA Inventory nevertheless does treat *some* of these as though they were polymers. For example, "laureth sulfate", which corresponds to the formula above where $R = C_{12}H_{25}$ and n = x, is on the Inventory (CASRN 9004-82-4). Variations in the number of ethylene oxide units, as long as n is *either* >10 *or* variable *or* represents an average value, will not produce a new (that is, non-Inventory) substance. Thus laureth sulfate with n averaging 7 is considered an existing substance, as is laureth sulfate with n = exactly 15. However, the case where n = exactly 7 is considered a discrete chemical substance, not a polymer, and would not be considered the same. It would have a different name and CASRN, and would be a new chemical if it is not already on the Inventory elsewhere. This has always been true, and is unchanged by the polymer exemption.

The "new" polymer definition does not affect the Inventory status of existing polymers or of new polymers submitted under the PMN rule. The polymer definition, which applies only to polymers manufactured under the polymer exemption, therefore does not have the effect of creating a set of "no longer polymers".

2. Would the following example count as a "polymer molecule?" (The longest straight chain is $1 + 1 + 2 = 3 + 1$.)

H(oxypropylene)-O-sorbitol-O-(propyleneoxy)$_2$-H

No. Sorbitol cannot be a repeating unit under the conditions of the relevant polymerization reaction (propoxylation), so it is considered an "other reactant." Therefore the longest sequence of monomer units (considered as derived from propylene oxide) is two. A continuous string of at least three monomer units is required, plus one additional monomer unit or other reactant.

3. How do you apply the molecular weight distribution requirement of the polymer definition (i.e., <50 percent of any one MW) to highly cross-linked polymers of essentially infinite MW?

For polymers of "essentially infinite" MW, unless the entire mass of polymer produced were in one continuous phase, the actual molecular weight would be limited by the size of the individual droplets, beads, pellets, flakes, etc. No two of these would be likely to have exactly the same mass, and the distribution criterion would be met. For that matter, the molecular weight determination itself would produce a range of values because of the finite precision of the instrument.

ELEMENTAL EXCLUSIONS:

4. Regarding elemental limitations, why was fluorine not included in 723.250(d)(2)(B) but included in ii(C)?

Fluoride ion (F^-) has a high acute toxicity, and would therefore be unacceptable as a counterion in a substance that is supposed to present no unreasonable risk to human health. Fluorine covalently bound to carbon is either unreactive and thus not available in the form of F^-, or is part of a reactive functional group such as acyl fluoride (COF) and subject to the reactive functional group criteria.

5. Can you give an example of F^- (anion) that is not allowed?

Consider a cationic ion exchange resin that would otherwise have been eligible (because it meets the criterion of insolubility). If the counterion is fluoride (F^-), it will be ineligible.

6. Ammonium is not listed as an acceptable monatomic counterion. Does this mean a polymer may be made under the exemption, but not its ammonium salt?

No. Ammonium may be used as a counterion. It is not monatomic, and is not excluded under section (d)(2)(ii).

7. Are *only* monatomic counterions allowed? What about CO_3^{2-}, HCO_3^-, NO_3^-, etc.?

Monatomic counterions are allowed only if they are on a list of specifically allowed ones. All other monatomic counterions are excluded. The polymer exemption says nothing whatsoever about polyatomic counterions as such; they are permitted if they do not otherwise render the polymer ineligible. Carbonate $(CO_3{}^{2-})$ is allowed, for example; perchlorate $(ClO_4{}^-)$ is not, because the chlorine is neither a monatomic ion nor is it covalently bound to carbon; trichloroacetate $(CCl_3CO_2{}^-)$ is allowed.

    8. Are monomers that have CF_2 or CF_3 groups allowed?

Monomers that contain CF_2 or CF_3 groups are acceptable, provided that the groups are not part of a reactive functional group. $-CF_2-$ is not generally a monomer unit because it is not "the reacted form of the monomer in the polymer"; however, $-CF_2CF_2-$ groups derived from the polymerization of tetrafluoroethylene certainly could be monomer units.

EXCLUSION FOR DEGRADABLE POLYMERS:

    9. What is the time frame for "polymers that do not degrade, decompose or depolymerize?" Does EPA want us to synthesize polymers that bioaccumulate in the environment? Does the term "degrade" apply to biodegradation or other degradation in waste treatment systems?

This restriction is essentially unchanged from the 1984 polymer exemption. While EPA recognizes in principle the beneficial effects of biodegradability, it commented in the discussion section of that rule that the Agency "... has little experience reviewing the mechanism by which breakdown may occur, the decomposition products that may result, and the potential uses of such polymers.... Because of the complexity of review necessary for many of these polymers and the lack of EPA review experience, the Agency did not believe that an expedited review period was sufficient to adequately characterize risk."

The Agency acknowledged in that discussion that essentially all polymers degrade or decompose to a limited degree over time. It gave as examples the normal fate of polymers in landfills and the weathering of paint, and specifically stated that the exclusion was not intended to address such degradation. *Substantial* biodegradation in a waste treatment system would render a polymer ineligible for the exemption.

    10. How does EPA define "degrade," "decompose," and "depolymerize?" If these are by-product minor reactions of a polymer, can the polymer still be eligible for the exemption, assuming other criteria are met?

The definitions are provided at §723.250(d)(3), and read: "For the purposes of this section, degradation, decomposition, or depolymerization mean those types of chemical change that convert a polymeric substance into simpler, smaller substances, through processes including but not limited to oxidation, hydrolysis, attack by solvents, heat, light, or microbial action." Minor byproduct degradative reactions will not exclude a polymer from the exemption; see the answer to the previous question, for example.

11. Starch is a polymer that readily degrades in the environment. If it were not listed on the TSCA Inventory, would starch be eligible for the exemption?

No; polymers that readily degrade are excluded from the exemption.

12. What does the Agency mean by "substantially" in the phrase "substantially degrade . . ."? Does this refer to any specific conditions (e.g., sunlight, water, low pressure) or under normal environmental conditions?

By "substantially," the Agency means considerably; meaningfully; to a significantly large extent. The restriction refers to polymers that undergo considerable degradation, under normally anticipated conditions of use or disposal, and in a reasonable length of time.

13. Will a polymer that is designed to be pyrolyzed or burned when it functions as intended be excluded from the exemption by the "degrade, decompose or depolymerize" conditions?

Yes, if that is the normal way it is used. A polymer propellant or explosive would be excluded. However, a plastic used for (say) garbage bags would not be excluded merely because it might under some circumstances be incinerated.

14. A manufacturer produces a polymer that is otherwise eligible for the exemption. It is readily biodegradable by the OECD test. There are two uses for the product. In one use, the manufacturer can reasonably anticipate that the polymer will eventually find itself in aqueous systems where it may degrade. In the second use, the polymer will be formulated into articles at a low percentage such that the articles themselves would not be anticipated to degrade once they are disposed of in a landfill. Provided that the manufacturer could control customer sales to assure that the polymer would only be used in the second use, could the polymer exemption apply?

Yes. Provided that the use is restricted to conditions under which the polymer would not be expected to degrade, decompose, or depolymerize, it would not be excluded from the exemption.

15. Will EPA specify testing conditions for evaluating "degradation"? Will manufacturers using the exemption have to test to prove their polymers don't degrade? Can we rely on *intent* to degrade?

This guidance document does not specify test conditions for degradability; there is no testing requirement to establish nondegradability; and, as the rule says in section (d)(3), polymers are excluded "... that could substantially decompose after manufacture and use, even though they are not actually intended to do so." In other words, it is what can actually be expected to happen to the substance, rather than just the intent of the manufacturer, that determines whether this criterion is met.

16. Are Diels-Alder polymers (for example, dicyclopentadiene polymers) considered degradable?

There are no specific constraints on structure or method of polymerization. If Diels-Alder polymers are "designed or reasonably anticipated to substantially degrade, decompose, or depolymerize," they would be excluded; if not, they would be eligible if the other exemption criteria are met.

EXCLUSION OF WATER-ABSORBING POLYMERS:

17. How are water-soluble, water-dispersible, and water-absorbing polymers distinguished with regard to the polymer exemption? Are they treated the same? Is dispersibility considered degradation?

Water-soluble and water-dispersible (that is, self-dispersing or already dispersed) polymers are not considered to be water-absorbing substances. Only water-insoluble, non-dispersible water-absorbing polymers are excluded. The distinction is based on an assumed mechanism for lung damage by water-absorbing polymers, which involves a failure of the lungs to clear particles of these materials. Water-soluble or water-dispersible materials are expected to be cleared, and are thus not excluded. Dispersibility is not considered to be degradation.

A water-absorbing polymer that is manufactured or imported in water and is sold in water at concentrations allowing full water-absorption is not excluded from exemption provided that it meets all other criteria of the exemption and is not otherwise specifically excluded.

18. Why are high MW water-absorbing polymers excluded from the polymer exemption?

EPA excluded this category of polymers from the exemption based on TSCA section 8(e) inhalation study, designated 8(e)-1795 and FYI-470, on a water-absorbing polyacrylate polymer with a MW in excess of 1 million daltons that indicated a potential cancer concern for this type of high MW water-absorbing polymer. The Agency concluded that exposure to respirable fractions of these polymers might present an unreasonable risk to human health. (For a discussion of this issue, also see pages 16319–16320 in the rule which this document compliments.)

19. If a polymer is partly ionized on use by a pH change which increases its water absorption to greater than 100 percent by weight, is the polymer no longer eligible for the exemption? What if the neutralizing agent is less than two weight percent of the polymer? Does the so-called "(h)(7)" pH neutralizer exclusion apply to polymers > 10,000 MW that absorb more than 100 percent of their weight of water upon neutralization? Does the "(h)(7)" exclusion take precedence over the polymer exemption, and vice versa, or what?

If the polymer becomes water-absorbing upon use in neutral water, it is a water-absorbing polymer, whether or not ionization is involved.

If it is deliberately converted to a water-absorbing polymer by neutralization, that constitutes manufacture for commercial purposes as a chemical substance *per se*, rather than processing. The resulting substance would be a *different* polymer that would be considered water-absorbing and consequently not eligible for the exemption. (Even if the neutralizing agent used is less than or equalt to two percent, a polymer must still meet the eligibility requirements in order to be exempt.) The unneutralized starting polymer could still be eligible for the exemption, if it met the other exemption criteria.

On the other hand, if the neutralization results in a substance excluded from reporting under 40 CFR §720.30(h)(7) (which basically covers processing rather than manufacture), that substance remains excluded from reporting even if it would have been ineligible under the polymer exemption. (See the Agency's published clarification on this issue available through the TSCA Assistance Information Service (202) 554–1404: the package from Joseph Carra, Deputy Director, Office of Pollution Prevention and Toxics, to the regulated community, dated June 29, 1994.) If an *exempt* polymer is converted into a water-absorbing substance as a result of a chemical process or reaction that produces a substance excluded from reporting under (h)(7), the starting polymer remains exempt. Both the polymer exemption and 40 CFR §720.30(h)(7) apply, independently, to the respective substances.

20. I have an acidic resin that is eligible for a polymer exemption. Would the salt of this resin automatically be eligible for exemption?

A salt of an exempt polymer would not automatically be eligible for exemption. However, if the conversion of the resin to its salt introduces no properties (for example, water-absorption) or constituents (for example, certain elements in amounts greater than permitted) that would cause it to be excluded from the exemption, the resulting polymer salt should also be eligible for the exemption. The manufacturer must ensure that the polymer salt does in fact meet all requirements for exemption and that the reaction making the salt has not caused a change in the polymer which could exclude it from exemption. (Bear in mind that the conversion of a polymer to its salt does not always produce a reportable substance; see the answer to question 19.)

LIMITATION ON CATIONIC PROPERTIES:

21. If you have a very "non-basic" amine (such as dialkyl aniline) is it anticipated to become cationic in the environment? Suppose you can calculate from the pK_a of the amine and the concentration of amine groups in the polymer that the functional group equivalent weight of the protonated form of the amine will be >5,000 in a natural aquatic environment. Could the polymer be eligible for the exemption?

If a manufacturer or importer can establish by pK_a data, or otherwise, that the amine groups in a polymer are "non-basic" and therefore would not become cationic in the environment, the polymer would not be excluded from exemption on the basis of potentially cationic character. However, amine groups are still considered reactive functional groups whether they are protonated or not. In other words, neither pK_a nor the "non-basic" character of amines affects the calculated reactive functional group equivalent weight. See the discussion under that section.

22. Does the phrase "used only in the solid phase" mean end use, as opposed to processing where the polymer may be melt extruded, injection molded, etc.?

"Used only in the solid phase" does refer to end use; a solid material melted during the course of processing does not have to be considered a liquid if it is solidified at the end of the processing step.

23. A polymer contains a potentially cationic group. The polymer is neither water soluble nor water dispersible but is manufactured by emulsion polymerization and therefore exists as particles dispersed in water. Is the polymer ineligible for the exemption?

Cationic polymers and potentially cationic polymers (see definitions in section 4.2.1 of this manual) are excluded from the exemption except for two types: 1) those that are solids, are neither water soluble nor dispersible in water, are only used in the solid phase and are not excluded by other factors; and 2) those that have low cationic density and are not excluded by other factors. If your polymer is neither water soluble not water dispersible, manufacture by emulsion polymerization alone would not render it ineligible. See also the answer to question 17.

24. What exactly is meant by water-insoluble with respect to cationic polymers that qualify for exemption? Does the phrase "[T]he polymer is a solid material that is not soluble or dispersible in water" relate to a specific test? Is this a drop in water test or formulating test?

The phrase in section (d)(1)(i) does not relate to a specific test, and the Agency has not prescribed any specific test for water-solubility of polymers. Whatever standard is used, however, should be applied to the commercial material as manufactured and sold. If an aqueous emulsion is the commercial form of the substance, the solubility criterion should be applied to that, rather than to a dried film of the final, end-use product. (An aqueous emulsion is a water-dispersed material, and a substance in that form would be considered to be soluble or dispersible; it therefore would not qualify.)

REACTIVE FUNCTIONAL GROUPS:

25. Please confirm that amine salts are permitted, as well as confirming that sulfonic and sulfuric acids ($-SO_3H$ and $-OSO_3H$) and their salts are considered non-reactive.

Amine *counterions* are permitted for anionic polymers. Sulfonate salts are not considered reactive. However, sulfonic and sulfuric acids are considered reactive (they were specifically designated as such in the 1984 polymer exemption rule, and the interpretation has not been changed in the new rule).

26. Regarding (e)(1) criteria, what are a few examples of "high concern" and "low concern" functional groups. Would acrylate, epoxide or isocyanate groups be considered "high" or "low" concern?

Epoxides are listed in (e)(1)(ii)(B), the list of "moderate concern" groups for which concerns exist at a functional group equivalent weight of 1,000 or less. Acrylate and isocyanate are not listed either in (e)(1)(ii)(B) or in (e)(1)(ii)(A), the "low concern" group list; they are therefore considered "high concern" groups and fall under (e)(1)(ii)(C), for which the functional group equivalent weight concern level is 5,000 or less. Sections (e)(1)(ii)(A) and (B) contain lists of all the "low concern" and "moderate concern" groups, respectively. Any reactive group not on either list is considered to be "high concern."

27. The nitro group does not appear on the low- or moderate- concern list of reactive functional groups. Does this mean that nitro would fall into the high-concern category by default? This is counter-intuitive, because I wouldn't consider the nitro group to be very reactive and of much concern.

Numerous groups were not listed because they were not considered to be reactive functional groups (for example, ester and ether groups). Nitro groups are also not considered to be reactive functional groups, unless they are specially activated (certain aromatic nitro groups are readily displaced by nucleophilic substitution reactions).

28. Is the amine group considered a high-concern reactive functional group? It is not listed specifically at either 40 CFR §723.250(e)(ii)(A) or (B), which would by default place it in category (C). However, because the criteria for a substance that "may become cationic in the environment" appears to address the concerns that EPA would have for amine groups in limiting the amount of amine in a polymer to one in 5,000 daltons, it does not seem that the amine group, in and of itself, should be regarded as a reactive functional group. Would the amine group be used in the calculation for FGEW$_{combined}$?

The amine group is considered a high-concern reactive functional group and therefore should be used in the calculation. It is reactive in undergoing condensation reactions to form polyamides and polyimides and, unlike the aliphatic hydroxyl group, was not identified as a low-concern functional (category (A)) group. The Agency has concern for this group as a reactive functional group unrelated to considerations of its aquatic toxicity. For polymers that are not water-soluble or -dispersible and that will be used only in the solid phase, the limitation on cationic functional groups (such as quaternary ammonium) would not apply; but the limit on amine groups *as reactive groups* would still apply.

29. Regarding FGEW of high concern groups vs. low concern groups, does one need to combine all high concern groups and separately combine all low concern groups—or add both together?

If any "high concern" (that is, (e)(1)(ii)(C)) groups are present, one needs to calculate the combined functional group equivalent weight of any "moderate concern" (that is, (e)(1)(ii)(B)) and "high concern" groups together. To meet the criterion, the resulting FGEW must be no less than 5,000. "Low concern" (that is, (e)(1)(ii)(A)) groups are not included in the computation.

30. If a polymer with a number-average molecular weight >10,000 meets the reactive functional group and oligomer content criteria of (e)(1), but not the more stringent oligomer content criterion of (e)(2), it seems to fall into a gap between (e)(1) and (e)(2). Is it therefore not eligible for the exemption? If it isn't, does the Agency plan to amend the (e)(1) criterion to omit the phrase "and less than 10,000 daltons"?

The (e)(1) and (e)(2) exemptions are indeed mutually exclusive. Polymers with molecular weight of more than 10,000 are eligible only for the (e)(2) exemption, which has lower allowable concentrations of oligomer than does (e)(1). A polymer like the one described would not be eligible for either the (e)(1) or (e)(2) exemption. The Agency received no comment on this issue from the time the rule was proposed on February 8, 1993 until after the final rule became effective on May 30, 1995. A modification of the criteria seems reasonable, but additional rulemaking rather than a simple correction would be required. The issue is under discussion, and Agency resource constraints may rule out near-term action.

THE TWO PERCENT RULE (AND NON-INVENTORY REACTANTS):

31. Please explain the changes in the "Two Percent Rule" for polymers.

The "Two Percent Rule," which has been in effect since 1977, allows manufacturers and importers of polymers to add monomers or other reactants to an Inventory-listed polymer at levels of two percent or less (based on the dry weight of the manufactured polymer) without making a polymer with a different chemical identity than the Inventory-listed polymer. It also serves as a basis for determining the identity of a polymer. Before May 30, 1995, the effective date of the PMN Rule amendments, the monomer content of a polymer was always calculated based on the weight percentage of monomer or other reactant "charged" to the reaction vessel. The 1995 amendments allow persons greater flexibility in determining the percentage composition and whether monomers and other reactants are present at more than two percent. In addition to being able to use the "charged" method, the 1995 amendments allow persons to use an alternative method, i.e., to determine the amount of monomer or other reactant that is present "in chemically combined form" (incorporated) in a polymer and to report the minimum weight percent of that monomer or reactant that is needed in theory to account for the amount incorporated. A manufacturer is free to use either method to determine a two percent level; however the "incorporated" method, while providing more flexibility, also requires supporting analytical data or theoretical calculations.

This change in the "Two Percent Rule" applies to all polymers under TSCA, including Inventory listings, PMN submissions, and polymer exemptions.

32. If I use the "chemically combined" method and claim that two percent or less of a reactant is incorporated in my polymer even though I charge a higher level to the reaction vessel, what records am I required to maintain to support this claim?

Your records must contain analytical data or appropriate theoretical calculations, if such an analysis is not feasible, to demonstrate that the minimum weight of monomer/reactant required to account for the monomer/reactant fragments chemically incorporated is two percent or less. Your records should take into account potential batch-to-batch variation.

33. It appears from the polymer exemption rule and the technical guidance manual that a person does not have the option of including a reactant/ monomer at less than or equal to two percent in the polymer identity. Is this true?

Yes, this statement is true. Polymers covered by a polymer exemption do not have a formal name. The "identity" is established by the percentages of monomers/reactants charged or incorporated in the polymer, as cited in the exemption-holder's records. If a polymer has less than or equal to two percent of a monomer/reactant, the identity does not contain that monomer/reactant. If an otherwise identical polymer is made, and the same monomer/reactant is a greater than two percent, the identity of the second polymer is different from the first. Two exemptions would have to be claimed to cover both polymers.

For polymers for which a PMN is submitted, the submitter does have the option of including a reactant/monomer at less than or equal to two percent in the polymer identity.

34. Does a manufacturer need to test every batch of polymer to prove that less than two percent is incorporated, or would one documented test on a typical batch be sufficient?

A company is not required to test every batch but is required to maintain in its records analytical data or theoretical calculations to demonstrate compliance with the "Two Percent Rule" when using the "incorporated" method. If the amount normally incorporated is expected to be close enough to two percent that occasional batches might exceed that level, either more frequent testing, or always considering the reactant to be part of the chemical identity, or manufacturing a *separate* exempt polymer with the reactant present at greater than two percent and included in the polymer identity, might be appropriate.

35. I use a prepolymer that is on the Inventory to make my polymer. The prepolymer contains a non-Inventory monomer, and the final polymer contains greater than two percent of that monomer. Will my polymer be ineligible for the exemption?

Not on the basis of the non-Inventory monomer; §(d)(4) bars the use of "monomers and/or other reactants ... that are not already included on the TSCA Chemical Substance Inventory ...", but the prepolymer is a reactant that is on the Inventory. The identity of the final polymer will probably include the non-Inventory monomer, though; see the answers to related questions in the section on Inventory Status of Reactants (questions 45–49).

36. If an initiator is incorporated at no more than two percent, does it have to be on the TSCA Inventory?

An initiator or other reactant present at no more than two percent does not have to be on the Inventory for a polymer to be eligible for the exemption. However, if the reactant is not on the Inventory, it cannot be *used* for commercial manufacture in the United States. Consequently, this provision will for all practical purposes be applicable only to imported polymers.

37. Can I use less than or equal to two percent of any monomer that is on the Inventory?

Yes, as long as that monomer doesn't introduce elements, groups or properties that would render the polymer ineligible at the concentration of monomer used. Note, though, that for the (e)(3) "polyester" exemption, *all* components of the polymer must be on the list of allowable reactants. In this case the use of non-listed monomers, even at two percent or less, would render the polymer ineligible for the (e)(3) exemption.

38. I wish to import a substance containing greater than two percent of a reactant not on the public TSCA inventory, but which may be on the confidential inventory. On what do I file a Bona Fide, if all I plan to import is the final polymer, to know whether it now qualifies for the new polymer exemption criteria or if I need to file a PMN for the polymer?

There is really no way to find out whether a substance is on the Inventory unless you intend to import or manufacture that substance itself. You may not file a Notice of Bona Fide Intent to Manufacture ("Bona Fide") on the reactant unless you have a *bona fide* intent to manufacture or import it. (Your supplier, if in the U.S., could file a Bona Fide on the monomer, however.) Therefore, the only substance for which *you* can file a Bona Fide is the final polymer. If the polymer is on the Inventory, no PMN will be needed. If not, you will need to file a PMN for the polymer; unless you have a real intent to import or manufacture the monomer, you cannot file a PMN or an exemption for the monomer. If the monomer is on the Inventory, your polymer may be eligible for exemption. If it is not, completing the review process for the monomer and commencing its manufacture or import will allow it to be used in an otherwise exemptible polymer.

39. Can the polymer exemption be used for import of a polymer made with a non-TSCA listed chemical? If not, why?

The polymer cannot be imported under the polymer exemption if the non-TSCA reactant is used at greater than two percent. The reason is that the Agency cannot make the determination that no unreasonable risk will be incurred by a polymer that contains residual amounts of a monomer or other re-actant that it has never reviewed. If the reactant is present at less than or equal to two percent, and if its presence does not otherwise render the polymer ineligible, the polymer may be imported (if eligible). A polymer may not be manufactured domestically if any reactant is not on the Inventory.

40. If a polymer is on the Inventory but contains a non-Inventory monomer, can you import it?

Yes. If the polymer is on the Inventory, it is an existing chemical, and no PMN or other notice or exemption is required. The exclusion of non-Inventory monomers and other reactants applies only to the polymer exemption. As in the answer to the previous question, you may not manufacture it domestically unless all the reactants are on the Inventory.

41. What if the non-Inventory-listed monomer is charged or incorporated at less than or equal to two percent?

A polymer containing a non-Inventory-listed monomer at less than or equal to two percent may be eligible for the exemption provided that the monomer does not "introduce into the polymer elements, properties, or functional groups that would render the polymer ineligible for the exemption." Language at §(g)(1) says that such reactants are not allowed "at any level"; but to the extent that below certain levels they do not render the polymer ineligible, they are not such reactants when used below those levels. Note again that a non-Inventory-listed monomer that is not on the list of permitted reactants for the (e)(3) exemption will render it ineligible for that exemption. There are in fact reactants on that list that are not on the Inventory. These are not subject to the two percent limitation, since they have already been reviewed by the Agency and are considered to be not of concern; see the answer to Question 50. However, if a monomer or other reactant is not on the Inventory or otherwise excluded from reporting or exempted from section 5 requirements, it cannot be used for domestic manufacture, regardless of its concentration in the product polymer.

42. Can polymers that utilize less than or equalt to two percent of non-Inventory listed monomers be eligible for the exemption?

Such polymers would be eligible for exemption as long as they meet all the other exemption criteria. However, a monomer used at any concentration must be on the Inventory or exempt before it can be used in the domestic manufacture of the polymer.

43. If a polymer contains any amount of a component that is not on the TSCA inventory, it cannot be manufactured domestically under the polymer exemption. Does that mean that a PMN for the polymer is

necessary, or does it mean the reactant must be put on the Inventory first before the polymer exemption can be used?

To use a substance domestically for any reason, it must be on the Inventory, excluded from reporting, or exempted under an applicable section 5 exemption (for example, low volume exemption, low release and exposure exemption, pre-1995 polymer exemption, current polymer exemption). Therefore, a PMN (or applicable section 5 exemption) is required for the new reactant, and the reactant must be on the Inventory or exempt before it can be used in the domestic manufacture of the polymer. Once the reactant is on the Inventory, a polymer containing it would not be automatically excluded from the exemption, as long as it was otherwise eligible.

44. If you have a TSCA-listed brominated flame retardant mixed at greater than two percent in a polymer base, is the polymer subject to PMN requirements or is it exempt?

The material is considered to be a mixture of polymer and the flame retardant. Mixtures are not subject to reporting under TSCA, provided that there is no intended reaction between the components of the mixture. The components of the mixture are separately subject to reporting if they are not on the Inventory. If they are both on the Inventory, no reporting is required. If the polymer is eligible for the exemption, the presence of the other component will not render it ineligible.

45. Are all of the exclusions under 40 CFR §720.30 ("Chemicals not subject to notification requirements") applicable to the polymer exemption?

Yes; however, a manufacturer must comply with the conditions of the exclusions even though the substances are being used in connection with the polymer exemption. For example, a substance subject to the low-volume exemption could be used as a monomer for an eligible polymer, but only if the supplier is a holder of the exemption and if the appropriate production ceiling is adhered to.

INVENTORY STATUS OF REACTANTS; CHEMICAL IDENTITY OF POLYMERS:

46. How do I find out whether:
    (a) my polymer is on the confidential TSCA Inventory?
    (b) a reactant in my polymer is on the confidential Inventory?

You can determine the Inventory status of your polymer by filing a Notice of Bona Fide Intent to Manufacture (or a PMN). You may not file a Bona

Fide on the reactant unless you have a *bona fide* intent to manufacture or import it. It is the responsibility of the manufacturer or importer (your supplier, in this case) of the reactant to determine the Inventory status of the reactant.

47. When a prepolymer is one of the precursors of a polymer, what should be considered to be the constituents of the final polymer: the ultimate reactants from which the prepolymer was manufactured, the prepolymer itself, or what?

The choice should follow Chemical Abstracts (CA) nomenclature rules and conventions for its Ninth Collective Index (9CI). In general, polymers are named on the basis of their ultimate monomers. Thus the name of a prepolymer derived from dimethyl terephthalate and 1,4-butanediol would be based on those reactants. However, there are some exceptions to this generalization. For example, although polyethylene glycol may be thought of as a homopolymer of ethylene oxide, it is not named as a homopolymer under CA naming practices, but rather according to the structural repeating unit (SRU) and end groups present: α-Hydro-ω-hydroxy-poly(oxy-1,2-ethanediyl). Similarly, polydimethylsiloxane is named on the basis of its SRU: di-Me Siloxanes and Silicones (and is considered to be end-capped with trimethylsilyl groups). If a prepolymer is named so as to represent a certain structural feature or definite repeating unit, its name cannot be decomposed into ultimate monomers for the purpose of naming the final polymer. The Agency's conventions for representation of polymeric substances are discussed in greater detail in a 1995 paper, "Toxic Substances Control Act Inventory Representation for Polymeric Substances," available from the TSCA Hotline: phone (202) 554–1404; fax (202) 554–5603.

48. Does the "Two Percent Rule" apply to the actual reactants used, or to the ultimate or putative reactants?

Consistent with the answer above, the ultimate reactants should be the basis of the chemical identity of the polymer. Thus, if a new polymer is made from the polymer in the answer above, plus additional dimethyl terephthalate and ethylene glycol, the final polymer name would be based on three constituents, and the total amount of dimethyl terephthalate would be the sum of the separate contributions. Ultimate reactants that contribute no more than two percent by weight to the final polymer may be omitted from the identity. If a homopolymer is used as a prepolymer constituent, the identity of the derived polymer should be based on the ultimate monomer, except where CA practice differs due to the applicability of SRU nomenclature (see the paper referenced in the answer to the previous question). Although *calculation* of the percentage composition of a polymer may be based on analysis (that is, "incorporated"), the *identity* should be based on the ultimate precursors.

49. In light of the modified "Two Percent Rule," which now allows reporting of polymers as incorporated as well as charged, can all polymer listings on the Inventory now be read either as incorporated *or* as charged?

Yes; polymers on the Inventory can be interpreted either as incorporated or as charged. Remember that "incorporated" means the minimum amount that theory requires to be charged in order to account for the amount monomer or reactant molecules or fragments found in the polymer itself.

50. If I import a polymer that is described as a sodium salt and I can determine analytically that sodium is present at two percent or less, can I assume that sodium hydroxide was the neutralizing agent used to produce that material, and should I use the sodium hydroxide molecular weight in determining the percent incorporated (and hence the chemical identity)?

Yes, in the absence of information about the source of the sodium ion, sodium hydroxide should be used as the default source and the calculations should be based on the molecular weight of sodium hydroxide. The hydroxides of magnesium, aluminum, potassium and calcium should also be used as the default sources of the respective ions.

POLYESTER CRITERION:

51. Some of the reactants on the polyester list are not on the TSCA Inventory. Am I allowed to use these to manufacture a polyester under the polymer exemption?

Yes, for imported polymers. Under the 1984 exemption those reactants were placed on the polyester ingredients list, even though they were not on the Inventory, because there was no exclusion for non-Inventory reactants. The Agency is continuing to allow these specific reactants, because the Agency has already made the determination that no unreasonable risk will be incurred by a polymer that contains residual amounts of these reactants.

For domestic manufacture, you may use only substances that are on the Inventory or are otherwise exempt or excluded from reporting.

52. If a monomer in my polyester is used at less than or equal to two percent and is not on the (e)(3) list, is the polymer eligible for the exemption if it meets all the other criteria and is not otherwise excluded from the (e)(3) exemption?

No, the polyester would not be eligible for the exemption. Only monomers and reactants on the (e)(3) list may be used for this category of polymer regardless of the percentage charged or incorporated.

53. Is there to be a mechanism to add new reactants to the polyester re-actants list? If so, what is expected to be required?

The list of permissible ingredients in the present exemption has already been enlarged since the 1984 version. To quote from the Agency's response to a comment addressing this specific issue in the preamble to the final rule, "The Agency believes that it would be appropriate in the future to propose amendments to this section to allow expansion of the list of eligible precursors, when additional candidates have been identified. To support requests for additional reactants, petitioners should provide health and environmental effects information on the candidate reactants, which must be already on the Inventory." No specific mechanism has yet been put in place. The Agency would prefer not to deal with such reactants piecemeal, but rather as part of a systematic process, perhaps initiated by trade organizations or consortia of interested companies.

OTHER ISSUES:

54. If a polymer contains a gel fraction (presumably high MW > 10,000) of 10 to 20 percent and the MW of the soluble fraction is <10,000, is it no longer exempt? Or is the gel fraction an impurity? Or by-product?

Since the two polymeric fractions have the same chemical identity and are not separately prepared, they would usually be considered as a single substance, for which one (not two) number-average molecular weight would be measured. However, impurities are not considered part of a polymer composition; if the 10–20 percent gel portion is undesirable, it may be considered an impurity. In that case, the appropriate number-average molecular weight would be for the portion below 10,000, and the polymer would have to meet the (e)(1) criteria. Whether the gel portion is considered an impurity does not depend upon whether it is a minor component; it depends upon whether it is not intended to be present.

55. Are inventory-listed monomers which have allowed groups, and a 5(e) order attached, eligible for the new polymer exemption?

Yes, as long as the use of the monomer is in accordance with the conditions of the 5(e) order.

56. There is no guidance on measurement of oligomer content. Is accumulated weight fraction on a GPC trace an adequate determination? In the absence of GPC, how can this be done?

Cumulative weight fraction is a commonly accepted method. The Agency has not prescribed any analytical methodology; others may be acceptable, depending on circumstances.

57. Do polymers made by "reactive processing" of two or more other polymers (both on TSCA) fall under the polymer exemption?

If not otherwise excluded, yes, as long as they meet the necessary criteria. There is no exclusion for polymers made from other polymers, nor is there any restriction on method of preparation.

58. What are the analytical requirements with respect to insoluble polymers? Can inference from melt flow data and comparison to other polymers be adequate? Can I use Monte Carlo simulation methods (such as Oligo 5) to estimate the MW of an insoluble polymer theoretically?

The Agency does not require any specific analytical methodology. Inference from physical behavior, from comparison to close analogues, and from theoretical calculation is acceptable where appropriate or where other methods are inapplicable. Monte Carlo methods, while widely used, have not been subjected to much experimental verification; if your polymer is expected to have values of MW or oligomer content near the allowable thresholds, you should probably not rely too strongly on such methods. For a discussion of analytical methods in general, see the relevant section of this guidance manual.

59. For persons who choose to use the "chemically combined" method for determining the amount incorporated in a manufactured polymer, does EPA prescribe a specific analytical method for this determination?

No. The rule does not specify any particular method. Guidance on this issue is found in this guidance manual.

60. If you make a new polymer in the laboratory which meets the exemption rule, do you need to send a research and development letter to the customer?

Substances considered to be research and development (R&D) chemicals are subject to the Research and Development Exemption, and must follow the conditions of that exemption. Polymers should be handled according to the R&D requirements until they reach the stage of being commercial products eligible for the polymer exemption. When the commercial activity is no longer R&D, provisions of that exemption no longer apply.

8. REFERENCES

1. OECD. 1994. (May). Organization for Economic Co-operation and Development. *OECD Guidelines for the Testing of Chemicals, Determination of the Low Molecular Weight Polymer Content (Draft Proposal)*.

2. OECD. 1994. (May 10). Organization for Economic Co-operation and Development. Chemicals Group and Management Committee. *Chairman's Report, Third Meeting of OECD Experts on Polymers, Tokyo, 14–16 April 1993*.

3. OECD. 1994. (May). Organization for Economic Co-operation and Development. *OECD Guidelines for the Testing of Chemicals, Determination of the Low Molecular Weight Polymer Content (Draft Proposal)*.

4. IUPAC Physical Chemistry Division, Engl. *Pure Appl. Chem.* **1976**, *48(2)*, 241–6.

5. Glover, C. A. *Tech. Methods. Polym. Eval.* **1975**, *4, Pt. 1*, 79–159.

6. Tung, L. H.; Runyon J. R. *J. Appl. Polym. Sci.* **1973**, *17(5)*, 1589–96.

7. Wagner, H. L.; Verdier, P. H. *J. Res. Natl. Bur. Stand. (U.S.)* **1978**, *83(2)*, 179–84.

8. Glover, C. A. *Advan. Chem. Ser.* **1973**, **Volume date 1971**, *No. 125*, 1–8.

9. FEDERAL REGISTER REFERENCES

TSCA. 1976. The Toxic Substance Control Act, 15 U.S.C. §§2601–2629 (1982 & Supp. III 1985).

USEPA. 1983a. (May 13). U.S. Environmental Protection Agency. Premanufacture Notification; Premanufacture Notice Requirements and Review Procedures; Final Rule and Notice Form. (48 FR 21742).

USEPA. 1983b. (September 13). U.S. Environmental Protection Agency. Premanufacture Notification; Revision of Regulation and Partial Stay of Effective Date. (48 FR 41132).

USEPA. 1984. (November 21). U.S. Environmental Protection Agency. Premanufacture Notification Exemptions; Exemptions for Polymers; Final Rule. (49 FR 46066). See also 40 CFR part 273.

USEPA. 1986. (April 22). U.S. Environmental Protection Agency. Toxic Substances; Revisions of Premanufacture Notice Regulations; Final Rule. (51 FR 15096–15103). See also 40 CFR part 720.

USEPA. 1991. U.S. Environmental Protection Agency. Premanufacture Notice for New Chemical Substances. EPA Form 7710–25.

USEPA. 1993a. (February 08). U.S. Environmental Protection Agency. Premanufacture Notice; Revision of Exemption for Chemical Substances Manufactured in Quantities of 1,000 Kg or Less Per Year; Proposed Rule. (58 FR 7646–7661). See also 40 CFR parts 721 and 723.

USEPA. 1993b. (February 08). U.S. Environmental Protection Agency. Premanufacture Notification; Revision of Notification Regulations; Proposed Rule. (58 FR 7661–7676). See also 40 CFR part 720.

USEPA. 1993c. (February 08). U.S. Environmental Protection Agency. Toxic Substances; Significant New Use Rules; Proposed Amendment to Expedited Process for

Issuing Significant New Use Rules; Proposed Rule. (58 FR 7676–7679). See also 40 CFR part 721.

USEPA. 1993d. (February 08). U.S. Environmental Protection Agency. Premanufacture Notification; Exemptions for Polymers; Proposed Rule. (58 FR 7679–7701). See also 40 CFR part 723

USEPA. 1995. (March 29). U.S. Environmental Protection Agency. Premanufacture Notification Exemptions; Revisions of Exemptions for Polymers; Final Rule. (60 FR 16316–16336).

Appendix 7.10

The PMN Review Process

(Reproduced from:
www.epa.gov/opptintr/chem-pmn/chap1.pdf)

Chapter 1

The Premanufacture Notification (PMN) Review Process

1.1 Introduction

Prior to the promulgation of the Toxic Substances Control Act (TSCA) in 1976 (TSCA 1976), there was no statutory requirement that required either risk assessment of new chemical substances prior to their commercial introduction or testing of substances suspected of being harmful. Unlike other federal statutes that regulate risk *after* a chemical is in commerce, TSCA requires the Environmental Protection Agency (EPA) to assess and regulate risks to human health and the environment *before* a new chemical substance is introduced into commerce. Section 5 of TSCA requires manufacturers and importers to notify the Agency before manufacturing or importing a new chemical substance.[1] EPA then performs a risk assessment[2] on the new chemical substance to determine if an unreasonable risk may or will be presented by any aspect of the new substance. Finally, EPA must make risk management decisions and take action to control any unreasonable risks posed by new chemical substances.

---

1. As discussed in the Appendix, these provisions apply to substances that are either manufactured within the U.S. or imported into the U.S. In the following discussion, the words manufacture or manufacturer include import or importer.

2. Risk assessment is the characterization of the potential for adverse health or ecological effects resulting from exposure to a chemical substance. Risk management is the weighing of policy alternatives and selecting the most appropriate regulatory (or non-regulatory) action after integration of risk assessment with social and economic considerations. Risk, in either case, is the probability that a substance will produce harm under specified conditions, and is a function of the intrinsic toxicity of a substance and the expected or known exposure to the substance. In practical situations, the critical factor is not the intrinsic toxicity of a substance, but the risk associated with its use.

TSCA implies that EPA will develop a review process for evaluating chemicals before they enter the marketplace. Other Acts, such as the Federal Food, Drug, and Cosmetic Act (FFDCA 1982) and the Federal Insecticide, Fungicide, and Rodenticide Act (FIFRA 1972), have led to the development of similar processes within the FDA's New Drug Application Program and EPA's Pesticide Registration Program, respectively.

TSCA, however, departs from FDCA and FIFRA in several significant ways in its treatment of new substances. First, under TSCA, the Agency only receives the data that are available (if any) and must then determine whether there may be an unreasonable risk associated with the chemical. Second, TSCA does not require toxicity testing of a new chemical substance prior to submission of a Premanufacture Notification (PMN) to EPA. Third, under TSCA, EPA is allowed only 90 days to review each substance (extendable to 180 days under certain conditions; see Appendix).

Currently, the EPA receives approximately 2,500 PMNs annually. The Agency must assess the risks posed by each of these new substances, regardless of the quantity or quality of data submitted or available. Charged with the difficult task of rapidly forecasting the environmental behavior and toxicity of chemical substances for which very little or nothing is known, EPA has developed the **PMN Review Process**. This process utilizes several general approaches to fill in data gaps so that the Agency can make rapid risk assessment and risk management decisions for new chemicals as prescribed by TSCA.

The PMN review process is used for "standard" PMNs as well as PMN exemption notifications (Appendix; USEPA 1986a; USEPA 1995b; USEPA 1995c). In this chapter, the terms "PMN submission" or "PMN" refer to all new substance submissions, unless one type of submission is mentioned explicitly. The types of submissions and their respective review periods are shown in Table 1-1.

Numerous acronyms are used to describe Divisions or Branches within the Office of Pollution Prevention and Toxics (OPPT) as well as to identify scheduled meetings and types of scientific reviews. Table 1-2 contains a list of frequently-used acronyms. This list is current as of December 1996. OPPT is scheduled to be reorganized in 1997 and some of these acronyms will change. The PMN review process, however, will remain essentially the same.

1.2 The PMN Review Process

The PMN Review Process consists of four distinct, successive technical phases: the chemistry review phase, the hazard (toxicity) evaluation phase, the exposure evaluation phase and the risk assessment/risk management phase. These phases are structured to "drop" substances of low-risk from review and to focus more sharply on, and explore more deeply, those substances of greater risk as the review progresses. Thus, the resource-intensive efforts of the later review phases are conserved by eliminating many PMN chemicals from con-

TABLE 1-1 Types of Submissions and Their Designators

Submission Type	Review Period	Designator	Reference: TSCA Section
PMN and Exemption Submissions:			
Standard Premanufacture Notification (PMN)	90 days	P	5(a)(1)
Low Volume Exemption (LVE)	30 days	L	5(h)(4)
Low Release and Exposure Exemption (LoRex)	30 days	X	5(h)(4)
Test Market Exemption (TME)	45 days	T	5(h)(1)
Polymer Exemption[1]	None	Formerly Y	5(h)(4)
Non-PMN Submissions:			
Correction Case[2]	varies	C	N/A
Enforcement Case[3]	varies	I	N/A

[1] Polymers meeting the conditions of the Agency's most recent Polymer Exemption Rule no longer need to be submitted to the Agency (USEPA 1995a). See text for details.

[2] Those correction cases that go through the PMN review process arise from requests by industry to revise a previous PMN chemical name. Inventory corrections, which are requests to correct chemical identity in initial Inventory reporting forms, do not go through the PMN review process.

[3] Enforcement cases arise from EPA investigations into potential TSCA violations.

TABLE 1-2 Acronym List: Organizational and Meeting Acronyms

Organizational Acronyms:*	
Office of Pollution Prevention and Toxics	OPPT
Economics, Exposure, and Technology Division	EETD
Industrial Chemistry Branch	ICB
Chemical Engineering Branch	CEB
Exposure Assessment Branch	EAB
Regulatory Impacts Branch	RIB
Health and Environmental Review Division	HERD
Health Effects Branch	HEB
Environmental Effects Branch	EEB
Information Management Division	IMD
TSCA Information Management Branch	TIMB
Confidential Business Information Center	CBIC
Chemical Control Division	CCD
New Chemicals Branch	NCB
Chemical Screening and Risk Assessment Division	CSRAD
Analysis and Information Management Branch	AIMB
Meeting Acronyms:	
Chemical Review and Search Strategy	CRSS
Structure-Activity Team	SAT

*This list is current as of December 1996. OPPT is scheduled to be reorganized in 1997 and some of these acronyms will change.

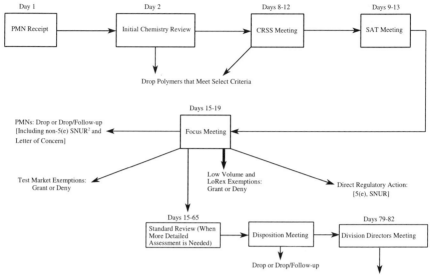

¹ See Appendix for additional information on EPA's authority under TSCA.
² SNUR stands for Significant New Use Rule.

Figure 1-1 *Office of Pollution Prevention and Toxics New Chemical (PMN) Review Process*[1]

sideration early in the process and by focusing only on those specific aspects of a few PMN substances for which there is the greatest concern. It is important to note that although a chemical substance may drop from review because of low risk, the 90-day review period still applies.

The PMN Review Process is designed to accommodate the large number of PMNs received, to assess the risks posed by each substance adequately within the strict timeframe prescribed by TSCA (whether or not toxicity data are available), and to maximize the efficiency of staff resources. Figure 1-1 provides an overview of the process as it exists today. Although some changes have taken place over the years, the process illustrated in Figure 1-1 is quite similar to the original PMN review process that began in 1979.

Table 1-3 contains historical information on the amount of test data submitted with PMNs; although the information is several years old, the amount of data submitted has not changed significantly. From Table 1-3, it is apparent that over half of all submitted PMNs have not contained any hazard or fate test data. More recent studies show that: less than 5% of PMN submissions contain ecotoxicity data (Zeeman et al. 1993); less than 4% contain at least one measured physicochemical property value (Lynch et al. 1991); and less than 1% contain biodegradation data (Boethling and Sabljic 1989).

For the vast majority of PMN substances, the Agency is unable to reach a decision based on the submitted data alone. The Agency utilizes a number of technical approaches to overcome the lack of data during risk assessment.

TABLE 1-3 Test Data Submitted with PMNs (1979–1985)[1,2]

Type of Data	Percent of PMNs Containing the Specified Data		
	All	Non-polymers	Polymers
Toxicologic Data (some)	44	55	28
Acute Toxicity (oral)	38	50	22
Acute Toxicity (dermal)	21	27	13
Acute Skin/Eye Irritation	34	45	21
Mutagenicity	13	18	6
Sensitization	8	12	5
Other	8	11	3
Ecotoxicological Data (some)	9	11	5
Acute Toxicity (vertebrate)	6	9	3
Acute Toxicity (invertebrate)	3	3	2
Environmental Fate Data (some)	9	11	5
Biodegradation	6	8	2
Log P	3	5	1
No Test Data	54	41	70

[1] These data are based on the receipt of approximately 5,500 PMNs. Current trends in test data submissions are similar. See text for additional details and references.

[2] Source: DiCarlo et al. 1986.

These approaches include, for example, chemistry review, analysis of structure-activity relationships (SARs), analysis of quantitative structure-activity relationships (QSARs), and the use of physicochemical properties to assess the likelihood of absorption in exposed individuals; the various approaches are discussed in greater detail in this chapter and in Chapter 2. The remainder of this chapter discusses the PMN Review Process, including the purpose and function of each phase, with particular focus on the technical approaches used by the Agency to assess the risks of new chemical substances. Other Agency publications are available to assist the reader in understanding the general PMN review process (USEPA 1986a) and in filing a PMN (USEPA 1991).[3]

1.2.1 Receipt of the PMN (Day 1)

PMN submissions are received at the Confidential Business Information Center (CBIC) where they are time- and date-stamped. Here, appropriate security management of any submissions containing TSCA Confidential Business Information (CBI) is initiated. The TSCA Information Management Branch (TIMB) performs an administrative review of each submission to verify that all of the required information, other than specific chemical information, is present in the PMN. This review includes submitter and chemical information, generic chemical name and use (if chemical name and use information are claimed as

---

3. These, and other useful documents for PMN submitters, are available through the TSCA Assistance Information Service at (202) 554–1404.

CBI), projected production volume, and the presence of any submitted health or environmental hazard studies in the sanitized version (i.e., the version that does not contain CBI). The submissions must also contain the English translations for any submitted studies originally written in a foreign language. Next, TIMB checks the user tracking sheets received from EPA's Financial Management Division to confirm that the appropriate fees have been paid.

The submission is then forwarded to the Industrial Chemistry Branch (ICB) of the Exposure, Economics, and Technology Division (EETD) where chemists check the adequacy of the submitted chemical name, molecular formula, and chemical structure diagram to describe the new substance. As of the effective date (May 30, 1995) of the Revisions to PMN Regulations (USEPA 1995c), EPA requires the submission of a correct Chemical Abstracts (CA) name that is consistent with listings of chemical names for similar substances already on the TSCA Inventory. A correct molecular formula and chemical structure diagram, where appropriate, are also required.

If the name is determined by EPA to be inadequate or incorrect, the Agency will declare the notice incomplete unless the submitter used Method 1 (USEPA 1995c) to determine chemical identification and submitted exactly the same substance information to EPA and the Chemical Abstracts Service (CAS) Inventory Expert Service. Only in this situation will EPA allow the PMN review period to continue while the problem is resolved. If the submitter did not use the CAS Inventory Expert Service (which solely constitutes Method 1) the Agency will not begin the review period until the problem is resolved by the submitter. (See USEPA 1995c for details.)

If no problems are identified during the administrative and nomenclature prescreening reviews, the first day of the 90-day clock[4] for PMN review is the day that the PMN submission was received at EPA Headquarters. If very minor problems are identified that would not constitute an incomplete notice, and the information is believed to be readily available, the submitter is contacted for this information by telephone. If the notice is incomplete, the submitter is given a list of the problems in writing so that the submitter will know what is needed to complete the notice and start the review period. If the submitter has not responded to EPA's request for additional information within 30 days, EPA terminates the notice and returns the PMN user fee. When all required additional information is received from the submitter, the first day of the review period is assigned as the day EPA receives this information.

Following the resolution of any minor problems with administrative information and chemical identification, the CBIC staff assign a case number to the PMN. Case numbers are assigned in sequential order using a one-letter designator to indicate the type of submission (see Table 1-1). The CBIC staff assign

---

4. The phrase "90-day clock" refers to standard PMN submissions. In the interest of brevity, the reader should note that this phrase will be used for the amount of time in which the Agency must complete its review; the actual time for exemption notices is less than 90 days, as indicated in Table 1-1.

document control numbers and log each submission (and copy) into a computerized document tracking system designed for TSCA CBI documents. Using established procedures to protect CBI (USEPA 1993), the CBIC staff forward copies of each case to technical staff in EETD and the Health and Environmental Review Division (HERD) as well as to program management staff in the Chemical Control Division (CCD) for their respective reviews.

1.3 Chemistry Review Phase (Days 2–12)

The first technical phase of PMN review by EPA scientists is the **chemistry review phase**, which is performed by the Industrial Chemistry Branch (ICB). This phase establishes a chemistry profile for each new substance and establishes the essential foundation for the review by other OPPT scientists in subsequent phases of PMN review. The chemistry review phase has four components: initial review, preparation of the Chemistry Report, Inventory review, and discussion at the Chemical Review and Search Strategy (CRSS)[5] meeting.

1.3.1 Initial Chemistry Review (Day 2)

The initial chemistry review is a rapid assessment by ICB chemists of each new chemical submission. The first step is to establish the technical completeness of the submission. The chemists check the reported Chemical Abstracts (CA) name, molecular formula, and chemical structure against the reactants and feedstocks used in its manufacture to determine quickly whether the PMN substance is identified correctly, as well as consistently, and check the generic chemical name (if provided) to verify that it is appropriate.

If the submission is an exemption notice, the chemist checks for compliance with the exemption guidelines.[6] For all submissions, an in-house electronic database is searched to establish if an identical substance has been submitted previously.[7] This check for previous exemptions is a rapid screening process, not to be confused with the definitive determination performed during the Inventory review (see below).

Based on its experience during the review of thousands of new chemical substances, EPA has identified a group of polymers (see below) that it believes poses no unreasonable risk of harm to human health or the environment. When a PMN substance in initial chemistry review falls within this group, the ICB

---

5. The CRSS meeting is the first meeting of the PMN review process.

6. Since the effective date of the Agency's revised Polymer Exemption Rule (USEPA 1995a), no notifications have been required for exempt polymers. Manufacturers must, however, follow the Agency's requirements for all polymers exempt under this rule.

7. In a change from the previous low volume exemption regulation, more than one low volume exemption may now be granted for any substance (USEPA 1995b), but the Agency will assess the risk of the total production volume if there is more than one exemption notification for the same substance.

chemist labels the case a "pre-CRSS drop" and the Agency performs no further review. As a general practice, the Agency does not notify the submitter that a PMN submission has been dropped from further review; by law, manufacture of a new substance cannot commence before the normal review period has expired, even for PMN cases that have been dropped from further Agency review.

For a polymer to be considered a pre-CRSS drop, it must satisfy all six of the following criteria:

(1) It must belong to one of twelve (12) acceptable polymer classes: polyesters, polyamides and polyimides, polyacrylates, polyurethanes and polyureas, polyolefins, aromatic polysulfones, polyethers, polysiloxanes, polyketones, aromatic polythioethers, polymeric hydrocarbons, and phenol-formaldehyde copolymers;

(2) The levels of oligomer present in the polymer must be less than or equal to (a) 10 weight percent of polymer molecules with molecular weight less than 500 daltons and (b) 25 weight percent of polymer molecules with molecular weight less than 1,000 daltons;

(3) It must have no more than the level of ionic character permitted by the polymer exemption rule (generally a functional group equivalent weight for ionic groups greater than or equal to 5,000);

(4) It must have (a) no reactive functional groups, (b) only reactive functional groups specifically excluded based on OPPT's risk assessment experience (e.g., blocked isocyanates), or (c) a reactive functional group equivalent weight no less than a defined threshold (e.g., for pendant methacrylates, the equivalent weight threshold is 5,000);

(5) The lowest number-average molecular weight of the polymer must be less than 65,000 daltons but greater than 1,000 daltons; and

(6) the polymer must not swell in water.

These criteria have been developed for use by EPA, although they can by useful to submitters interested in developing low risk polymers. These criteria should not be confused with the criteria stated in the Polymer Exemption Rule (USEPA 1995a), which specifically exempt certain polymers from PMN submission. (The criteria above were used, however, in the development of the Polymer Exemption Rule).

It has been the Agency's experience that polymers meeting these criteria have a low risk for causing adverse environmental and human health effects. Both the group of acceptable polymer classes and the reactive functional group criteria are being updated and expanded as OPPT's experience in risk identification and assessment continues to grow. The actual figure varies from time to time, but, in general, many of the PMNs for polymers meet these criteria and are dropped from further review. (Many of these polymers also qualify for exemption and need not be reported at all.)

Another important function of the initial chemistry review is to identify PMN cases for which pollution prevention opportunities may exist. For example, ICB has developed a PMN screening methodology known as the Synthetic Method Assessment for Reduction Techniques (SMART). The purpose of the SMART review is to identify pollution prevention opportunities (e.g., alternative syntheses, in-process recycling, etc.) and to encourage the PMN submitters to take advantage of these opportunities, if possible, during production of their new chemical substances. The SMART review of PMN cases takes place simultaneously with the chemistry review. PMN cases that are judged appropriate candidates for SMART review are assigned to staff chemists with expertise in identifying pollution prevention opportunities as they relate to the manufacture of the substance (see Chapter 3 and USEPA 1995e).

The next step of the initial chemistry review is to assign each PMN case (except those already dropped) to a chemist for preparation of a Chemistry Report. Generally, each PMN is assigned to a staff member with particular expertise in that chemical class. For example, a submission for a new dye would be assigned to an organic chemist with experience reviewing this class of substances. Substances submitted simultaneously that are closely related or that comprise a synthetic pathway are typically assigned as a group to an individual chemist for review.[8]

At this stage, the senior chemist also assigns each PMN case for presentation at a specific CRSS meeting. The CRSS meetings are held twice a week, on Monday and Thursday mornings. A routine CRSS meeting has between 10 and 30 PMN cases; frequently, some of the cases are grouped for review and are presented together. This twice-weekly bundling of cases for review greatly increases the efficiency of the PMN review process. Unless any unforeseen problems delay the review of individual cases, the cases bundled for review at this point will go through the review process together.

1.3.2 Inventory Review (Days 3–11)

The Inventory review is an extremely important component of the PMN review process, from both legal and technical standpoints. The Inventory review, performed by chemists within ICB, has two major functions. The first is to establish a complete and accurate chemical name for the new substance. The chemist compares the chemical structure, molecular formula, the reactants, and the reaction scheme for consistency with the CAS name submitted in the PMN; if a CAS Registry Number is provided, the chemist verifies it as well. The name must be consistent with CAS nomenclature policies and with how similar substances have been named previously for the TSCA Inventory. If inconsistencies

---

8. PMNs for closely related new chemical substances submitted at the same time by one manufacturer are frequently grouped into what is called a consolidated submission. Each new substance gets a unique case number, however. A consolidated submission must have prior approval by the EPA. See USEPA 1991.

are found, the chemist declares the notice incomplete, and review of the notice is terminated, unless the submitter used Method 1 to develop the name (See USEPA 1995c for details).

The second function of the Inventory review is to determine definitively that the new chemical substance is not (or is) on the TSCA Chemical Substance Inventory. For this search, the Agency uses the continually updated computer database of the Inventory, known as the Master confidential and non-confidential listings. The Agency maintains a separate list of low volume and LoREX exemptions on the Master Inventory File, in light of the special status of exempt substances.

If the Inventory review establishes that a PMN substance is currently on the TSCA Inventory or the intended use of the substance is a non-TSCA use (e.g., pesticide, pharmaceutical, pharmaceutical intermediate), the substance is excluded from PMN reporting.[9] If the review establishes that the same manufacturer had submitted the identical substance in an earlier PMN and that this submission was not withdrawn, the new notice is declared not valid. In either circumstance, Agency staff terminate the review and notify the submitter.

1.3.3 Preparation of the Chemistry Report (Days 3–11)

It is essential that all of the chemical aspects of PMN substances are thoroughly explored and understood, because the Agency's hazard and risk assessments are based largely on the chemistry of these substances. The chemistry information is summarized in the Chemistry Report, prepared for each PMN. In preparing the Chemistry Report, the chemist verifies the chemical identity information, researches the chemistry of the PMN substance, and examines and/or estimates the physicochemical properties that are critical for Agency risk assessment.[10]

Chemists frequently contact the PMN submitter to clarify information submitted or to discuss an apparent error. Most such problems are resolved over the telephone (at the submitter's discretion and with confidentiality preserved, as appropriate), allowing the PMN review to continue on its normal schedule. The manufacturer is required, however, to submit correction pages for the Agency's records. EPA may request a suspension of the 90-day clock from the submitter if obtaining the necessary information from the submitter is expected to be delayed. Examples of frequent chemistry problems with PMN submissions are given in Table 1-4 (helpful advice regarding these issues is also included). For answers to questions about procedural, technical, or regulatory

---

9. If the substance is already on the Inventory, the submitter is free to manufacture it, subject to any SNUR, section 4 test rule, or other rule that the Agency may have promulgated for that substance.

10. Many of EPA's risk assessments of PMN substances are based on the physicochemical properties of these substances. A detailed discussion of the use of physicochemical properties during risk assessment of PMN substances is provided in Chapter 2.

TABLE 1-4   Technical Problems Frequently Encountered in PMN Submissions

Page of PMN Form	Description of Problem
4	*Chemical Identity Problems*

Chemical name and structure do not agree because:
 (1) degree of specificity is different in name vs. structure, (e.g., the name indicates no specific isomer, but the structure is specific for a particular isomer);
 (2) submitter incorrectly drew the structure (i.e., the number of bonds or atoms is incorrect; the location of bonds or atoms is incorrect);
 (3) submitter did not draw a representative or partial structure of a complex/ variable/multi-component PMN substance (e.g., the appropriate form of a sulfur dye: leuco or oxidized).
CAS Registry Number (CASRN) and chemical name or structure do not agree because:
 (1) submitter made a typographical error, or
 (2) submitter is trying to cover a choice of alternative counterions with one PMN (e.g., using either Na or Li or Mg), or
 (3) submitter is trying inappropriately to cover multiple, class 1 chemicals with one PMN. The EPA allows a single PMN to cover multi-components if submitter is making only one product. For multi-component submissions, each unique substance should be drawn within a single PMN.
CASRN and reactant name(s) do not agree, for the same reasons.
Chemical name and molecular formula do not agree, for the same reasons.
Reporting two or more substances as a mixture when they should be considered collectively as a Class 2 substance.

| 5 | *Molecular Weight Values* |

The lowest number-average (NAVG) molecular weight is supposed to be measured for the complete polymer mixture from a series of reactions or an average of multiple analyses of a particular reaction; often it is submitted as the lowest peak in an individual run.[1]
Although submitters are not required to report values for *typical* number-average molecular weights for their polymers, this would be useful, especially if the typical and lowest molecular weights are far apart.
For polymers that cannot be analyzed by GPC (these polymers typically are high molecular weight and are solvent-insoluble), the molecular weight (in grams/ mole) can be estimated using Avagadro's number (6.02×10^{23}) multiplied by the mass of a typical particle.
Molecular weight values given as "greater than" some number are not helpful unless the base number is fairly close to the actual molecular weight. For example, MW > 10,000 is often listed; it might be more accurate, for example, to list MW > 30,000 or > 100,000 or > 1,000,000.

| 5 | *Monomer Composition of Polymers* |

If the submitter does not know the identity of one or more monomers because the identity is the proprietary information of a supplier, a letter of support from the supplier of the proprietary monomer(s) is required to complete the chemical identity information. The notice submitter must ensure that the supplier sends the letter of support directly to EPA, referencing the PMN submitter and the PMN user fee number. Often, these letters are missing.

TABLE 1-4 *(Continued)*

Page of PMN Form	Description of Problem
5	*Structural Diagram of Polymers*

The structural diagram for polymers often fails to show at least the most likely bond types (i.e., the chemical bonds of the polymer) expected to be present, or a representative arrangement of monomers and other reactants in the polymer. Submitters are expected to provide as much structural information as known to or reasonably ascertainable by them.

| 6 | *Impurities and By-products* |

Unreacted feedstocks and reactants are not listed when they should be. The description of impurities and byproducts/coproducts is incomplete.

| 6 | *Generic Names* |

Submitted generic names often are much more general than they should be, and are sometimes improperly deceiving. The degree of masking of specific parts of a name should be minimal, just enough to hide true proprietary details. (For guidance, see USEPA 1986c.)

| 6 | *Synonyms and Generic Names* |

Both of these need to be consistent with the chemical structure. For example, since polyethylene terephthalate is an aromatic polyester, it should not be described as an aliphatic or olefinic polyester.

| 7 | *Use Information* |

At least one use must be reported that is covered under TSCA. For example, a substance used for coatings on eyeglasses would be excluded from TSCA reporting, as it is part of a medical device covered under another statute, but the same substance used also for telescope lens coatings would be subject to reporting.

For substances with both TSCA and non-TSCA uses, submitters need to specify the percentage of each use. The production volume to be reported is the total amount manufactured for all uses.

If the use is given as "chemical intermediate," it would be useful to know the ultimate use of the final product. The ultimate use may determine whether the intermediate is even subject to TSCA. Further, unreacted chemical intermediate remaining in the final product may present risk issues.

| 8 | *Process Description* |

Weights of reactants and other starting materials charged and of product formed are often missing.

A simple diagram showing only the reaction vessel and a list of reactants and other starting materials doesn't reflect critical intermediate steps and separations. For example, a simple process flow diagram for polyurethane condensation polymers may show an alcohol in the reagent list as if the alcohol were capping the polymer; however, it could be a solvent in the formulated product.

TABLE 1-4 *(Continued)*

Page of PMN Form	Description of Problem
	Sometimes the diagram shows that both the free acid and its salt are formed and isolated, but the PMN reports only one of these. Both may be separately subject to reporting under TSCA.
	Submitters who are planning to import a chemical(s), but contemplating domestic manufacture should provide a prospective manufacturing process diagram. They should know and describe how the substance is made or how they plan to make it. A diagram of the processing or formulation of the PMN substance after import should not be substituted for the manufacturing process diagram.
	Releases of non-PMN substances, such as solvents, from the chemical reaction should be indicated. Mass or weight balance information would be helpful to tie in with pollution prevention information on page 11.
13	*Physical and Chemical Properties*
	The physical form of the neat substance would be very helpful and often is not stated.
	Physicochemical properties should be measured and reported for the neat substance, whenever possible. If data are available for mixtures, solutions, or formulations containing the PMN substance, the percent of the individual components should be specified. (Note that MSDS sheets, by law, reflect the formulated product, whereas the PMN physicochemical property sheet should reflect the neat substance.)
	Upon occasion, physicochemical properties that exist in the literature are inconsistent with those measured by the submitter.
	Physicochemical properties are used by Agency toxicologists; toxicologists usually consider water solubility or vapor pressure to be significant at lower levels than do submitter chemists. For example, vapor pressures given in PMNs as "<0.1 torr" are often significant for Agency reviews and should be measured more exactly. Further, estimated values expected to be less than 0.01 torr, for example, should be reported as <0.01 torr and not simply <0.1 torr. The terms "negligible" and "soluble" are not useful.
	For all submitted test data, the Agency requires submission of copies of the actual data; a summary of the data is not considered to meet this requirement.

[1] See Chapter 2 for methodology and discussion.

requirements prior to submitting a PMN, submitters are invited to telephone a PMN Prenotice Coordinator at (202) 260–1745, (202) 260–3937, or (202) 260–8994.

OPPT utilizes an electronic database on its own local area network (LAN) that captures and rapidly disseminates information on the PMN case to the various staff participating in the PMN review process. This database, as well as the LAN, is designed to protect CBI data. A portion of this electronic database contains the Chemistry Report data.

In establishing the chemical structure, EPA recognizes two classes of chemical substances (USEPA 1986b; USEPA 1991). Class 1 substances are single compounds composed of molecules with particular atoms arranged in a definite, known structure. Class 2 substances typically have variable or unknown compositions or are composed of complex combinations of different molecules and, hence, do not meet the criteria for Class 1 substances.

For Class 1 substances, there is only one molecular entity to review. For Class 2 substances, however, the chemist usually identifies a representative molecule(s) for review purposes. For example, a PMN substance may be the reaction product of an alcohol with a fatty acid feedstock having a carbon chain length ranging from 2 to 18 atoms. The various esters in this reaction product will differ somewhat in their physicochemical properties and will likely differ in potential health hazard, ecological hazard, and/or exposure. The chemist is responsible for deciding how this substance is best represented for Agency review.

Once a Class 2 substance is placed on the TSCA Inventory, the manufacturer may have some limited compositional freedom in the make-up of the substance. Given this freedom, the Agency concentrates its review on the composition with the greatest potential for harm to health or the environment (i.e., the worst case). Typically, the chemist chooses the component that is the lowest molecular weight, the most water soluble, the most volatile, or the most prevalent to represent the whole Class 2 substance, although all reasonable components are identified during the chemistry review. Thus, the review is representative of a very complex substance, but focuses on the worst-case scenarios.

The chemist next considers the synthesis of the PMN substance. He or she reviews the feedstocks to establish that they are identified correctly, that the PMN substance can be synthesized from them, and that they are individually listed on the TSCA Inventory.[11] This aspect of the chemistry review is critical. One of the most frequent errors in PMN submissions is that the named PMN substances cannot be synthesized from the listed feedstocks; either the feedstocks or the PMN substances are not identified correctly. For example, a straight-chain octyl group is frequently listed in PMNs, whereas a 2-ethylhexyl group is the actual feedstock moiety. Although each group contains eight carbons and there are not large differences in physicochemical properties, there may be significant differences in toxicity. The Agency anticipates that the most recent PMN rule revision (USEPA 1995c) will decrease the number of problems in this area through the requirement of CAS nomenclature for naming PMN substances. Regardless of the effect of the rule, however, careful review will remain an important function of Agency chemists.

Chemists also review the chemical synthesis to identify (or confirm) impurities or byproducts that may be present in the PMN substance. If present in substantial quantities, impurities may pose even greater risks than those of the PMN substance itself.

Chemists review the uses, production volumes, and manufacturing methods of the PMN substance. They determine whether the chemical nature of the

---

11. This is a quick check of the Inventory; more definitive searches of the Inventory are done as required.

PMN substance is consistent with its intended use and also identify other potential commercial and consumer uses to be included in Agency assessments of potential exposure to the PMN substance from these other uses.[12]

During the chemistry review of a PMN substance, chemists frequently identify closely-related or congeneric substances for which physicochemical and toxicity data are available. These structural analogs are used as surrogates for risk assessment of the PMN substance. EPA chemists also identify previous PMN cases with chemical structures analogous to the case under review (structural analogs). This allows EPA staff to compare the current assessments with earlier ones, promoting consistency and aiding in relative risk comparisons.

Chemists also identify "use analogs," which are other substances that have been or are known to be used for the same purpose as the intended use of the PMN substance. Use analogs allow the Agency to compare the risk of the PMN substance to that of other commercial substances intended for the same use.

Those physicochemical properties of the PMN substance that are important to risk assessment are also determined during the chemistry review. These typically include molecular weight, physical state, melting point, boiling point, water solubility, vapor pressure, and octanol/water partition coefficient. Chemists develop a value for each of these properties for every PMN in the review process at this point; they may also add values for other properties as warranted by the specific PMN substance. Chemists confirm submitted values (if provided), locate experimental values from the literature, or derive estimated values using appropriate techniques. Chapter 2 provides a detailed discussion of physicochemical properties, their measurement or estimation, and their subsequent use in risk assessment.

Most PMNs contain few physicochemical data. Consequently, the majority of physicochemical properties used for risk assessment of PMN substances are obtained by EPA scientists, usually by estimation. Any chemical estimation technique possesses some degree of uncertainty. In the absence of data, it is the practice of the Agency to select the estimation method that, within reasonable limits, maximizes the exposure or hazard potential. The Agency's aim is to estimate physicochemical properties to result in somewhat higher exposure and risk, so that a margin of safety results. Therefore, actual exposures and risks will not be under-estimated due to lack of data. For this reason, it is in the submitter's best interest to provide reliable experimental values in the PMN, if these can be measured. Even accurately measured (reliable) values for close analogs of a PMN substance are likely to be helpful for accurate estimation of exposure and risk. A more detailed discussion of the importance of accurate

---

12. The exposure to a chemical substance that has more than one use can vary substantially from one use to the next. Thus, depending upon use, the overall risk of such a chemical can vary substantially. If there are known uses (i.e., in the case of an imported substance, commercial uses outside of the U.S.) or potential new uses that would be of concern for unreasonable risk, the Agency may choose to develop a SNUR. See Appendix.

physicochemical property data in the risk assessment of PMN substances is provided in Chapter 2.

For polymers, EPA chemists review additional data, including the number-average molecular weight of the polymer, how it was determined, and what percentages of the molecules in the polymer have a molecular mass of less than 500 daltons and 1,000 daltons (USEPA 1995d). This is a result of the Agency's findings that lower weight oligomers may pose a greater degree of risk than their corresponding higher weight polymers, all else being equal. Finally, chemists determine the equivalent weight of any reactive functional group(s) and charged species.

In rare cases, the chemist may determine, during the more thorough chemistry review, that a polymer fulfills the requirements for a pre-CRSS drop (even though the initial chemistry review did not reach that conclusion). When this occurs, the Agency drops the PMN from further review.

1.3.4 Chemical Review and Search Strategy (CRSS) Meeting (Days 8–12)

As stated earlier, the Agency's ability to assess the potential hazards and risks of a given PMN substance is based largely on the chemistry of the substance. The chemistry of each PMN substance, summarized in the form of a Chemistry Report, is presented at the CRSS meeting. The CRSS meeting is thus an extremely important meeting within the PMN process: it is at this meeting that the chemistry needed for subsequent hazard and risk assessments is discussed and evaluated. The CRSS meeting is chaired by one of the senior chemists in ICB and attended by approximately 20 Ph.D.-level scientists. The key participants are ICB chemists, but representatives of most other groups involved in the PMN review process also attend. Typically, these include toxicologists, chemical engineers, and chemists from other branches in OPPT.

The CRSS chairman follows a defined agenda to initiate discussion of each new chemical submission that is in active review at that point. (Pre-CRSS drops, invalid, delayed, withdrawn, or incomplete submissions are not discussed.) Cases that previously had been delayed while the submitter resolved problems are presented first. Second are low volume cases.[13] Finally, all test market exemptions and regular PMN and SNUN cases are discussed in the order in which they were received at EPA headquarters. Occasionally, corrections to PMN or exemption notices are discussed at CRSS meetings, as are enforcement cases. (Those enforcement cases discussed at CRSS meetings are usually PMN submissions for substances already in commerce in violation of TSCA.)

The chemist who performed the review presents the PMN case at the CRSS meeting rapidly, but comprehensively, using standardized visual aids to

---

13. Polymer exemption cases had been discussed here as well; however, under the revised polymer exemption rule, the Agency no longer reviews polymer exemption notifications.

facilitate understanding. He or she starts with the case number (which indicates the submission type), chemical name, manufacturer, production volume, and method of manufacture, then continues with specific uses of the substance, focusing on the structure and functional group(s) that impart the characteristics of the PMN substance. Next, the chemist discusses the values of the physicochemical properties, along with the methods used for their estimation or, in the case of measured values, the literature sources and measurement methods used. These values are closely scrutinized by meeting attendees, as they form a basis for subsequent risk assessments. The chemist compares and contrasts any structure or use analogs from previous PMN cases to the new submission.

Chemists also scrutinize PMN submissions for pollution prevention opportunities. This is discussed in detail in Chapter 3. When applicable, the chemist will discuss known or potential alternative syntheses that appear to offer greater pollution prevention opportunities than the synthesis intended to be used by the PMN submitter. If a Synthetic Method Assessment for Reduction Techniques (SMART) review (see Chapter 3, section 2.2) was undertaken, the chemist presents these results, concentrating on any less polluting alternative syntheses that he or she may have identified.

Finally, the chemist initiates a discussion of any unique, interesting, or important information regarding the new chemical substance. These additional comments may range from the curious (e.g., an unexpected shade of red displayed by a new dye) to the serious (e.g., it appears that the synthesis will form a particularly toxic byproduct that was not identified in the PMN), and may include information needed by others in the PMN review process. The chemist may discuss other potential uses of the new substances (based on use data of analogs or the substance itself) and the anticipated production volumes.

Following the Chemistry Report presentation, another ICB chemist presents the proper chemical name for the PMN substance; he or she also states whether it is present on the TSCA Inventory. This chemist further identifies any feedstocks or other reagents that are not on the Inventory. If the PMN substance is declared to be on the TSCA Inventory, all review stops, as the chemical is excluded from reporting.

Typically, during the presentation of a case, attending staff members ask questions and provide comments in informal, round-table peer review. These discussions draw on the combined experience (both academic and industrial) and scientific expertise of all participants to evaluate the chemistry of the PMN substance. Attendees also suggest ways to resolve any problems that have arisen. If, following all this discussion, the CRSS meeting participants feel they do not have sufficient information to be comfortable with the technical quality and reliability of the chemistry for the PMN substance, they will delay further Agency review of the case until additional information can be gathered. The vast majority of cases, however, proceed to the next step.

After the case is presented, ensuing discussions are completed, and a consensus is reached, the meeting chairman records the status of each case using one or more identifiers (shown in Table 1-5). The case number, the chemist

TABLE 1-5   Notations Used for CRSS Meeting Notes

Notation	Description
BT	Biotechnology Case: The PMN substance is a biotechnology case.
CP	Consolidation Problem: The different substances contained in a consolidated submission are not sufficiently similar in nature or use.
DE	Delayed: Indicates that the case could not be discussed at its initially-scheduled CRSS meeting and will be delayed to the next meeting. Typically due to missing, ambiguous, inconsistent, or incorrect information that could not be obtained, clarified, or corrected prior to the meeting. The review period clock (between 30 and 90 days) does not stop for delayed cases.
DR	Dropped: Indicates a polymer that was dropped from further review, i.e., a pre-CRSS drop or a drop decision made during the CRSS Meeting.
ER	Excluded from Reporting: Indicates a substance that is specifically excluded from TSCA §5 reporting requirements (i.e., the chemical substance is listed on the TSCA Inventory, is not subject to TSCA reporting, or does not meet the definition of "chemical substance" under TSCA).
EL	Eligible: The new chemical meets the requirements for exemption. Only substances submitted as PMN exemptions may be declared eligible.
IC	Incomplete: The submission does not contain mandated information.
ID	Chemical Identity: The correct identity of the new chemical substance is not accurately described or cannot be ascertained.
MC	Multi-component Case: A reaction product combination reported in one submission (one PMN case number) that is represented as a mixture under TSCA Inventory policy.
MX	Mixture: The substance is a mixture of chemical substances and thus is excluded as a whole entity under TSCA; the individual substances are, however, subject to PMN notification if they are not already on the Inventory.
NE	Not Eligible: The PMN substance is not eligible for the type of exemption filed.
NV	Not Valid: The submission is identical to an earlier one submitted by the same manufacturer. (Previously, only one low volume exemption was allowed per substance and any subsequent exemption requests were declared not valid; see the revised exemption, USEPA 1995b.)
NX	Not Exposure-Based: The substance is a polymer produced at greater than 100,000 kg/yr that does not meet certain criteria for inhalation toxicity. It is exempted from a human and environmental exposure review.
SP	Suspended: Review of the substance is suspended at the submitter's request, although this process is usually initiated by EPA phoning the submitter; the review clock stops.
SR	Suspension Requested: A significant problem affecting the review of the case was found; the suspension request is transmitted to the CCD manager who contacts the submitter to request a suspension.
UF	User Fee: A problem with the fee payment must be resolved before the review (and the review clock) can be started.
WD	Withdrawn: The submitter withdrew the submission.
YX	Exposure-based: The new chemical substance is produced at greater than 100,000 kg/yr, is not a polymer (unless it meets certain criteria for inhalation toxicity) and is, therefore, subject to a section 5(e) exposure review.

responsible for the case, and the identifier(s) are entered into the CRSS meeting notes. These notes are physically posted in a central location and on the CBI LAN. The CRSS notes are used by subsequent reviewers for scheduling purposes.

Following the CRSS meeting, the chemist who presented a specific case makes any necessary changes to his or her Chemistry Report and files the report electronically on the CBI LAN and in hard copy in the CBIC. Subsequent reviewers at EPA use this report as a source of validated chemical information for the next steps in the PMN process: hazard identification and risk assessment. The report is especially critical to the hazard determinations performed by the Structure-Activity Team (SAT); correct structure, presence of impurities, and physicochemical properties identified during the chemistry review are key to the accuracy of the SARs used by the Agency to predict human and environmental hazard, especially in the absence of toxicological test data.

1.4 Hazard Evaluation

The second phase of the PMN review process is the **hazard evaluation phase.** The term "hazard," in the vernacular of PMN review, is synonymous with toxicity. The purpose of this phase, as the name implies, is the identification of possible hazards (toxic properties) of PMN substances to human health and the environment; this phase includes analyses of the likelihood of absorption and metabolism in humans, human toxicity, toxicity to environmental organisms, and environmental fate. During this phase, OPPT convenes a team of scientists who specialize in organic chemistry, biochemistry, medicinal chemistry, pharmacokinetics, metabolism, toxicology, genetics, oncology, environmental toxicology, and environmental fate. It is the responsibility of this multidisciplinary team to assess the potential hazards and risks of each new substance within the narrow time constraints of TSCA, using the sparse data available for most of the substances. During the hazard identification phase, these EPA scientists strive to elucidate the probable human toxicity, environmental fate, and environmental hazards posed by each new chemical substance. The hazard identification phase begins at approximately the same time as the Inventory review and preparation of the chemistry report and continues after the CRSS meeting.

1.4.1 Human and Ecological Hazard Identification (Days 2–12)

For any case that is not a pre-CRSS drop, scientific staff from the EAB, the Health Effects Branch (HEB), and the Environmental Effects Branch (EEB) of HERD initiate reviews in the areas of environmental fate, human toxicity, and ecological effects, respectively, at approximately the same time as the Chemistry Report is being prepared by the ICB. The first step is to evaluate submitted test data and to search the scientific literature for published information on the PMN substance. As previously stated, however, PMNs seldom contain enough measured toxicity data to perform a complete hazard assessment (see Table 1-3). In addition, because PMN substances are "new" substances, there are seldom any data available on them in the scientific literature.

The paucity of human, animal, and aquatic toxicity data for most PMN substances has led OPPT scientists to use several different approaches for hazard identification. These approaches include: consideration of the likelihood of absorption from the lung, gastrointestinal tract, and skin; consideration of the expected products of metabolism and their toxicity; structure-activity relationships (SARs); and consideration of the presence of structural groups or substituents that are known to bestow toxicity. SARs are the comparison of the substance under review with structurally analogous substances for which data are available.[14] In SARs, a series of structurally similar chemicals for which a measured toxicological or environmental endpoint (the "activity") is available is used as a basis for qualitative estimation of the same endpoint for an untested chemical of the same structural class. The underlying assumption in using SARs is that the toxicological properties of substances belonging to the same chemical class are related or attributable to the general structure (or some particular portion thereof) of the class. Logically, any substance that has the same general structure is likely to have the same toxicological properties. Using SARs, for example, one can be alerted to the possibility of a new, untested chemical sharing the same toxic effect(s) with structurally similar chemicals that are known to produce the effect(s). On the other hand, SARs can be used to mitigate a health concern for a substance if an analog is identified with data showing that the analog is nontoxic.

HEB scientists qualitatively estimate human acute and chronic toxicity of PMN substances, including: oncogenicity; mutagenicity; developmental toxicity; neurotoxicity; reproductive toxicity; and systemic toxicity, irritability, and sensitization. Again, the Agency's findings of the likelihood of these effects occurring in humans are seldom based on measured animal data on the PMN substance. Rather, they are usually based on structural comparison of the PMN substance with closely-related substances for which toxicity data are available (SARs). To use SARs during PMN review, OPPT scientists try to identify structural analogs of PMN substances from the literature or from in-house sources, including PMN structural databases, TSCA section 8(e) toxicity databases, and other in-house substructure-searchable databases of substances for which toxicity data are available.

Subtle differences in molecular structure within a congeneric series of substances can greatly change the relative toxicity. Knowledge of the biochemical mechanisms of toxicity can help to explain why such structural differences affect toxicity. OPPT scientists utilize their knowledge of toxic mechanisms, whenever possible, to improve the predictive quality of SARs. In cases where analogs

---

14. For most chemical substances, toxicity data are almost always derived from animal studies. It is the policy of the EPA to assume that chemicals that are capable of causing toxic effects in animals will cause the same toxic effects in humans.

TABLE 1-6   The Role of Pharmacokinetics in Predicting Health Hazards

Metabolic Process	Role in Human Health Risk Assessment
Absorption	If a substance is not absorbed, its toxic expression is limited to topical effects such as skin and eye irritation, and to unfavorable effects on nose, mouth, respiratory tract, and gastrointestinal tract membranes. Qualitative estimation of the rate and extent of absorption is based on lipophilicity and water solubility. The susceptibility of the substance to (and the likely products of) degradation by microorganisms in the gastrointestinal tract is important for assessing absorption following oral exposure.
Distribution/Redistribution	Tissue distribution and redistribution determine the potential for a substance to reach a site where toxicity can be expressed. These assessments require knowledge of blood flow rates, the octanol/water partition coefficient, and the dissociation constant of the PMN substance.
Biotransformation	The rate of degradation as well as the nature and reactivity of the metabolites are required for this assessment. Although the body frequently uses biotransformation to detoxify absorbed xenobiotics, in some cases toxic metabolites are created.
Excretion	If a compound is absorbed, its capability to express a biological effect is generally limited by the amount of time it remains in the body. Thus, a rapid rate of excretion will limit the potential for an adverse effect.

closely related to the PMN substance are equally good but vary greatly in toxicity and for which mechanistic data on the chemical class are unknown to EPA, it is the general practice of EPA to assume that the PMN substance is as toxic as the most toxic analog. If, however, mechanistic data are available and such data lead OPPT scientists to believe that the PMN substance is less toxic than other analogs, then EPA will assume that the PMN substance is less toxic. Although not required under TSCA, it would be extremely helpful if PMN submitters would provide analogs of the PMN substance for which toxicity data are available in their PMN submissions, particularly if mechanistic data for the chemical class are known to the submitter. Such information would greatly enhance EPA's ability to make more accurate hazard assessments of PMN substances and lessen the likelihood that OPPT scientists will over-estimate the toxicity of PMN substances.

HEB scientists also estimate the probable human pharmacokinetics of the PMN substance, evaluating absorption, distribution and redistribution, metabolism (biotransformation), and excretion of the substance. Special attention is given to the possible formation of toxic metabolites. (The role of pharmacokinetics in predicting health hazards is illustrated in Table 1-6 and described further in DiCarlo 1986.) Estimation of absorption is a particularly important component of hazard identification in that a PMN substance may appear toxic (based on SARs), but it may have other characteristics that will lead HEB scientists to believe that the substance will not be significantly absorbed through the gastrointestinal tract, skin, or lungs of humans. A human toxicity concern for a PMN substance derived by SARs may be mitigated by EPA's belief that the substance will be poorly absorbed.

Although SARs are useful in estimating toxicity, the likelihood of absorption of a PMN substance through the skin, lung, and gastrointestinal tract may not be inferred easily from the structure without careful consideration of the physicochemical properties of the substance. The relationship between the physicochemical properties of a substance and its absorption is discussed in greater detail in Chapter 2. OPPT scientists use physicochemical properties extensively to predict the likelihood of absorption of a PMN substance.

Another approach used by EPA to identify the likely toxicity of PMN substances is quantitative structure-activity relationships (QSARs), which combine physicochemical properties with SARs. In QSARs, a particular biological (toxicological) or environmental property of a series of structurally analogous chemicals is mathematically correlated with one or more physicochemical properties of the chemicals using a regression equation. The goal of QSAR is to delineate a particular property or activity more precisely than is possible by intuition or SAR alone. Using QSARs, one can predict, for example, the acute toxicity (LD_{50}) value of an untested substance directly from a physicochemical property of that substance.

EEB scientists use QSARs to estimate chronic and acute toxicity values for fish (vertebrates), daphnids (invertebrates), and algae (plants) (USEPA 1994). Based on these values, EEB scientists determine a concentration of concern, the minimum concentration at which Agency scientists have concern about harm to these aquatic species. These QSARs most frequently utilize octanol/water partition coefficient as the physicochemical descriptor of toxicity. Some other physicochemical properties used by EEB scientists in QSARs include melting point, dissociation constant, and water solubility.

1.4.2 Environmental Fate

The environmental fate of PMN substances is assessed by EAB scientists. Environmental fate is a very important component of hazard identification; it predicts where a chemical will partition in the environment, which is useful in determining environmental and human exposure and, ultimately, long-term health and environmental effects of a substance. Information on the partitioning and environmental lifetime of a substance is important in determining levels, routes, and the likelihood of both human and environmental exposure. Environmental fate assessment includes the consideration of: relative rates of environmental biodegradation, hydrolysis, and photolysis; adsorption to soils and sediments; treatability (generally in publicly-owned treatment works (POTWs)); and half-lives in the atmosphere, surface waters, soils, and sediments.

Because fewer than 10% of submitted PMNs contain environmental fate data, EAB scientists typically must estimate the environmental fate of new substances. EAB scientists estimate the environmental fate of a new chemical substance utilizing the substance's water solubility, octanol/water partition coefficient, soil adsorption coefficient, vapor pressure, Henry's Law constant, absorption spectra, and bioconcentration factor (BCF). Utilizing the physico-chemical properties obtained not only from the Chemistry Report, but also from their own preliminary review, EAB scientists estimate the potential for a substance to adsorb onto soils and sediments, pass into streams, rivers, and groundwaters, and to volatilize into the atmosphere. EAB scientists also estimate the environmental lifetime of a PMN substance by determining the percentage of the substance removed by wastewater treatment plants and the speed of hydrolysis, primary and ultimate biodegradation, and destruction by sunlight (photolysis) or atmospheric oxidants.

It is readily apparent from the preceding paragraphs of this section that physicochemical properties play an important role in estimating the likelihood of human exposure and absorption, environmental fate, ecological toxicity, and thus, risk of chemical substances. A more comprehensive discussion of physicochemical properties, including their measurement, estimation, and use in estimating absorption, environmental fate, QSARs, and exposure is provided in Chapter 2. It is important to stress here, however, that when PMN submitters do not submit accurately-measured physicochemical property data to EPA, OPPT scientists will estimate such data if they are unavailable from the literature or other sources. The estimated values may not always be accurate and may vary greatly from one estimation method to another because of the limitations of the estimation methods. As a general practice during physicochemical property estimation, OPPT scientists will use those estimated values that indicate significant exposure, absorption, or toxicity. The importance of OPPT possessing, and consequently utilizing, accurately-measured physicochemical property data for hazard identification cannot be overstated.

1.4.3 Structure-Activity Team Meeting (Days 9–13)

Because of the strict time constraints imposed by TSCA for PMN review, the OPPT scientists involved with assessing the potential hazards posed by PMN substances must have their hazard and environmental fate evaluations completed by the time the PMN substances are to be discussed at the designated SAT meetings. For most PMN substances, this allows only two weeks or less for the chemistry review, environmental fate, *and* hazard evaluation by OPPT scientists.

The SAT is a multidisciplinary team composed of approximately twenty OPPT scientists who specialize in disciplines that include organic chemistry, biochemistry, medicinal chemistry, pharmacokinetics, general toxicology, neurotoxicology, reproductive and developmental toxicology, genetics, oncology, aquatic toxicology, and environmental fate. These scientists are the same scientists who perform the hazard identification for PMN substances. The purpose of the SAT meeting is for these scientists to make a critical judgement on the likely hazard(s) posed by each PMN substance to human health and the environment, so that subsequent risk assessments and risk management decisions regarding these substances can be made.

The SAT meetings are held twice a week, on Tuesday and Friday mornings. In general, the PMN cases discussed at the CRSS meeting the day before (Monday or Thursday, respectively) are discussed at the SAT meeting. Exceptions are those cases for which technical problems at CRSS delay the review or those cases dropped from review at CRSS. Each PMN substance is discussed separately, and each SAT member individually discusses his or her findings and opinions, as well as the scientific basis for those opinions.

The discussion of a PMN submission begins with a summary of the chemistry of the substance by the CRSS chairperson, including: synthesis; byproducts or products from side reactions that may be present as a result of the synthesis; intended use; and physicochemical properties. The environmental fate specialist then summarizes the potential for the substance to adsorb onto soils and sediments, pass into streams, rivers and groundwater, and volatilize into the atmosphere; the percentage removed by wastewater treatment plants; rates of hydrolysis; primary and ultimate biodegradation; and destruction by sunlight (photolysis) or atmospheric oxidants. Following the environmental fate discussion, the pharmacokinetic specialist discusses the extent to which the substance is expected to be absorbed through the skin, lung, and gastrointestinal tract and the expected metabolites of the substance following absorption. The other SAT members then individually discuss their findings and judgements regarding the case being presented. The discussion may include, for example, the toxicity of analogs, previous related PMN cases, the significance of functional groups, and toxic mechanisms. These discussions culminate in deliberations that lead to establishing separate, overall ratings of the level of concern for human health effects and for ecological effects of each PMN substance using the following scale: low, low to moderate, moderate, moderate to high, or high.

1.5 Exposure Evaluation (Days 13–15)

The third phase of PMN review involves **exposure evaluation.** Following the SAT meeting, other OPPT scientists and engineers estimate the degree of human exposure (occupational and general population) and environmental exposure for those PMN substances that receive a SAT score of at least "low to moderate" for either health or ecological effects. Like hazard identification, exposure evaluation is a critical component of risk assessment; it consists of establishing the likelihood and magnitude of occupational, consumer, general population, and environmental exposure of a substance through careful consideration of the substances's physicochemical properties, expected environmental releases, known commercial or consumer use(s), potential commercial or consumer use(s) (identified during the chemistry review), and environmental fate.

Substances that receive "low" SAT scores for both human health and environmental effects may also undergo an exposure analysis if their production volumes are greater than 100,000 kg per year, because high production volumes such as these may lead to significant exposure and risk. Substances that receive "low" SAT scores for both human health and environmental effects and that have production volumes below 100,000 kg per year are generally not reviewed further.

The initial part of an exposure review of a PMN substance is performed by the Chemical Engineering Branch (CEB) of EETD, two to four days prior to the Focus meeting where the substance will be discussed. CEB engineers utilize the physicochemical properties of the PMN substance, most notably vapor pressure and molecular weight, to establish the importance of both dermal and inhalation exposure. For example, volatile substances and powder are typically evaluated for their potential for inhalation exposure.

CEB relies on the process flow and unit operations to identify potential release and exposure points. Using physicochemical property data and identified release and exposure points, CEB evaluates the potential for occupational exposure and for releases to the environment expected to result from manufacturing, processing, and commercial or industrial use of the substance. In addition, CEB may apply exposure and release data available on chemical substances analogous to the PMN substance, that are produced or used in similar circumstances as the PMN substance, to further evaluate occupational exposure and environmental release.

Using models that take into account the physicochemical properties of the PMN substance as well as unit operations, number of workers performing each operation, and industry-specific worksheets to fill remaining gaps, CEB engineers estimate the number of workers potentially exposed, their activities, their duration of exposure, and potential dose rates.

Emissions to the environment are obtained by evaluating data contained in the PMN and industry-specific worksheets to establish the potential for releases from manufacture, processing, and use of the PMN substance. Releases may be process-related, such as equipment vents and container residual. For example, losses to waste by a component of a photoresist pattern are expected to be relatively high (since most of a photoresist washes away during the developing stage), whereas those from a site-limited synthetic intermediate are expected to be relatively low. The physicochemical properties of the PMN substance may also be important at this stage; for example, water solubility is sometimes used along with information in the PMN to estimate potential releases to water, and vapor pressure could be used to estimate emissions to air.

EAB staff then receive data generated by CEB staff, allowing them to estimate levels of consumer and general population exposure as well as the resulting environmental concentrations that arise from emissions. For example, a component of a new spray coating designed for the household market might be expected to have higher levels of consumer exposure (through inhalation) during use than a new additive for motor oil (through dermal contact). To estimate exposure to the general population, EAB scientists consider the level of emissions into each environmental medium and the expected rate of removal. For releases to water, EAB will consider the percentage removed in a POTW (using the actual facility expected to receive that waste as indicated in the submission), the rates of biological and chemical degradation, and the degree of partitioning between water and sediment. For releases to air, EAB uses the rates of oxidation and photolysis to determine probable fence-line concentrations at the manufacturing facility. EAB uses the rates of biodegradation, volatilization, and percolation through soils to derive the concentration of the PMN substance in groundwater following its release to land (including landfills). The concentrations derived through this process are then compared to the ecological concentrations of concern developed prior to the SAT meeting to establish the potential for ecological effects that may result from environmental emissions. Estimations of yearly human intake from drinking water and fish consumption (if bioaccumulation is expected) are used to evaluate the potential for health effects.

As in hazard identification, physicochemical properties play a very important role in estimating occupational, population, and environmental exposure to PMN substances. The quality of these exposure estimates is obviously dependent on the accuracy of the physicochemical property data. Measured data are always preferred over estimated data because estimation methods, even the very good ones, do not take into account all of the intra- and intermolecular interactions responsible for given physicochemical properties. Estimated physicochemical properties, therefore, generally contain errors, which may vary widely. Estimated physicochemical properties that contain significant errors obviously affect the reliability of the exposure and hazard estimates derived from them. In cases where physicochemical property data are not available to EPA, the Agency estimates such data using several methods. It is the policy of the Agency to use those estimated values which lead to greater hazard and greater exposure. It behooves PMN submitters, therefore, to submit accurately measured physicochemical properties whenever possible.

An economist from the Regulatory Impacts Branch (RIB) assesses the validity of the production volume data submitted in the PMN by comparing the reported values to the historical median for similar chemical substances.

1.6 Risk Assessment/Risk Management Phase (Days 15–82)

The fourth phase of the PMN review process is the **risk assessment/risk management phase.** As stated earlier in this chapter, risk is the probability that a substance will produce harm under specified conditions. Risk is a function of the inherent toxicity (hazard) of a substance and the expected or known exposure to the substance. Risk assessment is the *characterization* of the potential for adverse health or ecological effects resulting from exposure to a chemical substance.

Risk management refers to the way in which the risks posed by a chemical substance are minimized. This involves the weighing of policy alternatives and selecting the most appropriate regulatory (or non-regulatory) action after integration of risk assessment with social and economic considerations. It is in the risk assessment/risk management phase of PMN review that the results of the hazard and exposure evaluation phases are used to assess the risk of PMN substances and make the necessary decisions to manage any unreasonable risks that may be posed by PMN substances.

1.6.1 Focus Meeting (Days 15–19)

The general purpose of the Focus meeting is to allow EPA staff and management to discuss the hazard and exposure evaluations of PMN substances and to make risk assessment and risk management decisions. More specifically, the purposes of the Focus meeting are to: (1) characterize (assess) the risks posed by each PMN substance; (2) decide which PMN substances will not present an unreasonable risk and drop them from further review; (3) identify the PMN substances that may present an unreasonable risk but for which risk management decisions can be made without additional review; and (4) identify the PMN substances that may present an unreasonable risk but require additional review for risk characterization.

Focus meetings are held twice weekly, on Monday and Thursday afternoons. Focus meetings are chaired by representatives from CCD; they are attended by the chairpersons of the CRSS and SAT meetings, and representatives from the groups that performed the economic analysis, environmental fate, and exposure assessments.

The discussion of a PMN substance at the Focus meeting begins with a summary by the CRSS chairperson of its chemistry, intended use, potential uses identified by EPA, and any remarkable attributes of the substance, as claimed by the submitter or identified by EPA. Next, the SAT chairperson summarizes the human health and ecological hazards identified by the SAT. This is followed by a summary of the occupational, population, and environmental exposures expected to occur from the intended or potential uses of the PMN substances by the people who made these estimates. A RIB economist will discuss the validity of the production volume estimates.

From the information presented, the Focus meeting participants assess and characterize the risks posed by the PMN substance to human health and the environment, and carefully consider these risks along with the expected or potential societal benefits of the substance. Often, EPA may identify significant risks of a PMN substance that also has significant benefits to society (e.g., the PMN substance will supplant an existing chemical substance that poses a greater risk). In such instances, it is the practice of EPA to balance these factors in making risk management decisions regarding the PMN substance. It is the policy of EPA's PMN Review Program to encourage creative thinking by chemical manufacturers and producers to design and produce efficacious substances, and not make risk management decisions (e.g., over-regulation) that stifle creativity. Almost 90 percent of the PMNs submitted to the EPA complete the review process without being restricted or regulated in any way (USEPA 1995f).

There are eleven possible outcomes for a PMN substance at the Focus meeting (Table 1-7). These range from dropping a regular PMN from further review (or granting an exemption) to pursuing a regulatory ban on the production, use, or disposal of the new substance. Approximately 80% of all PMN submissions are dropped between pre-CRSS and the end of the Focus meeting.

Some of the remaining 20% fall into one of approximately 46 Chemical Categories (USEPA 1996b) that have been identified to date by the New Chemicals Program. These categories were developed as an administrative aid to facilitate reviews by grouping chemicals into categories with similar hazard concerns and testing requirements. For each Category, the Agency has developed a standard regulatory response, often involving a section 5(e) order to limit chemical production (and, thus, exposure) pending a certain pertinent test. This categorical approach is continually evolving as EPA's experience increases.

For PMNs outside of the Categories that the Focus group characterizes as possessing significant risks, the chairman of the Focus meeting will recommend a specific regulatory response to mitigate the concerns of the Agency's risk assessment. For example, the meeting chairman may decide to pursue regulation under an exposure-based section 5(e) order if a high production volume substance has high predicted levels of worker, consumer, and environmental exposure and a long environmental lifetime. For another substance that is expected to be released to the environment in moderate amounts and is similar in structure to a substance of known chronic aquatic toxicity, the chairman may decide to pursue a risk-based section 5(e) order. Finally, the chairman may decide to drop from further review a substance expected to be released to the environment in moderate amounts yet expected to have a very short environmental lifetime.

For low volume exemptions and LoRex exemptions, the Focus meeting usually serves as the final regulatory decision meeting because of the short review period for these exemptions.

If a question concerning a PMN arises that cannot be answered during the meeting, but may be answered quickly with further investigation, the chairman may delay a regulatory decision until the next Focus meeting. If more substantial questions remain or if closer examination of the chemical is deemed necessary, the chairman may put the PMN into Standard Review (see section 1.6.2, below).

If a Focus meeting decision on a PMN is to pursue regulation, the Program Manager for a PMN case (from CCD staff) will contact the manufacturer and describe the reasons for the Agency's concern as well as the regulatory controls that EPA intends to impose.[15] Often, the manufacturer may disagree with

TABLE 1-7   Possible Outcomes of the Focus Meeting

Outcome	Description
Grant	A PMN exemption is granted.
Deny	A PMN exemption is denied; the submitter is free to submit the substance as a regular PMN.
Drop	A regular PMN case is dropped from further review.
Standard Review	Further review of the substance is required before a regulatory decision can be made; this review is often targeted to answering one or more specific questions.
Letter of Concern	A concern for harm to health or the environment exists for the substance although the risk is relatively low due to low production, exposure, or release. After the meeting, the Agency will send a letter to the manufacturer explaining the expected risk and suggested (i.e., voluntary) controls to reduce human and environmental exposure. Letters of concern may be appropriate for routine PMNs, exemption cases, enforcement cases, or corrections.
Non-5(e) SNUR (Significant New Use Rule)	EPA will begin to draft a non-5(e) SNUR, which prohibits manufacture of the substance for any use other than that contained in a regular PMN submission; manufacturers who wish to use a substance for such a prohibited use must submit a Significant New Use Notification (SNUN) to the Agency. Non-5(e) SNURs are used for those PMNs in which the intended use is judged *not* to be an unreasonable risk, whereas uses other than the intended use may lead to unreasonable risk.
5(e) SNUR	In conjunction with a 5(e) order, EPA will begin to draft a SNUR to restrict the uses of a routine PMN substance. This is often necessary because 5(e) orders apply only to the original submitter, whereas SNURs apply to all manufacturers of that specific substance.
5(e) Consent Order	EPA will begin to negotiate with the submitter to prepare a written agreement under section 5(e) that specifies testing required to determine the risk of a routine PMN substance. The negotiated 5(e) order will restrict the production, distribution, use, or disposal of the substance until EPA has received and acted upon the required test data. Consent orders are used for those regular PMNs whose intended use, manufacture, processing, etc. may lead to an unreasonable risk unless certain conditions are met to reduce exposure.
5(e) Exposure-Based Authority	This is not a risk-based finding. The Agency begins to prepare a 5(e) order requiring testing based on exposure only.
Unilateral 5(e) Order	The Agency begins to prepare a unilateral order restricting a PMN substance under section 5(e) until specified tests have been carried out.
5(f) Order	The Agency begins to prepare an action to initiate a order under section 5(f) restricting or banning a PMN substance because unreasonable risk has been established.

the Agency's concern, and may ask the Agency to suspend the review period to allow the manufacturer time to conduct the appropriate tests[16] that the manufacturer feels will mitigate the EPA's concern and lead the Agency to reverse its regulatory controls. The Agency will then use these measured data in preference to estimated data or worst-case assumptions. In some cases, the real data mitigate the risk sufficiently and the Agency drops the case (or grants the exemption, as appropriate) without the manufacturer having to contend with the potential effects of EPA regulation on the substance's marketability. Discussions of the Agency's regulatory mandate are available elsewhere (Appendix; USEPA 1986a).

1.6.2 Standard Review (Days 15–65)

If it is decided at the Focus meeting that a PMN substance may present significant risk(s), but either the hazard or exposure information identified prior to the meeting is inadequate to characterize the risk fully at the Focus meeting, a more detailed review may be necessary for adequate risk characterization, and the PMN submission will be put into Standard Review. The purpose of a Standard Review is to explore further the potential or known hazards and exposures posed by a PMN substance, so that an adequate risk assessment may be made. Currently, approximately 5% of all PMN submissions go into Standard Review.

All of the scientists and other PMN review personnel who have participated in the regular review of the PMN substance before the Focus meeting typically participate in the Standard Review. In Standard Reviews, individual detailed reports on the chemistry, environmental fate and exposure, worker and consumer exposure, and health and ecological effects of the PMN substance are prepared. Considerable effort is devoted to identifying related analogs, performing comprehensive literature searches on these analogs, and retrieving and analyzing toxicity data on these analogs.

In addition, RIB staff perform an economic assessment of the PMN substance that includes comparing the PMN substance to other commercial products that are used for the same purposes. The economic analysis identifies alternative uses (if any) of the PMN substance, evaluates the markets for the PMN substance and their potential for growth, and estimates the selling price of the substance. The economist may also perform specialized financial studies to evaluate claims in the PMN including market limitations due to cost of the PMN substance and the feasibility of process and input modifications.

---

15. The detailed regulatory process itself is outside the scope of this document and the reader is referred to other documents for further information (USEPA 1986a).

16. EPA is developing final test guidelines; for status, contact the TSCA Assistance Information service at (202) 554–1404 or access the guidelines on the Internet at *http://fedbbs.access.gpo.gov/ epa01.htm* See also USEPA 1996a.

These detailed, individual reports are used by a designated technical integrator to prepare a single report that summarizes the findings of the Standard Review. In addition to summarizing the findings of the review team, the technical integrator writes a risk characterization of the PMN chemical, including recommendations for testing. The information contained in this report is then used by the review team and the senior risk assessors of OPPT to make a more complete risk characterization and to decide on the most appropriate risk management option(s). These findings are then presented at the Division Directors' meeting for a risk management decision. For PMN substances that go into Standard Review, the Division Directors' meeting is the final phase of the PMN review process and takes place between days 79 and 82. This meeting is attended by the Directors of the seven divisions participating in the PMN review process and is chaired by the Director of CCD, or his designee. It is the role of the Division Directors at this meeting to discuss the risk assessment findings and to make risk management decisions.

Following the Division Directors' meeting, the PMN program manager takes the necessary steps to implement the risk management decision.

REFERENCES FOR CHAPTER 1

Boethling RS, Sabljic A. 1989. Screening-level model for aerobic biodegradability based on a survey of expert knowledge. Environ Sci Technol 23: 672–679.

DiCarlo FJ, Bickart P, Auer CM. 1986. Structure-metabolism relationships (SMR) for the prediction of health hazards by the Environmental Protection Agency. I. Background for the practice of predictive toxicology. Drug Metabolism Reviews 17(1–2): 171–184.

FFDCA. 1982. The Federal Food, Drug, and Cosmetic Act, 21 U.S.C. §§301–392.

FIFRA. 1972. The Federal Insecticide, Fungicide, and Rodenticide Act, 7 U.S.C., §136 et. seq.

Lynch DG, Tirado NF, Boethling RS, Huse GR, Thom GC. 1991. Performance of online chemical property estimation methods with TSCA premanufacture notice chemicals. Ecotoxicol Environ Safety 22: 240–249.

TSCA. 1976. The Toxic Substances Control Act, 15 U.S.C. §§2601–2629. (1982 & Supp. III 1985).

USEPA. 1986a. U.S. Environmental Protection Agency. New Chemical Review Process Manual. Washington, DC: U.S. Environmental Protection Agency. Office of Toxic Substances. EPA report number: EPA-560/3-86-002.

USEPA. 1986b. U.S. Environmental Protection Agency. Toxic Substances Control Act Chemical Substance Inventory. TSCA Inventory: 1985 Edition. Volume I. Eligibility Criteria for Inclusion of Chemical Substances on the Inventory. Washington, DC: U.S. Environmental Protection Agency. EPA report number: EPA-560/7-85-002a.

USEPA. 1986c. U.S. Environmental Protection Agency. Toxic Substances Control Act Chemical Substance Inventory. TSCA Inventory: 1985 Edition. Volume I. Appendix

B. Generic Names for Confidential Chemical Substance Identities. Washington, DC: U.S. Environmental Protection Agency. EPA report number: EPA-560/7-85-002a.

USEPA. 1991. U.S. Environmental Protection Agency. Instructions Manual for Premanufacture Notification of New Chemical Substances. Washington, DC: U.S. Environmental Protection Agency. EPA report number: EPA/7710-25(1).

USEPA. 1993 (January). U.S. Environmental Protection Agency. TSCA Confidential Business Information A Security Manual. Washington, DC: U.S. Environmental Protection Agency. Office of Pollution Prevention and Toxics. Document number 7700.

USEPA. 1994. U.S. Environmental Protection Agency. ECOSAR. A Computer Program for Estimating the Ecotoxicity of Industrial Chemicals Based on Structure Activity Relationships. User's Guide. Washington, DC: U.S. Environmental Protection Agency. Office of Pollution Prevention and Toxics. Document number 748-R-93-002.

USEPA. 1995a (March 29). U.S. Environmental Protection Agency. Office of Pollution Prevention and Toxics. Premanufacture Notification Exemptions; Revisions of Exemptions for Polymers; Final Rule. (60 FR 16316–16336).

USEPA. 1995b (March 29). U.S. Environmental Protection Agency. Office of Pollution Prevention and Toxics. Premanufacture Notification Exemption; Revision of Exemption for Chemical Substances Manufactured in Small Quantities; Low Release and Exposure Exemption; Final Rule. (60 FR 16336–16351).

USEPA. 1995c (March 29). U.S. Environmental Protection Agency. Office of Pollution Prevention and Toxics. Premanufacture Notification; Revisions of Premanufacture Notification Regulations; Final Rule. (60 FR 16298–16311).

USEPA. 1995d. U.S. Environmental Protection Agency. Office of Pollution Prevention and Toxics. The Polymer Exemption Manual, in preparation. For Document availability, call the Toxic Substances Control Act (TSCA) Assistance Information Service at (202) 554–1404.

USEPA. 1995e. U.S. Environmental Protection Agency. Office of Pollution Prevention and Toxics. Pollution Prevention Technical Manual, in preparation. For Document availability, call the Toxic Substances Control Act (TSCA) Assistance Information Service at (202) 554–1404.

USEPA. 1995f. U.S. Environmental Protection Agency. Office of Pollution Prevention and Toxics. New Chemicals Program. Washington, DC: U.S. Environmental Protection Agency. Document number EPA-743-F-95-001.

USEPA. 1996a (April 15). U.S. Environmental Protection Agency. Proposed Testing Guidelines; Notice of Availability and Request for Comments. (61 FR 16486–16488).

USEPA. 1996b. U.S. Environmental Protection Agency. Office of Pollution Prevention and Toxics. TSCA New Chemicals Program (NCP). Chemical Categories. Available from the Toxic Substances Control Act (TSCA) Assistance Information Service at (202) 554–1404.

Zeeman MZ, Nabholz JV, Clements RG. 1993. The development of SAR/QSAR for use under EPA's Toxic Substances Control Act (TSCA): An introduction. In: Gorsuch JW, Dwyer FJ, Ingersoll CG, LaPoint TW (eds.). Environmental Toxicology and Risk Assessment, 2nd Volume; SAR/QSAR in the Office of Pollution Prevention and Toxics. Philadelphia: American Society for Testing and Materials. pp. 523–539.

LIST OF SELECTED READINGS FOR CHAPTER 1

For more information on the use of Structure-Activity Relationships in predicting health and environmental hazards see:

DiCarlo FJ, Bickart P, Auer CM. 1985. Role of structure-activity team (SAT) in the Premanufacture Notification process. In: QSAR in Toxicology and Xenobiochemistry. Tichy M, ed. Amsterdam: Elsevier. pp. 433–449.

DiCarlo FJ, Bickart P, Auer CM. 1986. Structure-metabolism relationships (SMR) for the prediction of health hazards by the Environmental Protection Agency. II. Application to teratogenicity and other toxics effects caused by aliphatic acids. Drug Metab Rev 17(3–4): 187–220.

Lipnick RL. 1990. Narcosis: Fundamental and baseline toxicity mechanism for non-electrolyte organic chemicals. In: Practical Applications of Quantitative Structure-Activity Relationships (QSAR) in Environmental Chemistry and Toxicology. Karcher W, Devillers J, (eds). Dordrecht, The Netherlands: Kluwer Academic Publishers. pp. 129–144.

Lipnick RL. 1995. Structure-activity relationships. In: Fundamentals of Aquatic Toxicology: Effects, Environmental Fate, and Risk Assessment—2nd Edition. Rand G, (ed). Washington, DC: Taylor and Francis. pp. 609–655.

Lipnick RL, Pritzker CS, Bentley DL. 1987. Application of QSAR to model the toxicology of industrial organic chemicals to aquatic organisms and mammals. In: Progress in QSAR. Hadzi D, Jerman-Blazic (eds). New York: Elsevier. pp. 301–306.

USEPA. 1980 (July 29). U.S. Environmental Protection Agency. Office of Pesticides and Toxic Substances. Availability of TSCA Revised Inventory. C. Corrections to Previous Inventory Reporting Forms. (45 FR 50544–50545)

Wagner PM, Nabholz JV, Kent RJ. 1995. The new chemicals process at the Environmental Protection Agency (EPA): Structure-activity relationships for hazard identification and risk assessment. Toxicol Lett 79: 67–73.

Woo Y-T, Lai DY, Argus MF, Arcos JC. 1995. Development of structure-activity relationship rules for predicting carcinogenic potential of chemicals. Toxicol Lett 79: 219–228.

Woo Y-T, Lai DY, Argus MF, Arcos JC. 1996. Carcinogenicity of organophosphorous pesticides/compounds: An analysis of their structure-activity relationships. Environ Carcino & Ecotox Revs C14(1): 1–42.

Woo Y-T, Lai DY, Argus MF, Arcos JC. 1996. Carcinogenic potential of organic peroxides: Prediction based on structure-activity relationships (SAR) and mechanism-based–short-term tests. Environ Carcino & Ecotox Revs C14(1): 63–80.

Zeeman M, Auer CM, Clements RG, Nabholz JV, Boethling RS. 1995. U.S. EPA regulatory perspectives on the use of QSAR for new and existing chemical evaluations. SAR & QSAR in Environ Res 3(3): 179–201.

For more information on the Chemical Categories see:

Moss K, Locke D, Auer C. 1996. EPA's new chemicals program. Chem Hlth Saf 3(1): 29–33.

Appendix 7.11

Chemical Information Needed for Risk Assessment

(Reproduced from:
www.epa.gov/opptintr/chem-pmn/chap2.pdf)

Chapter 2

Chemical Information Needed for Risk Assessment

2.1 Introduction

EPA requests various types of chemical information from companies submitting PMNs, including information on the physicochemical properties, synthesis, purity, and use of PMN substances. EPA receives approximately 2,000 PMN submissions annually and many of these do not contain all of the information necessary for a good screening-level risk assessment of the PMN substance (some contain no useful information other than the chemical name and structure). PMN submitters are required to provide certain information whereas other information is optional. This optional information is, nonetheless, important in EPA's review of chemicals, and its inclusion in PMN submissions improves the basis for EPA's evaluation and facilitates the review process. Such information can also be very helpful in avoiding misunderstandings leading to additional but unnecessary EPA review.

Chapter 1 addressed the process that EPA uses in its evaluation of PMN substances. The present chapter (Chapter 2) discusses the chemical information considered by EPA in its review process, how this information is used, and EPA's strategy when pertinent information is omitted from PMN submissions. The chemical information requested in a PMN submission is very important because it forms the underlying basis for risk assessment and risk management decisions made during the PMN review process.

The first section of this chapter discusses each of the different types of chemical information that EPA uses in its evaluation of PMN substances and the importance of this information to risk assessment. Definitions of physicochemical properties are included, and methods of measuring or estimating properties are described. EPA depends very heavily upon physicochemical properties of chemical substances for estimating their transport, environmental fate, exposure, and toxicity to mammalian and aquatic species. The use of this information in risk assessment is presented briefly graphically and is discussed.

The final section of this chapter describes EPA's methods for obtaining or estimating values for physicochemical properties essential in the review of PMN substances, but often not included in PMN submissions. Although accurately-derived empirical data are preferred over estimated data, if such data are not provided in a PMN submission, EPA will first search the literature for data on the PMN chemical, then search for data on analogous substances, and, finally, estimate the required data. Data sources and methods used by EPA in this process include reference books, on-line databases, and computer estimation programs.

This chapter is intended to provide submitters with an understanding of the basis for EPA's requests for certain chemical information. The solicited information is important in EPA's review of PMN chemicals. In all cases, EPA prefers accurate empirical data. If such data are not provided by the submitter and EPA is unable to find data on the PMN chemical, it is EPA's policy to make conservative assumptions and use credible worst case scenarios in its evaluations. Worst case scenarios may, in some cases, lead to overestimating the exposure and risk of a chemical. By providing as much physicochemical property data as possible in PMN submissions, submitters can aid EPA in assessing exposure and risk more accurately.

2.2 Important Chemical Information

To many people, properties such as physical state, melting point, boiling point, vapor pressure, water solubility, lipophilicity (octanol/water partition coefficient), molecular weight, etc., seem to have little to do with toxicity and environmental fate, although the relevance of some of these properties to exposure assessment may be clear. The main purpose of this chapter is to show how these and other physicochemical properties are used extensively by EPA for risk assessment of new chemical substances during PMN review.

Other factors such as intended use, other uses, and synthesis as they relate to risk assessment are also discussed. The intent is not to describe all aspects of risk assessment and associated physicochemical properties. Comprehensive treatises on risk assessment, physicochemical properties, and their measurement, estimation, and use in predicting environmental fate, exposure, toxicity, and pharmacologic response are listed under the Suggested Readings heading at the end of this chapter.

Figure 2-1 illustrates the physicochemical properties most commonly used during risk assessment of PMNs. The important lesson to be learned in this chapter is that essentially all forms of risk assessment of new chemical substances are largely dependent upon physicochemical properties. When mea-

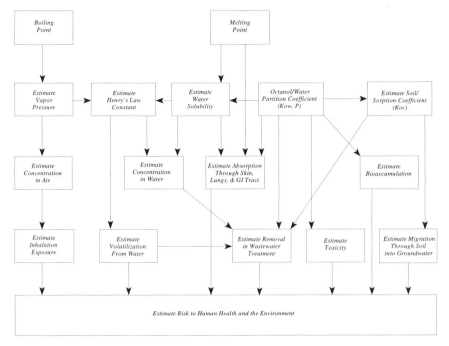

Figure 2-1 *Important Physicochemical Properties, Their Interrelationships, and Their Uses in Risk Assessment*

sured physicochemical properties of chemicals are not available, they must be estimated. Although many reliable estimation methods are available, in any estimation a certain degree of error is always present. Thus, estimation of physicochemical properties should never supplant actual measurement. This section discusses the chemical data that are most important to EPA in reviewing PMNs and how EPA uses these data in risk assessments.

2.2.1 Melting Point

Melting is the change from the highly ordered arrangement of molecules within a crystalline lattice to the more random arrangement that characterizes a liquid. Melting occurs when a temperature is reached at which the thermal energy of the molecules is great enough to overcome the intracrystalline forces that hold them in position in the lattice. As a solid becomes a liquid, heat is absorbed, and the heat content (enthalpy) increases. In other words, the enthalpy of a substance in the liquid state is greater than the enthalpy of the same substance in the solid state. The entropy (a measure of the degree of molecular disorder) also increases as substances change from solid to liquid.

Melting point is an important property used by EPA in the evaluation of PMN substances. The melting point of a pure substance is characteristic of that substance. Melting point, therefore, can be used in the identification of an unknown substance (theoretically, a substance has a single melting point value; however, several substances can coincidentally have the same melting point). The melting point also provides information about the purity of a substance. A sharp melting point or narrow melting range is a good indication that the substance is pure. A fairly wide melting point range generally indicates the presence of impurities. Some substances may decompose or sublime rather than melt. Decomposition and sublimation are also characteristic properties and, hence, are useful for identification purposes.

Melting point is a function of the crystal lattice of a solid, which in turn is dictated primarily by three factors: molecular forces, molecular symmetry, and the conformational degrees of freedom of a molecule (Dearden 1991). Most ionic substances have very high melting points because the forces that hold the ions together are extremely strong. For organic substances, the most important force influencing melting point is intermolecular hydrogen bonding. A substance that has less intermolecular hydrogen bonding and more intramolecular hydrogen bonding will have a lower melting point than a structural isomer of the same substance that has more intermolecular and less intramolecular hydrogen bonding. Melting point also tends to increase with molecular size, simply because the molecular surface area available for contact with other molecules increases, thus increasing the intermolecular forces (Dearden 1991).

Melting point can provide information about the water solubility of non-ionic organic substances. Both melting point and water solubility of non-ionic organics are affected by the strength of the intermolecular forces in the substance. If the intermolecular forces are very strong in a solid, the melting point is likely to be high and the solvation of the individual molecules by water is likely to be low. The melting point of a non-ionic solid, therefore, may be used as an indicator of water solubility. The water solubility of a non-ionic solid depends largely on the temperature of the water, the melting point, and the molar heat of fusion of the solid (Yalkowsky and Banerjee 1992). Abramowitz and Yalkowsky (1990) have reported the use of melting point with total molecular surface for the accurate, quantitative estimation of water solubility for a series of PCBs. Melting point has also been used with K_{ow} (i.e., octanol/water partition coefficient) for an accurate, quantitative estimate of water solubility of liquid or crystalline organic non-electrolytes (Yalkowsky et al. 1979, 1980). Melting point may also be used with other physicochemical properties to derive quantitative estimates of water solubility for non-ionic solids; some of these methods have been summarized by Yalkowsky and Banerjee (1992).

Because the melting point can provide an indication of a substance's water solubility, it can also serve as a tool for estimating the distribution of the substance in aqueous media. If a chemical substance is poorly soluble in water, its concentration in aqueous media may be too low for significant exposure; however, if a substance is highly soluble in water, its concentration in aqueous media is higher, thus increasing exposure potential. In general, high-melting non-ionic solids are likely to have low water solubility and exposure, whereas low-melting, non-ionic solids are likely to have higher water solubility and exposure.

For non-ionic organic substances, melting point can provide an indication of the likelihood of human exposure to a chemical via absorption through the skin, lungs, or gastrointestinal tract. In general, low-melting substances are more likely to be absorbed than substances that melt at higher temperatures, because, for a substance to diffuse through biological membranes, the molecules must be in their greatest state of molecular disaggregation (i.e., in solution). Non-ionic substances that melt at lower temperatures have less energy within their crystalline lattice, are more water soluble, and will be absorbed more readily than compounds that melt at higher temperatures. Substances that are liquids at ambient temperature are generally much better absorbed than solids (USEPA 1992).

Although reasonably accurate methods for the quantitative estimation of melting point have been reported for certain classes of substances (Abramowitz and Yalkowsky 1990; Dearden 1992), estimation of melting point is generally very difficult because the property depends upon a significant number of complex interactions including crystal packing and symmetry, molecular size, and hydrogen bonding (Yalkowsky et al. 1980; Yalkowsky and Banerjee 1992). While melting point may be roughly estimated by analogy with other chemicals that have similar structures, it is well known that even subtle changes within a homologous series of compounds can greatly affect melting point. Accurate estimation of a substance's melting point by comparison to similar substances, therefore, is not always feasible. Melting point is easily measurable for most organic substances (Shriner et al. 1980).

EPA chemists routinely estimate melting points if submitters do not provide them, but measured values are preferable. There is little justification for a PMN submitter to omit melting points for solids since melting point is easy and inexpensive to measure; in many cases, the submitter's analytical laboratory will have measured melting points during research and development activities. These data are considered health and safety data and must be submitted with the PMN. For known substances, the melting point is often available in the scientific literature, but literature values, of course, have no bearing on the purity of the submitter's chemical. Submitters should so indicate when they use literature values in PMN submissions.

When reviewing a PMN substance for which the melting point has been omitted by the submitter, EPA chemists search the literature for an empirical (measured) value. If an empirical melting point is not available, it is the general policy of EPA to estimate a more conservative, relatively low melting point in its risk assessment for that substance. As a consequence, EPA may conclude that the substance may be absorbed more readily through the skin, lung, or gastrointestinal tract than is actually the case and, thus, may predict that the substance will be toxic to humans. Likewise, in the absence of data, EPA will make the assumption that the substance has relatively high water solubility and may be toxic to aquatic life. These reasonable worst-case estimation scenarios can be avoided or mitigated if the submitter provides EPA with empirical melting points.

2.2.2   Octanol/Water Partition Coefficient (K_{ow}, P)

A partition coefficient describes the equilibrium ratio of the molar concentrations of a chemical substance (the solute) in a system containing two immiscible liquids (the solvents). The partition coefficient is not simply a comparison of the solubility of a substance in one immiscible solvent with that in another such solvent. The most common partition coefficient is the octanol/water partition coefficient, expressed as either K_{ow} or P, in which the two immiscible solvents are n-octanol and water. The equation for K_{ow} (or P) is:

$$K_{ow} = \frac{[\text{chemical substance}] \text{ in } n \text{ octanol}}{[\text{chemical substance}] \text{ in water}}$$

where concentrations are in moles/liter.

For purposes of simplification, K_{ow} is usually reported as its common logarithm (log K_{ow} or log P). A large log K_{ow} value for a chemical (relative to other substances), indicates that the chemical has a greater affinity for the n-octanol phase and, hence, is more hydrophobic (lipophilic). A low or negative log K_{ow} value indicates that a chemical has a greater affinity for the water phase, and hence, is more hydrophilic. A chemical substance with a log K_{ow} of 1 has ten times the affinity for n-octanol that it has for water, whereas a chemical substance with a log K_{ow} of -1 has ten times the affinity for water that it has for n-octanol. A chemical substance with a log K_{ow} of 0 has equal affinity for n-octanol and water. Substances containing polar substituents (e.g., –OH, –SH, –NH$_2$, etc.) tend to have lower log K_{ow} values than substances that lack such substituents.

For practically any given non-ionic organic substance, it is possible to use the octanol/water partition coefficient to estimate other physicochemical properties and, in many cases, the distribution of the chemical within a living system or the environment. This is why octanol/water partition coefficients are extremely helpful and are used extensively during risk assessment of chemical substances. Specifically, octanol/water partition coefficients are often used by EPA and others to estimate water solubility, soil and sediment adsorption, biological absorption (following oral, inhalation, or dermal exposure), bioaccumulation, and toxicity.

A primary reason for the versatility of the octanol/water partition coefficient in risk assessment is that it serves as a model for the distribution of a chemical substance within both biological and non-biological systems. Biological membranes and systems (e.g., organs, cell membranes, capillaries, blood–brain barrier, skin, intestines) typically contain various combinations of lipid and aqueous components. For a chemical substance to gain entry into and distribute

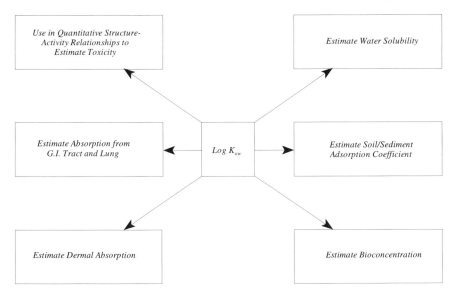

Figure 2-2   *Use of Octanol/Water Partition Coefficient (*Log *K*ₒw*) in Risk Assessment*

throughout a biological system, it must have a certain amount of both lipid and water solubility. The octanol and water phases of an octanol/water system are representative of the lipid and aqueous components of biological systems, respectively. Thus, the octanol/water partition coefficient is an important property influencing the biological activity of a chemical substance (Hansch and Dunn 1972; Hansch and Clayton 1973). For this reason, the octanol/water partition coefficient is used extensively by EPA and others in the quantitative prediction of toxicity (Blum and Speece 1990; Karcher and Devillers 1990; Hermens and Opperhuizen 1991; Grogan et al. 1992) and environmental fate (Lu and Metcalf 1975; Kenaga and Goring 1980; Swann et al. 1983). Pharmaceutical companies use the octanol/water partition coefficient for the quantitative prediction of pharmacological activity of many chemical substances (Martin 1978; Yalkowsky et al. 1980). Figure 2-2 illustrates the usefulness of log K_{ow}. Suggested readings, including the use of octanol/water partition coefficient in estimating bioavailability, toxicity, and pharmacological activity, are provided at the end of this chapter.

Substances with high (>5) log K_{ow} values are so hydrophobic that they partition very poorly into the aqueous components of biological systems, remain within the lipid components and are generally poorly absorbed following acute exposure. Chemical substances with high log K_{ow} values, although poorly absorbed, are more likely to bioaccumulate into fat tissue, whereas compounds with lower log K_{ow} values generally do not bioaccumulate because of their lower affinity for lipids (Lyman et al. 1982; Noegrohati and Hammers 1992). Substances with high log K_{ow} values that exist in the environment at sub-toxic levels may bioconcentrate to toxic levels within aquatic organisms, following sufficient exposure duration to achieve steady state partitioning. The ability of a very hydrophobic chemical to produce toxic effects may be limited by high melting point, resulting in both insufficient water and lipid solubility to reach toxic levels at the site of action within the aquatic organism (USEPA 1985). Generally, chemicals with good lipid and water solubilities are likely to be absorbed from all routes of exposure, including the skin (Shah 1990).

Substances with high log K_{ow} values tend to adsorb more readily to organic matter in soils or sediments because of their low affinity for water. Compounds with lower log K_{ow} values are not as likely to adsorb to soils or sediments because they will be more prone to partition into any surrounding water. Log K_{ow} is often used, in fact, by EPA to estimate quantitative soil/sediment adsorption coefficients, K_{oc} (Lyman et al. 1982) and qualitative removal of a substance during wastewater treatment.

Because the octanol/water partition coefficient is an equilibrium ratio of the molar concentrations of a chemical substance in n-octanol and water, it is often useful in estimating water solubility. Water solubility is often a difficult property to estimate; however, regression equations for the quantitative estimation of water solubility using log K_{ow} have been reported for organic chemical substances from several classes (Yalkowsky et al. 1979; Yalkowsky et al. 1980; Yalkowsky and Valvani 1980; Yalkowsky and Valvani 1979; and Yalkowsky and Banerjee 1992; Bowman and Sans 1983; Isnard and Lambert 1989; Kenaga and Goring 1980). As a general rule of thumb with non-ionic organic substances, the higher the log K_{ow} value, the lower is the water solubility. Estimation of water solubility is discussed in more detail later in this chapter. The EPA is currently developing guidelines for the selection of measured or estimated K_{ow} data. These will provide additional guidance to PMN submitters.

Measuring log K_{ow}

Several methods of measuring octanol/water partition coefficient are described in EPA's Test Guidelines (USEPA 1996), and newer methods continue to appear in the literature. Each of these methods has advantages and disadvantages; one must be very careful to select the best method for a particular chemical in order to obtain an accurate value. It is very important to state the method of measurement along with each log K_{ow} value, so that the reliability of the value is apparent.

The classical method for measuring log K_{ow} is the "shake-flask" method. In this method, the test chemical is mixed with an appropriate n-octanol/water mixture and shaken for some given period during which equilibrium between both phases is achieved. It is important for the n-octanol and water phases to be mutually saturated prior to shaking with the test chemical. After the phases separate, the concentrations of the test chemical in the octanol and aqueous phases are determined. The aqueous phase often needs to be centrifuged to remove any small octanol droplets.

The shake-flask method is widely used to measure the K_{ow} accurately for many chemicals. This method is not appropriate, however, for substances with high partition coefficients (log $K_{ow} > 4.5$). The shake-flask method is also inappropriate for (1) polycyclic aromatic substances lacking polar substituents, (2) halogenated hydrocarbons, and (3) large, non-polar chemicals, because large volumes of the aqueous phase are required for analysis and, in addition, the aqueous phase becomes contaminated with micro-emulsions formed during shaking. Although it may be possible to prevent or remove the emulsions formed during the shake-flask procedure, literature data for K_{ow} measured by this technique indicate that in many cases, the formation of emulsions has influenced the observed K_{ow} values. This may account for the high variance among literature values for rather hydrophobic chemicals whose K_{ow} values were determined by independent investigators using this method (Hansch and Leo 1979; Kenaga and Goring 1980).

Brooke and co-workers (1986) have described a "slow-stir" method for measuring octanol/water partition coefficients for hydrophobic chemicals. This method is similar to the shake-flask method, but differs in that the octanol and water phases are equilibrated under conditions of slow stirring rather than vigorous shaking. By careful stirring and rigid temperature control, the formation of emulsions can be prevented, and accurate partition coefficients can be obtained relatively easily for very hydrophobic substances. De Bruijn and co-workers (1989) found that for substances with log K_{ow} values ranging from 0.9 to 4.5, experimental data obtained by the slow-stir method were in good agreement with literature values based on the shake-flask method. For substances having log K_{ow} values of 4.5 and higher, there was reasonable agreement between data obtained using the slow-stir method and data obtained using either reversed-phase high performance liquid chromatography (HPLC) or the generator column method. Thus, the slow-stir method appears to be very useful for measuring log K_{ow} for hydrophobic as well as hydrophilic substances. In addition, the method is easy to use, relatively fast, and does not require expensive equipment. Detailed discussions of the slow-stir method in determining K_{ow} are available (Brooke et al. 1986; de Bruijn et al. 1989).

Another very versatile method for measuring log K_{ow} is the generator column method (USEPA 1985). In this method, a generator column is used to partition the test substance between the octanol and water phases. The column is packed with a solid support and is saturated with a fixed concentration of the test substance in *n*-octanol. The test substance is eluted from the octanol-saturated generator column with water. The aqueous solution exiting the column represents the equilibrium concentration of the test substance that has partitioned from the octanol phase into the water phase. The primary advantage of the generator column method over the shake-flask method is that the former completely avoids the formation of micro-emulsions. Therefore, this method is particularly useful for measuring K_{ow} for substances having log K_{ow} values over 4.5 (Doucette and Andren 1987, 1988; Shiu et al. 1988), as well as for substances having log K_{ow} values less than 4.5. A disadvantage of the

TABLE 2-1. **Methods of Measuring Octanol/Water Partition Coefficient (K_{ow})**

Method	Advantages	Disadvantages	References
Shake-Flask	Easy to use. Reliable for substances that have log K_{ow} values < 4.5. Doesn't require expensive equipment.	Generally not useful for measuring K_{ow} values for substances having log K_{ow} values > 4.5; shaking may form micro-emulsions, which lead to inaccurate measurement.	USEPA (1985); Kenaga and Goring (1980).
Slow-Stir	Easy to use. Relatively fast, doesn't require expensive equipment. Reliable for essentially all substances.	Requires careful stirring and close temperature control to avoid formation of micro-emulsions.	Brooke et al. (1986); de Bruijn et al. (1989).
Generator Column	Reliable for essentially all substances. Avoids formation of micro-emulsions.	Requires expensive equipment.	USEPA (1985); Doucette and Andren (1987, 1988); Shiu et al. (1988).

generator column method is that it requires sophisticated equipment. A detailed description of the generator column method is presented in USEPA 1985.

EPA encourages PMN submitters to provide accurately-measured log K_{ow} data in PMN submissions. For certain types of chemical substances, however, it is not necessary to do so. Substances that contain several aromatic rings, lack polar substituents, or are polyhalogenated most likely have log K_{ow} values greater than 7. Similarly, chemicals that contain long-chained (10 or more carbons) alkyl substituents with few polar groups (e.g., fatty acids) are also likely to have log K_{ow} values above 7. Such substances are so clearly hydrophobic that it is not necessary to have an accurately-measured K_{ow} value for risk assessment purposes. In addition, it is generally not necessary to measure K_{ow} values for substances that have strong surfactant properties. Measuring K_{ow} for surfactants (particularly ionic surfactants) is usually difficult because the surfactant causes the octanol and water phases to become miscible, preventing partitioning between the two solvents. EPA does not generally recommend measuring log K_{ow} for polymers or PMN substances that lack definite structure (class 2 substances). For most substances, especially class 1 compounds (i.e., those with defined structures), measured K_{ow} values are very helpful for properly and fairly characterizing risk potential. It is also helpful to provide EPA with the method used for measuring K_{ow}. Table 2-1 summarizes the methods used for measuring octanol/water partition coefficient.

Estimating log K_{ow}

Recognizing the importance of log K_{ow} in predicting absorption, biological properties, and environmental fate, scientists over the years have measured and recorded log K_{ow} values for thousands of substances, largely from the shake-flask method. These empirical data sets have served as a basis for developing techniques to estimate log K_{ow}. Numerous methods for estimating log K_{ow} accurately for many different classes of substances are now available. Some of these methods have recently been reviewed (Leo 1993; van de Waterbeemd and Mannhold 1996). Most of the log K_{ow} estimation methods are based upon one or more of the following approaches:

- fragment or substituent additivity (Hansch and Leo 1979; Leo 1990);
- correlations with capacity factors on reversed-phase HPLC (Lins et al. 1982; Brent et al. 1983; Garst 1984; Garst and Wilson 1984; USEPA 1985; Dunn et al. 1986; Minick et al. 1988; Yamagami et al. 1990);
- correlations with descriptors for molecular volume or shape such as molecular weight, molar refraction, parachor, molar volume, total molecular surface area and total molecular volume (Dunn et al. 1986; Doucette and Andren 1987; de Bruijn and Hermans 1990); and
- correlations with molar volume, solvatochromic (thermodynamic) parameters, or charge transfer interactions (Kamlet et al. 1988; Saski et al. 1991; Dunn et al. 1991; Moriguchi et al. 1992; Da et al. 1992).

A major problem in estimating log K_{ow} is that most methods work well for certain classes of substances, but not for other classes. Typically, originators of these estimation methods are quick to point out the shortcomings of other methods, but not the limitations of their own methods. Before using any method for estimating log K_{ow}, the user should become familiar with the theoretical basis of the method, its applicability, and its limitations. Estimation methods that have not been validated (i.e., tested against accurately-measured log K_{ow} values) should not be used. The remainder of this section briefly discusses the methods above for estimating log K_{ow} and attempts to provide some guidance with respect to their use. Table 2-2 summarizes the advantages and disadvantages of the methods. A detailed description of each estimation method is beyond the scope of this text; however, a comprehensive listing of references describing various estimation methods of log K_{ow} is provided in the Suggested Readings section at the end of this chapter.

The foremost method used in estimating log K_{ow} is that of Hansch and Leo (1979). This method uses empirically-derived fragment constants and structural factors to calculate log K_{ow} from a structure. Estimates are made from addition of fragment constants and structural factors, which are compiled for thousands of structural fragments and atoms stored in a database. The method has been validated by many investigators. A detailed description of how the method is used is available (Lyman et al. 1982). Using this method, one can estimate log K_{ow} for almost any substance. If an accurately-measured value of K_{ow} is available for a structurally similar or "parent" compound, this measured value can be used to estimate the log K_{ow} of the "derivative" by adding or subtracting the

TABLE 2-2.   Methods of Estimating Octanol/Water Partition Coefficient (K_{ow})

Method	Advantages	Disadvantages	References
Fragment Constant Additivity	Calculation of log K_{ow} for many substances can be accomplished directly from structure. Available as a computer program. Known to be very accurate for substances having log K_{ow} values less than 4.5	Inaccurate for substances with log $K_{ow} > 6$. Cannot estimate log K_{ow} for substances containing substituents that are not in the fragment constant database (except for the Meylan and Howard method).	Hansch and Leo (1979); Meylan and Howard (1995).
Correlation of Reversed-Phase HPLC Retention Times	Known to be very accurate.	Requires a dataset of accurately-measured log K_{ow} values and HPLC retention times of substances closely related to the test substance.	Garst (1984); Garst and Wilson (1984); USEPA (1985).
Correlation of Molecular Surface Area and Volume	Very accurate for certain non-polar hydrophobic substances.	Requires a dataset of accurately-measured log K_{ow} values and HPLC retention times of substances closely related to the test substance. Only accurate for non-polar hydrophobic substances such as halogenated and nonhalogenated benzenes and biphenyls.	Yalkowski and Valvani (1976); Doucette and Andren (1988); de Bruijn and Hermens (1990); Brooke et al. (1987).
"Three-Dimensional" Modeling	Calculation of log K_{ow} for many substances can be accomplished directly from structure. May be used for substances whose log K_{ow} values cannot be calculated by the fragment constant additivity method (due to missing fragment constants).	Requires knowledge of molecular modeling. Requires sophisticated computer hardware and software. Has not been thoroughly validated.	Sasaki et al. (1991); Moriguchi et al. (1992); Waller (1994).

appropriate fragment constant or structural factor. This approach is preferred whenever a reliable measured value of a parent compound is available because the solvent–solute interaction terms in the parent molecule are already accounted for. A major advantage of the Hansch and Leo method is that log K_{ow}

values can be estimated (calculated) directly from structure alone. This method is very accurate for many classes of chemical substances, but is known to over-estimate log K_{ow} for some substances with log K_{ow} values greater than about 6 (Lyman et al. 1982). A computer program (CLOGP) of the Hansch and Leo method is available.[17] A disadvantage of the method is that it cannot estimate log K_{ow} for substances that contain substituents whose fragment or structural factor contributions to log K_{ow} are unknown. Meylan and Howard (1995) have recently reported a variation of the Hansch and Leo fragment addition method for estimating K_{ow}. This variation uses atom/fragment contribution values and correction factors obtained from measured K_{ow} values of structurally diverse substances. Using the Meylan and Howard method, the K_{ow} of a substance is estimated by summing all atom/fragment contribution values and correction factors pertaining to the structure. The primary advantage of this method over the Hansch and Leo method is that it can calculate K_{ow} for substances for which K_{ow} cannot be calculated using the Hansch and Leo method. The Meylan and Howard method is easy to use, and reported to be very accurate. A computer program (LOGKOW) of the method is available.[18]

A great deal of effort has been directed toward estimating K_{ow} from retention times determined by reversed-phase HPLC. A detailed discussion of this method is available in USEPA 1985. In this technique, accurately-measured log K_{ow} values for a set of closely related substances are correlated to the reversed-phase HPLC retention times of the substances, and a regression equation is obtained. The log K_{ow} of a structurally similar substance can be estimated using its retention time and the regression equation. This method is semi-empirical since HPLC retention time must be measured.

The reversed-phase HPLC method is known to be very accurate for many chemical substances (Lins et al. 1982; Brent et al. 1983; Garst 1984; Garst and Wilson 1984; Minick et al. 1988; Yamagami et al. 1990). Obvious disadvantages of this method, however, are that it requires accurately-measured log K_{ow} values of analogous substances, sophisticated technical equipment, and a certain amount of technical expertise. Another disadvantage is that the linear regression equations cannot be extrapolated beyond the K_{ow} range for which the equations were derived. Also, log K_{ow} values for the reference chemicals are usually determined by the shake-flask method and, therefore, are not very reliable for hydrophobic substances. Leo (1990) has discussed other disadvantages to this approach. The reversed-phase HPLC method should only be used for chemicals and reference compounds whose chemical structures are similar.

Several investigators have reported exceptional correlations between log K_{ow} and molecular surface area or molecular volume for hydrophobic aromatic substances, such as halogenated benzenes and biphenyls (Yalkowsky and

---

17. The CLOGP computer program is available through the Pomona College Medicinal Chemistry Project, Claremont, California, 91711.

18. The LOGKOW computer program is available from Syracuse Research Corporation, Environmental Science Center, Merrill Lane Syracuse, NY, 13210.

Valvani 1976; Doucette and Andren 1987, 1988; Brooke et al. 1986, 1987; de Bruijn and Hermans 1990). Like the reversed-phase HPLC method, correlations with molecular surface area or volume require a data set of measured K_{ow} values for structurally similar substances. Molecular surface areas or molecular volumes are calculated for each chemical in the group and are then correlated with log K_{ow} to give a regression equation. Log K_{ow} of an analogous substance can then be estimated using the substance's calculated molecular surface area or volume in the regression equation. This method is not useful for estimating log K_{ow} for aromatic substances (or others) that contain polar substituents, since it does not take into account the effects that these substituents have on octanol/water partitioning.

An extension of this approach uses polarizability/dipolarity and hydrogen bonding terms in addition to molecular volume, and also has been found to predict log K_{ow} values accurately for PCBs and polycyclic aromatic hydrocarbons (Kamlet et al. 1988). Use of these descriptor terms in predicting log K_{ow} for more polar substances is presumably under investigation. A potentially serious drawback to this approach is that the descriptor terms may not always be available.

Recent advances in computer hardware and software have made estimation of log K_{ow} possible through consideration of three-dimensional intra- and intermolecular interactions (Sasaki et al. 1991; Moriguchi et al. 1992). This three-dimensional approach estimates log K_{ow} for organic substances through correlation with molecular surface area, electrostatic potential, charge transfer interactions, and other electronic and structural effects derived from three-dimensional molecular structures. Advantages to these methods are that log K_{ow} can be estimated directly from chemical structure and for substances to which Hansch and Leo's fragment constant approach has not been applicable. Although the three-dimensional methods for estimating log K_{ow} have not yet been completely validated, they appear to be very useful for rapid estimation of log K_{ow} for a wide variety of chemical substances. When in doubt regarding the applicability of a particular log K_{ow} estimation method, one should seek measured data on an analog and test the estimation. Alternatively, the analog can be used as the basis for estimation by subtracting and adding needed small fragments to obtain the PMN structure.

The octanol/water partition coefficient is very important in EPA's evaluation of PMN substances. EPA uses either measured or estimated log K_{ow} values in assessing approximately 50% of all PMN substances (which represents about 80% of all non-polymer PMN substances). As discussed above, octanol/water partition coefficients can be used to estimate other properties (e.g., solubility, bioaccumulation, toxicity); these other properties are then used to evaluate the potential risk of a chemical to human health and the environment. The submission of accurately measured octanol/water partition coefficients allows for the reliable prediction of the effects of a chemical on human health and the environment. Accurately estimated log K_{ow} values are also useful to EPA. If an accurately measured or estimated log K_{ow} value is not provided by the submitter, then the EPA will estimate K_{ow} using one of the methods discussed previously. In cases where it is not apparent to EPA as to which estimation method will provide the most accurate log K_{ow} value, EPA will select the method that provides a log K_{ow} value that results in the highest toxicity or exposure.

2.2.3 Water Solubility

Water solubility is defined as the maximum amount of a substance in its finest state of molecular subdivision that will dissolve in a given volume of water at a given temperature and pressure. For risk assessment, EPA is most interested in the water solubility of chemical substances given at environmental temperatures (20–30°C). Water solubility may be expressed in a number of units; EPA prefers water solubility data to be given in grams/liter (g/L). Most common organic chemicals have water solubilities that range anywhere from 0.001 g/L (1 part per million, ppm) to 100 g/L (100,000 ppm) at environmental temperatures. Solubilities for extremely hydrophobic substances (e.g., dioxins) have been measured below 1 part per billion, whereas some substances are infinitely soluble (completely miscible) in water.

Water solubility is one of the most important properties affecting bioavailability and environmental fate of chemical substances. Chemicals that are reasonably water soluble (that have low log K_{ow} values) are generally absorbed into biological systems because most of these systems contain a significant number of aqueous components. Such chemicals have relatively low adsorption coefficients for soils and sediments, and they bioconcentrate poorly, if at all, in aquatic species. Furthermore, highly water soluble substances tend to degrade more readily by processes such as photolysis, hydrolysis, and oxidation (Klopman et al. 1992). Water solubility also affects specialized transport pathways such as volatilization from solution and washout from the atmosphere by rain (Lyman et al. 1982). Water solubility, therefore, is a key element in the risk assessment of any chemical substance.

Measuring water solubility

The two most common methods for the experimental determination of water solubility are the shake-flask and generator column methods (Yalkowsky and Banerjee 1992; USEPA 1985; Lyman et al. 1982). Although these methods are not technically difficult, there can be considerable variation in the water solubility measured for the same substance using the same method, but in different laboratories. These discrepancies result primarily from the large number of experimental variables that are known to affect solubility measurements. These variables include properties of the water such as temperature, pH, presence of suspended solids, salt content, and organic content, and include properties of the chemical such as the physical state (especially particle size of solids), purity, and adsorption of the chemical onto the walls of the experimental apparatus

(Kenaga and Goring 1980; Yalkowsky and Banerjee 1992). It appears that discrepancies increase as hydrophobicity increases (USEPA 1979). The shake-flask method is acceptable for determining water solubilities for substances that have log K_{ow} values of 3 or lower. Disadvantages of the shake-flask method are: (1) the method requires considerable sample handling between saturation and analysis steps; (2) colloid formation may occur as result of the shaking; and (3) the method is inaccurate for hydrophobic substances. The generator column method does not have the shortcomings of the shake-flask method and, therefore, is the preferred method for measuring water solubility. In addition, it is very rapid, precise, and is applicable to substances with water solubilities ranging from 10 parts per billion to grams per liter (Yalkowsky and Banerjee 1992; USEPA 1985). The equipment used in the generator column method, however, is more sophisticated and, hence, more expensive. PMN submitters are encouraged to provide information on the method used to measure water solubility, as well as an estimate of systematic and random errors of the reported result.

Estimating water solubility

A considerable amount of effort has been devoted to understanding the mechanism of aqueous solubility and developing methods that enable accurate estimation. A comprehensive treatise on water solubility and methods for its estimation has been published (Yalkowsky and Banerjee 1992). To summarize the contents of the text, water solubility is governed by three major factors: (1) the entropy of mixing; (2) the differences between the solute-water adhesive interaction and the sum of the solute-solute and water-water adhesive interactions; and (3) the additional intermolecular interactions associated with the lattice energy of crystalline substances (Yalkowsky and Banerjee 1992; Klopman et al. 1992). In estimating the water solubility of liquid substances, only factors 1 and 2 need to be considered, whereas in estimating the water solubility of solids, factor 3 must be included as well.

Most estimation methods for water solubility consist of regression equations that contain K_{ow} data as descriptors of factors 1 and 2 (Lyman et al. 1982; Yalkowsky and Banerjee 1992). Generally, if K_{ow} data are not available, it is difficult to estimate water solubility accurately. Some estimation methods also incorporate atomic fragment constants, and have been moderately successful for certain types of substances (Lyman et al. 1982; Wakita et al. 1986; Yalkowsky 1988; Klopman et al. 1992). Methods for estimating water solubility have been more successful for liquids than for solids. This is largely because of the difficulty in incorporating descriptors of intermolecular interactions for solid substances into the regression equations of the estimation methods. Incorporation of melting point, entropy of fusion, or enthalpy of fusion as descriptors of factor 3 has met with limited success for only certain types of compounds and, thus, has limited applicability (Lyman et al. 1982; Yalkowsky and Banerjee 1992). In short, accurate estimation of water solubility is generally difficult, particularly for solid substances. As a general rule, non-ionic substances that are liquids at room temperature are usually more soluble than solids. Solid non-ionic substances with higher melting points or greater polarity tend to be less soluble than non-ionic solids that have lower melting points or lower polarity.

As noted earlier, when estimation of properties is difficult, EPA uses conservative values that ultimately tend to increase the Agency's overall concern for the chemical. EPA encourages the inclusion of reliably measured water solubility data in PMN submissions. By providing such information, the PMN submitter both eliminates the possibility that EPA will overestimate the water solubility of a chemical and ultimately assists EPA in making the most accurate risk assessment and risk management decisions.

It is not always necessary, however, for PMN submitters to provide EPA with measured water solubility data. For example, it is not necessary to measure the aqueous solubility of substances that are obviously very soluble, such as mineral salts of amines, metal salts of sulfonic acids, and quaternary ammonium compounds. For risk assessment purposes, EPA is not concerned with discerning the precise aqueous solubility for substances that are considerably water soluble. It is also, in general, not necessary for PMN submitters to determine water solubility for substances that are extremely water insoluble. Chemicals that are extremely hydrophobic (log K_{ow} greater than 7) are so poorly soluble that for risk assessment purposes, such substances are regarded as essentially insoluble. Finally, it is not necessary to measure water solubility for polymeric materials that are dispersible.

To decide whether water solubility should be measured, one should first determine or estimate the log K_{ow} of the substance. It is best to measure water solubility for substances whose log K_{ow} values are between -1 and 7. The generator column method is preferred for measuring water solubility for substances that have log K_{ow} values of 3 or greater. The shake-flask method is acceptable for measuring water solubility of substances having log K_{ow} values less than 3.

It is important that water solubility be determined for the substance itself, not for formulations of the substance. It is not uncommon for EPA to receive PMN submissions that include measured water solubility data for formulations of the PMN substance in co-solvents (e.g., alcohols, dimethylformamide, or dimethylsulfoxide). Such measured data are useless to EPA for risk assessment purposes.

Terms such as "insoluble" or "not very soluble" should not be used unless they are accompanied by data from attempted solubility measurements (such as "log K_{ow} is greater than 7"). A substance that is regarded as "insoluble" by a chemist may be sufficiently soluble to contribute to risk, as determined by a toxicologist or environmental fate specialist. Similarly, terms such as "soluble" or "very soluble" should not be used unless, again, they are accompanied by data from attempted solubility measurements (such as "water solubility is greater than 100 g/L").

2.2.4   Soil/Sediment Adsorption Coefficient

The soil/sediment adsorption coefficient, K_{oc}, is a measure of the tendency of a chemical to be adsorbed onto soils or sediments. K_{oc} is defined as the ratio of the amount of chemical adsorbed per unit weight of organic carbon (oc) in soils or sediments to the concentration of the chemical in solution at equilibrium:

$$K_{oc} = \frac{\mu g \text{ adsorbed/g organic carbon}}{\mu g/mL \text{ solution}}$$

Discussions on soil and sediment adsorption are available (Karickhoff et al. 1979; Means et al. 1982). Values of K_{oc} can range from 1 to 1×10^7 (Lyman et al. 1982).

K_{oc} is important in the assessment of the fate and transport of chemicals in soils and sediments. A chemical with a high K_{oc} value is likely to be adsorbed to soils and sediments and thus, is likely to remain on the soil surface. In contrast, a chemical with a low K_{oc} value is not likely to be adsorbed to soils and sediments but is likely to leach through these soils and sediments and, if not degraded, may reach ground and surface waters. Chemicals that adsorb tightly to soils and sediments may accumulate in soils, but will be less prone to environmental transport in the gas phase or in solution. Chiou and co-workers (1983) reported that the extent of a chemical's insolubility in water is the primary factor affecting its adsorption to soils and determines its degree of mobility in rivers, groundwater, and runoff. Also, a substance that is tightly adsorbed to soils is less likely to be subject to other fate processes (such as volatilization, photolysis, hydrolysis, and biodegradation) than a substance that tends to partition into water.

EPA's Toxic Substances Control Act Test Guidelines (USEPA 1985) describe an experimental method for determining the adsorption coefficient K, which can be used to calculate K_{oc}. The method involves equilibrating various aqueous solutions containing different concentrations of the test chemical and a known quantity of sediment or soil. After equilibrium is reached, the distribution of the chemical between the aqueous phase and the solid phase is determined. The coefficient, K, is determined from the following equation:

$$\frac{x}{m} = KC^{1/n}$$

where

$x/m = (\mu g \text{ of chemical absorbed})/(g \text{ soil or sediment})$

$C = (\mu g \text{ of chemical})/(mL \text{ of solution})$

$n = $ a parameter ranging from 0.7 to 1.1 (Lyman et al. 1982)

K_{oc} is determined from K and the percent of oc in the soil or sediment:

$$K_{oc} = \frac{K}{\%oc} \times 100$$

Several methods are available for the estimation of K_{oc} from empirical relationships with other properties (Lyman et al. 1982). Octanol/water partition coefficient (K_{ow}) is often used in regression equations for the estimation of K_{oc}. Other properties used to estimate K_{oc} include water solubility, bioconcentration factor (BCF) for aquatic life, and parachor. Swann et al. (1983) found that the retention times of chemicals in reversed-phase high-performance liquid chromatography (RP-HPLC) correlate well with measured K_{oc} values. Bahnick and Doucette (1988) and Sabljic (1984, 1987) have reported the use of molecular connectivity indices for estimation of K_{oc}. Meylan and co-workers (1992) have recently reported a model for K_{oc} estimations that uses molecular connectivity indices and fragment descriptors. This last method appears to produce more accurate estimates of K_{oc} than other models, is easier to use since measured or estimated K_{ow} or water solubility values are not needed, and is more comprehensive in its applicability to a variety of structurally diverse organic compounds.

K_{oc} provides a measure of a substance's distribution between soil and water. For practical reasons, EPA does not expect PMN submitters to measure K_{oc} values for substances submitted in PMNs. In fact, EPA has, to date, never received a PMN that included a K_{oc} value; however, EPA estimates K_{oc} values for practically every PMN substance submitted to the Agency because of the importance of this property in predicting environmental partitioning and distribution. This emphasizes the need for the inclusion of certain physicochemical property data (such as water solubility and K_{ow}) in PMNs, which EPA can then use in estimating K_{oc}. K_{oc}, used with the K_{ow}, BCF, and Henry's Law constant, can predict the environmental distribution of a chemical and, thus, is a measure of environmental risk (McCall et al. 1983).

2.2.5 Henry's Law Constant

A substance that is introduced into the environment by release to air, water, or land tends to diffuse through all environmental media in the direction of establishing an equilibrium between these media. Henry's Law describes the distribution of a chemical between water and air and states that when a substance is dissolved in water, the substance will have a tendency to volatilize from the water into the air above until an equilibrium is reached. Henry's Law constant (H) can be considered an air-water partition coefficient and is defined as the concentration of the chemical substance in air relative to the concentration of the chemical substance in water:

$$H = \frac{[\text{chemical substance}] \text{ in air}}{[\text{chemical substance}] \text{ in water}}$$

This equation is appropriate only for equilibrium conditions of dilute solutions (those typically observed in the environment). Chemicals that have high H values have a greater tendency to volatilize from solution and partition towards air, whereas relatively low H values indicate that the substances will tend to partition into water. Some groups of substances tend to partition significantly toward air despite possessing relatively low vapor pressures. These high H values are primarily the result of the poor solubility of these substances (hydrocarbons, for example) in water.

Henry's Law constant can be expressed as a ratio of the partial pressure of a substance in the vapor above a solution to the concentration of the substance in the solution:

$$H = \frac{\text{equilibrium vapor pressure}}{\text{solubility}}$$

where vapor pressure is in atmospheres and the solubility is in moles per cubic meter.

The vapor pressure of the pure substance, typically in units of atmospheres-cubic meters per mole (atm-m^3/mol), is often used as an approximation of the partial pressure (Lyman et al. 1982). This approximation is valid for substances with low water solubilities. If the solubility of a substance exceeds a few percent, then the dissolved substance's vapor pressure will be lower than that of the pure substance due to its dilution by water (Mackay and Shiu 1981). The thermodynamic principles that govern the relationships between vapor pressure, water solubility, and H for solid and liquid substances have been addressed in detail by Mackay and Shiu (1981). Also included in this discussion are experimental techniques for obtaining these properties. The inverse of the H value is also used by some investigators (McCall et al. 1983); therefore, the ratio H must be defined as being either air/water or water/air. The vapor pressure term can be expressed in other units (e.g., Pascals, torr), and the solubility term can be expressed in other concentration units (e.g., grams per cubic meter) or as a mole fraction.

The H value is often calculated from data for vapor pressure and water solubility that are measured independently (see the sections on these two properties for information on obtaining experimental measurements). As mentioned, this method may not be accurate for substances with water solubilities exceeding a few percent, but it is considered to be satisfactory for less soluble substances (Mackay and Shiu 1981). A second method for determining H involves measuring the water solubility and vapor pressure of a substance in a system that is at equilibrium (Mackay and Shiu 1981). This method is typically used for substances with high water solubilities. A third method described by Mackay and Shiu (1981) is most appropriate for substances with very low solubilities and vapor pressures. The method involves measuring the relative concentration changes in one phase during an equilibrium air–water exchange process. The H value is then determined from the slope of a semilogarithmic plot of concentration versus time.

EPA often estimates H using vapor pressure and water solubility data. Several methods are also available for estimating H from molecular fragments (Bruggemann and Munzer 1988; Hine and Mookerjee 1975) and bond contribution values (Meylan and Howard 1991).

Whereas the soil adsorption coefficient (K_{oc}) provides a measure of a substance's distribution between soil and water, H provides a measure of a substance's distribution between water and air. As with K_{oc}, EPA does not expect PMN submitters to measure H values for substances submitted in PMNs. EPA, however, does estimate H values for many PMN substances submitted to the Agency to describe the volatilization of a substance from water. This further emphasizes the need for the inclusion in PMNs of certain physicochemical property data (such as water solubility and vapor pressure, or at least boiling point), which EPA can then use for estimating H. The H value, water solubility, K_{ow}, K_{oc}, and BCF are all important properties used in determining the environmental distribution pattern of a substance and in assessing its environmental risk.

2.2.6 Boiling Point

Boiling point is the temperature at which the vapor pressure of a substance in the liquid state is equal to atmospheric pressure. A substance boils when it has absorbed enough thermal energy to overcome the attractive forces between the molecules of the substance. The heat required to overcome these forces is the latent heat of vaporization. Solid substances, of course, must first liquify (melt) before they can boil. Some solid chemicals *sublime*; they pass directly from the solid to the gaseous state without melting. Boiling points and sublimation temperatures, like melting points, are characteristic properties of pure substances and may be used for the purpose of identification. Boiling points can also provide an indication of the purity of a liquid. With the exception of azeotropes, a liquid that is a mixture of several substances will begin to boil at a temperature equal to the boiling point of its most volatile component. The temperature will then gradually increase as the vapor phase becomes more rich with the less volatile component(s), until the temperature equals the boiling point of the least volatile component.

Boiling point is an indication of the volatility of a substance. It is particularly important in EPA's assessment of PMN substances, because it can be used to estimate vapor pressure, a vital property in estimating exposure (see section on vapor pressure). Boiling points are easily measured; EPA's Toxic Substances Control Act Test Guidelines (USEPA 1985) describe five methods for measuring boiling points. These methods include: (1) determination by use of an ebulliometer, in which the substance is heated under equilibrium conditions at atmospheric pressure until it boils; (2) the dynamic method, in which the vapor recondensation temperature is measured by means of a thermocouple; (3) the distillation method, in which the liquid is distilled and the vapor recondensation temperature is measured; (4) the Siwolloboff method, which involves heating the sample in a heat bath and measuring the temperature at which bubbles escape through a capillary tube; and (5) the photocell method, in which a photocell is used with the Siwolloboff method to detect rising bubbles in the capillary tube. Boiling point should always be measured using a pure sample of the substance and should never be measured from a mixture or a solution containing the substance.

The boiling points of members of a homologous series of substances generally increase in a uniform manner with increasing molecular weight. Therefore, the boiling point of a substance may be estimated using its molecular weight, if boiling points for homologous substances are available. Boiling points measured or estimated at reduced pressure can be used to estimate boiling points at one atmosphere (760 mm Hg).

Lyman et al. (1982) discuss seven different methods for estimating boiling point. At the time of this writing, no other methods have been reported since. All of the methods discussed by Lyman are capable of estimating boiling point from structure alone. Each method has its own advantages and disadvantages with respect to applicability and, therefore, is typically used only for a particular class of substances. EPA chemists often use these methods to estimate boiling point when an experimental value is not included in PMN submissions and is not found in the literature. EPA chemists frequently have difficulty determining which method is the most appropriate for a chemical that has multiple functional groups and falls into several different chemical categories. In such cases, EPA usually selects the estimation method that results in the lowest boiling point, consequently maximizing exposure to the PMN substance. As with estimating water solubility, boiling points of liquid substances are easier to estimate than boiling points of solids, since the latter include intermolecular, intracrystalline forces (such as crystal packing) that are very difficult to estimate (see section on water solubility).

Experimental boiling points are known for many chemicals and are easily measured. PMN submitters, therefore, should be able to provide boiling point data for many new chemical submissions, provided that the substance does not decompose rather than melt or boil. It is not necessary, however, for PMN submitters to provide EPA with measured boiling point data for every PMN substance. EPA is concerned primarily with chemicals that melt below 100°C, since these substances are most likely to volatilize readily. High melting solids (> 150°C) typically have very high boiling points and, therefore, do not volatilize significantly. Polymers and other structurally large substances (solid or liquid) usually have low volatilities because of their high molecular weights, and often decompose upon heating. Salts also have low volatilities because of their strong ionic forces and very high melting points. Therefore, it is not necessary (or it may not be possible) for a PMN submitter to provide EPA with boiling point data for substances that have high molecular weights or very high melting points.

2.2.7 Vapor Pressure

Vapor pressure is the pressure at which a liquid substance and its vapor are in equilibrium at a given temperature. At this equilibrium, the rate of condensation of the vapor (conversion of gaseous substance to liquid) equals the rate of vaporization of the liquid (conversion of liquid substance to vapor); the vapor phase in this equilibrium is saturated with the substance of interest. Vapor pressure is characteristic of a substance at a given temperature, and is usually expressed in units of millimeters of mercury (mm Hg, or torr), atmospheres (atm), or Pascals (Pa); EPA prefers mm Hg or torr.

Because vapor pressure is an indication of the volatility of a substance, it can be used to estimate the rate of evaporation of that substance and is very important in the exposure assessment of chemicals. EPA uses the vapor pressure and molecular weight of PMN substances to estimate their concentrations in air and assess occupational exposure and potential environmental releases. Vapor pressure is also used in assessing potential exposure to consumers from products that contain the PMN substance. In the exposure evaluation of PMN chemicals, EPA is particularly concerned with substances that have vapor pressures greater than 10^{-3} mm Hg.

Vapor pressure is also an important property in the assessment of environmental fate and transport of a chemical substance. Volatilization is an important source of material for airborne transport and may lead to the distribution of a chemical over wide areas and into bodies of water far from the site of release (USEPA 1985). Chemicals with relatively low vapor pressure, high soil adsorptivity, or high solubility in water are less likely to vaporize and become airborne than chemicals with high vapor pressure, low water solubility, or low soil adsorptivity. Chemicals that do become airborne are unlikely: (1) to be transported in water; (2) to persist in water and soil; or (3) to biodegrade or hydrolyze. Such chemicals may undergo atmospheric oxidation and photolysis. Non-volatile chemicals, however, are of greater concern for accumulation in soil and water (USEPA 1985).

Several experimental procedures are available for measuring vapor pressure; two are described in EPA's Toxic Substances Control Act Test Guidelines (USEPA 1985). The first method, the isoteniscope technique, is a standardized procedure applicable to pure liquids with vapor pressures from approximately 0.75 to 750 mm Hg. The second method, the gas saturation procedure, involves a current of inert gas passed through or over the test material and can be used for solids or liquids with vapor pressures ranging from 7.5×10^{-8} to 7.5 mm Hg (USEPA 1985).

Lyman et al. (1982) discuss several methods for estimating vapor pressure. EPA often uses these methods when vapor pressure data for a substance are not included in a PMN and are unavailable from the literature. Theoretically derived equations are used to estimate the vapor pressures of solids, liquids, and gases from measured or estimated normal (760 mm Hg) boiling points or from boiling points obtained at reduced pressure. Vapor pressure data, either estimated or measured, are necessary to estimate other properties such as Henry's Law constant.

EPA encourages PMN submitters to provide vapor pressure data in PMNs whenever possible because of the importance of vapor pressure in determining human exposure and environmental fate. Vapor pressure data should be obtained for the pure PMN chemical and not for a formulation of the substance. A frequent problem in PMNs is that the vapor pressure data submitted were measured for the PMN substance dissolved in a solvent. In such cases, the vapor pressure data represent the solvent, *not* the PMN substance, and are, therefore, useless to EPA. If measured vapor pressure data are not supplied, then measured boiling point data may be used to estimate vapor pressure reliably. If measured boiling points are not available, estimated boiling points may also be used to estimate vapor pressure, but estimated boiling points can decrease accuracy and increase the possibility of error. As with other physicochemical properties, if EPA is uncertain about its estimated vapor pressure, it will most likely use a value that reflects a worst-case scenario, leading to greater exposure.

As with boiling point, PMN submitters do not necessarily need to provide EPA with measured vapor pressure data for every PMN substance. EPA is concerned primarily with chemicals that are liquids or gases at room temperature or solids that melt below 100°C, since these substances are most likely to volatilize readily, which can result in significant exposure during manufacture or use. High melting solids ($> 150°C$) are expected to have very high boiling points (and very low vapor pressures) and, therefore, are not expected to volatilize significantly. Polymers or other high molecular weight substances (solid or liquid) typically have low volatility because of their large size. PMN submitters do not need to provide EPA with vapor pressure data for such substances.

2.2.8 Reactivity

The reactivity of chemical substances within biological and environmental systems is crucial to EPA's risk assessment of PMN substances. Toxicity is often the result of a chemical's ability to interfere with normal biochemical processes at the cellular level. Many biochemical processes are enzyme-mediated reactions involving various organic molecules used to produce other organic molecules for a specific function that is vital to the organism. The mechanisms for these enzyme-mediated reactions are fundamentally identical to reaction mechanisms of organic chemistry. Biochemical reactions may involve, for example, nucleophilic attack, electrophilic substitution, loss of electrons (oxidation), gain of electrons (reduction), or hydrolysis.

A knowledge of organic reaction mechanisms is necessary in understanding how a xenobiotic (a chemical that is not part of a biological system or process) will behave or react with molecules that are part of a biochemical pathway. EPA chemists and toxicologists examine every PMN substance to ascertain how these substances may react following absorption into the human body.

For example, PMN substances that contain electrophilic substituents, such as acid chlorides, isocyanates, anhydrides, or α,β-unsaturated carbonyls (acrylates, acrylamides, quinones), may undergo nucleophilic attack by free amino (NH_2) groups present in proteins, thus perturbing the biochemical pathway. In fact, substances containing these functional groups are often quite toxic because of their susceptibility to nucleophilic attack by biological molecules (Anders 1985; De Matteis and Lock 1987; Gregus and Klaassen 1996). EPA does not automatically assume, however, that a PMN substance is toxic just because it contains a reactive functional group. Physicochemical properties must also be considered to assess exposure and bioavailability. Poor water solubility, for example, may mitigate EPA's concerns for the toxicity of a PMN substance containing a reactive functional group, because substances with poor water solubility are expected to be poorly absorbed. This example further illustrates the importance of physicochemical properties in EPA's risk assessment of PMN substances.

EPA chemists and toxicologists consider potential reactivity in predicting the toxicity of PMN substances that contain reactive functional groups and for which few or no toxicological and physicochemical property data are provided. However, it is often difficult to predict the reactivity of a functional group, especially if, for example, the group is hindered or otherwise chemically influenced by other substituents contained within the molecule. In such cases, EPA's policy is to assume reactivity, which may lead EPA scientists to predict a health concern. EPA chemists would prefer to have more information from the PMN submitter with respect to the relative reactivity of any functional groups in a PMN substance. EPA does not expect submitters to conduct extensive laboratory experiments investigating the reactivity of functional groups. EPA believes, however, that the opinions of the submitter's in-house chemists, with respect to chemical reactivity, would be very helpful.

2.2.9 Hydrolysis

Substances may also react in the environment to produce other substances with properties different from those of their precursors. A type of reaction of particular interest is hydrolysis, which is the decomposition of a substance upon reaction with water. Hydrolysis is often described using rate constants (the rate of disappearance of the substance) and half-lives (the time required for the concentration of the substance undergoing hydrolysis to be reduced to one-half its initial value). In addition to hydrolysis, reactions with water in the environment can include elimination of a chemical group, isomerization, and acid–base reactions. Hydrolysis is likely to be the most important reaction of organic substances in aqueous environments, although elimination reactions can also be significant (Lyman et al. 1982).

Chemicals released into the environment are likely to come into contact with water following direct release into surface water, soil, or the atmosphere. It is important to know whether a substance will hydrolyze, at what rate, and under what conditions. If a substance hydrolyzes rapidly, then the hydrolysis products may be more important than the original substance in assessing environmental fate and effects. For a substance that hydrolyzes slowly, however, both the parent substance and the hydrolysis products should be assessed.

Certain chemical groups (e.g., haloformates, acid halides, small alkoxy-silanes, epoxides) are very susceptible to hydrolysis, while others hydrolyze more slowly (e.g., alkyl halides, amides, esters). Water solubility can be a limiting factor in hydrolysis. Generally, the more soluble a substance is, the faster it will hydrolyze. Substances with very low water solubility that contain hydrolyzable substituents may hydrolyze very slowly, if at all. Half-lives (the time required for the concentration of the chemical to be reduced to half its initial value) for the hydrolysis of even reasonably similar chemicals can vary widely, from seconds to years, depending primarily on water solubility, but also on pH and temperature.

EPA's Toxic Substances Control Act Test Guidelines (USEPA 1985) describe a procedure for determining hydrolysis rate constants and half-lives at several pH levels. The method involves preparing solutions of a substance of known concentrations and then determining the changes in concentrations of these solutions at various time intervals. This method is also applicable to elimination reactions. The rate constants generated by this method can be used to determine the hydrolysis rates at any pH of environmental concern.

In the absence of experimental data, EPA makes qualitative and semi-quantitative estimates of hydrolysis rates based upon chemical structure, physicochemical properties, and comparison to similar substances with known rates of hydrolysis (Mabey and Mill 1978; USEPA 1986, 1987, 1988a, 1988b). This estimation approach is most reliable when measured physicochemical properties (particularly water solubility) for the substance of interest are available, as well as measured hydrolysis rate constants for analogous substances. Physicochemical properties for the substance and rate constants for analogous substances, however, are not always known. In such cases, EPA bases hydrolysis estimates on chemical structure and estimated physicochemical properties. In the face of uncertainty, EPA will rely on conservative assumptions (e.g., EPA will assume a slower hydrolysis if EPA has environmental concerns for the intact chemical; if EPA has concerns for the hydrolysis products, EPA will assume a faster rate of hydrolysis). EPA does not expect PMN submitters to provide measured hydrolysis data routinely along with their PMN submissions. However, providing EPA with any qualitative or quantitative information pertaining to hydrolysis would be very helpful. This information would make it possible for the EPA to make more accurate risk assessments and to avoid the use of credible worst-case assumptions.

2.2.10 Spectral Data

Many PMN submitters include spectral data in their submissions, which EPA finds helpful in verifying the identity of PMN substances. Spectral data are also helpful in identifying the presence of unreacted functional groups (e.g., isocyanate) and unknown, possibly toxic byproducts (e.g., dioxins, PCBs), especially if EPA suspects that such chemical species may be present. If EPA chemists suspect that unreacted functional groups or toxic byproducts may be present, given the synthesis of a PMN chemical, but no spectral data are provided, then their presence may be assumed by EPA. In actuality, EPA chemists often use spectral data provided by PMN submitters to rule out (rather than confirm) the presence of toxic byproducts or unreacted functional groups.

The spectral data that EPA finds most useful include mass spectra (MS), infrared (IR), hydrogen (^{1}H) and carbon (^{13}C) nuclear magnetic resonance (NMR), and ultraviolet (UV). Each of these spectral techniques provides unique information and collectively this information is extremely useful for structure elucidation (Pavia et al. 1979; Silverstein et al. 1981).

Ideally, EPA would like to have spectral data on a purified sample of the PMN substance; however, spectral data on a less pure commercial grade product are also helpful. It is not necessary for PMN submitters to provide spectral data for polymers (other than the data obtained from spectral techniques used to determine molecular weight) that were synthesized from monomer species with no reactive functional groups other than those necessary for the polymerization reaction.

2.2.11 Photolysis (Direct/Indirect)

Many chemicals released into the atmosphere or surface water undergo chemical transformation through absorption of sunlight. Photolysis is the decomposition of a substance as a result of absorbing one or more quanta of sunlight radiation; it can take place in water or in air. Rate constants (measurement of the rate of disappearance of the substance) and half-lives (the time required for the concentration of the substance undergoing photolysis to be reduced to one-half its initial value) provide information on photochemical transformation in water and the atmosphere. In direct photolysis, a substance absorbs solar radiation and undergoes a photochemical reaction. In indirect photolysis, one substance absorbs sunlight, then transfers the energy to another substance, thus initiating a chemical reaction. Absorption of light in photochemical reactions (direct and indirect) can result in intramolecular rearrangements, isomerization, homolytic and heterolytic cleavages, redox reactions, energy-transfer reactions, and reactions with water.

Photochemical processes in the atmosphere can produce reactive atoms and free radicals such as the hydroxyl radical ($\cdot$OH). Chemicals that do not absorb sunlight (i.e., do not undergo direct photolysis) may undergo indirect photolysis in the atmosphere by reacting with hydroxyl radicals or with ozone (Finlayson-Pitts and Pitts 1986). The oxygen present in water may participate in direct or indirect photochemical reactions as an acceptor of energy or electrons. Decaying vegetation in water may also absorb sunlight; energy is then typically transferred to another substance, thus initiating an indirect photochemical reaction (Leifer 1988).

Photochemical reactions in the atmosphere and water are important examples of chemical transformations that should be considered when assessing the environmental fate of chemical substances. The products of photochemical reactions and their resulting effects on human health and the environment are also important considerations in chemical evaluations.

Like K_{oc}, EPA has never received a PMN submission that included photolysis rate constants. EPA estimates photolysis rate constants, however, for essentially every PMN submitted. For practical reasons, the Agency does not expect PMN submitters to provide measured photolysis data in their PMN submissions, although it would be helpful to EPA if PMN submitters at least provided UV absorption data. UV data can be used by EPA to determine if a substance will undergo direct photolysis and, if it does, the data will then be used to estimate the relative rates of the direct photolysis of the substance (USEPA 1985).

EPA, in its Toxic Substances Control Act Test Guidelines (USEPA 1985), describes test methods for determining molar absorptivity and reaction quantum yield (the fraction of absorbed light that results in a photoreaction at a fixed wavelength) for direct photolysis of a substance in an aqueous solution. The Guidelines also discuss methods for determining the rate constant and half-life of a substance in an aqueous solution or in the atmosphere, as a function of latitude and season of the year in the United States.

Photolysis of chemicals in the atmosphere and water can be estimated by various methods. Computer programs are available that calculate rate constants and half-lives for reactions with hydroxyl radicals and ozone in the atmosphere (e.g., the EPI program described in Section 2.4.4 of this chapter). Lyman et al. (1982) describe several methods for estimating atmospheric residence time, which is related to half-life. Qualitative estimates of photolysis can be made based on the types of compounds that may be subjected to photolysis and the types of reactions they may undergo. Certain types of chemical groups are known to absorb light and undergo photolysis; therefore, the rate constant and half-life for a particular substance may be estimated qualitatively by analogy to known data on other compounds with similar structures.

2.2.12   Other Chemical Information

Use (Intended Use/Other Uses/Potential Uses). Information on the intended use(s) of a PMN substance and the percent of total production estimated for each use, both provided by the submitter, are important to EPA's review of the substance. EPA uses this information to trace a PMN chemical's life cycle and to estimate health and environmental exposures to the chemical. Use and disposal information also reveals which release scenarios are likely to be the most significant with regard to exposure to a substance, and could determine which physicochemical properties are most important during the review of the substance. In addition to evaluating the occupational exposure of workers to a chemical during its manufacture, EPA considers potential consumer exposure if the chemical is to be used in a commercial product. A substance with consumer use(s), for example, will most likely lead to a significantly greater number of exposures than a chemical with only industrial uses.

In addition to the listing of intended uses provided by the submitter, EPA identifies and evaluates other possible or potential uses of the chemical by searching the literature and EPA's in-house database of PMN submissions for structurally-analogous substances, particularly those that pose a potential risk to human health or the environment. The identification of other uses is important because anyone may market or use a PMN substance for any purpose once the substance is on the TSCA Inventory (unless the substance is restricted by a 5(e) consent order or a SNUR). If a substance is used for an entirely different purpose than originally stated in a PMN submission, then production volume, environmental releases, and human exposures could be significantly different than those estimated from the initial PMN. A new use for a substance, therefore, could pose a threat to human health and the environment. The potential for other uses, especially those involving high exposure or release, leads EPA to restrict the future uses of some PMN substances through SNURs. The manufacturer of a chemical may not always be aware of other potential uses for a substance or may not be planning to pursue other uses because of the substance's marketability or the company's interests. It would be helpful to EPA, however, if submitters would provide known potential uses of a substance even if they are not planning to pursue them.

Synthesis. EPA requests information on the synthesis of PMN substances, including data on feedstocks, solvents, catalysts, other reagents used in the synthetic process, and byproducts (chemicals produced in the synthetic process without a separate commercial intent). This information is supplemented by process and operation descriptions and is utilized during several stages of EPA's evaluation of PMN substances.

Information on the synthesis of a chemical is important in several ways. Review of the synthetic process helps EPA to verify the identity of the PMN substance. From a review of reaction conditions, EPA may also be able to predict the existence of impurities and by-products, including toxic reaction products (e.g., PCBs, dioxins or nitrosamines), that are unknown to the submitter because, for example, such substances may be present only in very small concentrations.

EPA scientists also review the synthetic processes for selected, potentially higher-risk PMN substances with respect to pollution prevention. EPA investigates whether any modifications could feasibly be implemented in this synthesis that would limit or avert the use of hazardous substances (including solvents and all reactants) or that would reduce or prevent the production, not just of hazardous waste, but of all waste. In a few cases, EPA scientists may also identify alternative synthetic sequences that would at least reduce the production of toxic byproducts or the use of high-risk solvents and feedstocks.

Submitters may demonstrate to EPA on the Optional Pollution Prevention page of the PMN [page 151 of Appendix 6.10] any pollution prevention strategies that they plan to implement. Some companies provide detailed descriptions of synthetic pathways that incorporate pollution prevention (e.g., processes that give high yields and use few or no organic solvents). EPA would like to see more companies do the same. For PMN submissions that do not contain synthetic information (synthetic data are not required for imported substances), pollution prevention information voluntarily supplied by submitters can assist EPA in its review of the PMN substance. For example, if a synthetic scheme is not given for a PMN substance, EPA may be concerned about the possible existence of toxic byproducts and impurities, based on information known about the synthetic scheme of similar substances. If the submitter, however, includes pollution prevention information explaining how their synthesis has improved upon known methods, then EPA would not need to assume a worst-case scenario.

Purity/Impurities. The purity of a PMN substance, as well as the identities, concentrations, and hazards of all impurities are considered in the evaluation of every PMN substance. During review, EPA investigates whether any reported physicochemical properties submitted for a PMN substance (especially melting point and boiling point) coincide with any data previously recorded in the literature. Discrepancies between literature values and the data contained in the PMN submission may be attributable to impurities. EPA will contact the submitter if it is not clear in the PMN what the identities of impurities are, especially if impurities are predicted from EPA's analysis of the synthetic process. The presence of hazardous impurities (such as dioxins, PCBs, or nitrosamines) is cause for concern and, if present at significant levels, such impurities would lead EPA to predict potential risk to human health and the environment, especially if the PMN substance is intended for consumer use.

Molecular Weight. The molecular weight of a substance is the sum of the atomic weights of all the atoms in a molecule. For a simple molecule, the molecular weight is easily determined if the structure is known. Polymers, however, are typically comprised of a variable number and sequence of monomer units that may themselves also have varying chain length and molecular weight. The molecular weight of a polymer is frequently reported as a number-average weight (the sum of the molecular weights of the molecules divided by the number of molecules).

Very large molecules are unlikely to be absorbed and, therefore, may be of little concern to EPA unless, of course, they contain reactive functional groups. EPA consequently exempts under TSCA section 5(h)(4) certain polymers (those with number-average molecular weights greater than 1,000 and certain polyesters, for example) from some of the PMN requirements. EPA does have concerns, however, for certain polymers with average molecular weights of 10,000 daltons or greater. These concerns are largely for lung toxicity (USEPA 1995).

2.3 Use of Chemical Information in Assessment of PMN Chemicals

Each physicochemical property discussed in this chapter is important in EPA's evaluation of the potential risks posed to human health or the environment by PMN substances. Refer back to Figure 2-1, which illustrates some of the physicochemical data used, their interrelationships, and their importance in risk assessment. Because of the large volume of data that EPA uses in its evaluation of PMN substances, Figure 2-1 does not attempt to include all of the types of chemical information used or to describe all of their functions in risk assessment.

2.4 How EPA Obtains Physicochemical Information

2.4.1 General Approach

When physicochemical property data required for chemical evaluation are not reported in a PMN submission, EPA finds or estimates values for the missing data. EPA's general approach for obtaining physicochemical property data is first to search for data on the PMN substance by following a sequence of literature and database sources. If data on the PMN substance cannot be found, EPA scientists may identify close structural analogs and use the same search strategies to find property data for those analogs. EPA scientists then use professional judgment to extrapolate property values for the PMN substance from the data available for the analogs. If the required properties for structural analogs cannot be found, EPA scientists estimate the properties needed for the PMN substance using the best estimation method available to EPA (Lynch et al. 1991). If properties for structural analogs are found, EPA scientists may still estimate the same properties for the PMN substance. EPA scientists then analyze and compare both sets of data to determine which set is most reasonable. A flowchart illustrating EPA's procedure for obtaining physicochemical properties is presented in Figures 2-3 and 2-4. The sources EPA uses for searches and the programs used for estimating property values are discussed below. Additional information on the on-line databases, reference books (e.g., Verscheuren 1983), and computer programs EPA uses to obtain property data is provided below.

2.4.2 Methods of Searching for Measured Physicochemical Properties

CAS On-Line Search. The American Chemical Society's Chemical Abstracts Service (CAS) On-Line Database includes several files that can be searched for chemical information. EPA first conducts a CAS On-Line search on the Registry File by CAS Registry Number (CAS RN), chemical name, or molecular formula. The easiest search to perform uses the CAS RN, if it is available.

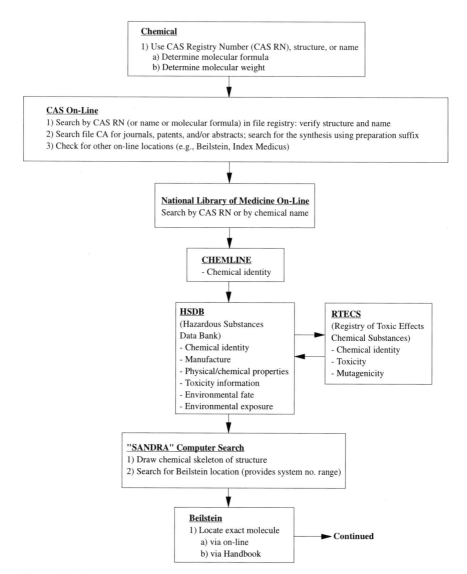

Figure 2-3 *Methods for Obtaining Measured Physicochemical Property Values on Exact Structures*

If EPA does not have a CAS RN for the PMN substance, then an accurate chemical name or molecular formula is used for searching.

Linking a molecular formula in a search with a chemical name or name fragments can also be useful for finding the exact substance or a closely related analog. The CAS Registry File provides, among other information, the most recent CAS Registry chemical name, molecular formula, the chemical struc-

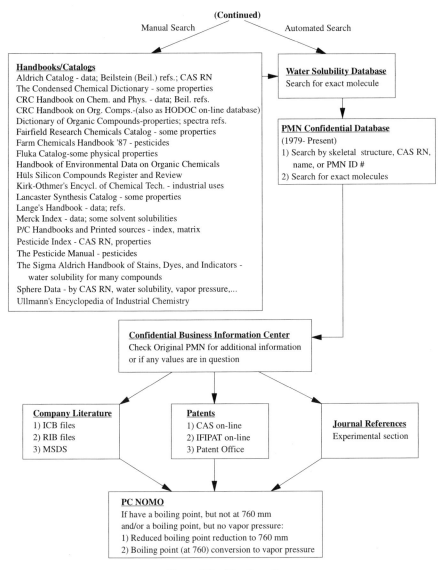

Figure 2-3 (*Continued*)

ture, other on-line sources where the substance may be found (e.g., Beilstein On-Line, discussed below), and abstracts of the literature references to that substance. This information can be used to verify any name and structural information already provided.

Information on the synthesis of a substance can be obtained by searching the Chemical Abstracts file using the CAS RN. This file provides references (usually scientific journal citations or patents) and may contain physicochemical

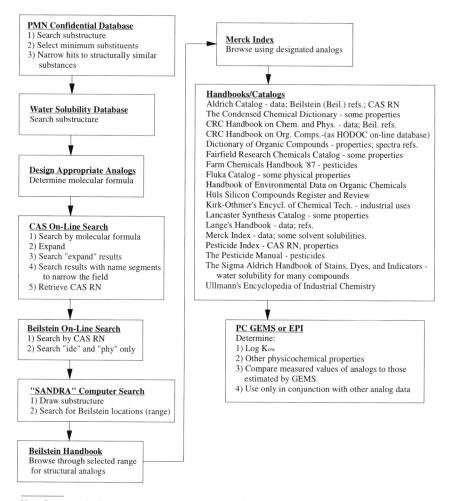

PMN Confidential Database
1) Search substructure
2) Select minimum substituents
3) Narrow hits to structurally similar
   substances

Merck Index
Browse using designated analogs

Handbooks/Catalogs
Aldrich Catalog - data; Beilstein (Beil.) refs.; CAS RN
The Condensed Chemical Dictionary - some properties
CRC Handbook on Chem. and Phys. - data; Beil. refs.
CRC Handbook on Org. Comps.-(as HODOC on-line database)
Dictionary of Organic Compounds - properties; spectra refs.
Fairfield Research Chemicals Catalog - some properties
Farm Chemicals Handbook '87 - pesticides
Fluka Catalog - some physical properties
Handbook of Environmental Data on Organic Chemicals
Hüls Silicon Compounds Register and Review
Kirk-Othmer's Encycl. of Chemical Tech. - industrial uses
Lancaster Synthesis Catalog - some properties
Lange's Handbook - data; refs.
Merck Index - data; some solvent solubilities.
Pesticide Index - CAS RN, properties
The Pesticide Manual - pesticides
The Sigma Aldrich Handbook of Stains, Dyes, and Indicators -
   water solubility for many compounds
Ullmann's Encyclopedia of Industrial Chemistry

Water Solubility Database
Search substructure

Design Appropriate Analogs
Determine molecular formula

CAS On-Line Search
1) Search by molecular formula
2) Expand
3) Search "expand" results
4) Search results with name segments
   to narrow the field
5) Retrieve CAS RN

Beilstein On-Line Search
1) Search by CAS RN
2) Search "ide" and "phy" only

PC GEMS or EPI
Determine:
1) Log K_{ow}
2) Other physicochemical properties
3) Compare measured values of analogs to those
   estimated by GEMS
4) Use only in conjunction with other analog data

"SANDRA" Computer Search
1) Draw substructure
2) Search for Beilstein locations (range)

Beilstein Handbook
Browse through selected range
for structural analogs

Note: Once an analog has been found, further data can be searched using Figure 2-3.

Figure 2-4 *Methods for Identifying Analogs of PMN Substances and Their Physicochemical Properties*

property data (in the experimental sections of scientific papers) or potential uses.

GMELIN On-Line Database. For information on organometallic or inorganic compounds, EPA searches the Gmelin on-line database which contains the critically reviewed and evaluated data from the *Gmelin Handbook of Inorganic and Organometallic Chemistry*. Useful information includes structural data, structural images, chemical and physical properties, and bibliographic data.

National Library of Medicine (NLM) On-Line Databases. This inexpensive on-line system contains individual databases that include information on chemical identification, physicochemical properties, manufacturing processes, and uses. These databases are, therefore, useful for obtaining a variety of information on many chemicals or on analogous substances. NLM databases include the Hazardous Substance Data Bank (HSDB), the Registry of Toxic Effects of Chemical Substances (RTECS), and Chemline.

HSDB entries contain information and data on chemical identity (name, CAS RN, synonyms, molecular formula), methods of manufacture (including impurities and formulations), manufacturers, major uses, and chemical and physical properties (such as color, physical state, odor, boiling point, melting point, molecular weight, density, dissociation constant, heat of combustion, heat of vaporization, octanol/water partition coefficient, pH, solubility, spectral properties, surface tension, vapor density, and vapor pressure). Toxicity, environmental fate, and exposure data may also be provided.

RTECS is primarily a database of toxicological data and references, including information on acute and chronic toxicity, mutagenesis, and skin and eye irritation. The database also includes chemical identity information such as chemical name, CAS RN, synonyms, molecular formula, and molecular weight.

Chemline is an interactive chemical dictionary file containing approximately one million chemical substance records. The data elements consist of CAS RN, molecular formula, synonyms, ring information, and a locator to other on-line databases that might contain further information on a compound.

Beilstein On-Line Database. The Beilstein On-Line Database is an on-line version of the Beilstein Handbook of Organic Chemistry (see below), an extensive compilation of information on organic compounds comprised of a multi-volume Home Register and five supplements. Information includes synthetic methods, measured physicochemical properties, and references. If the CAS On-Line search (described above) identifies a compound as listed in Beilstein, then a Beilstein On-Line search can be performed to provide physical data quickly, particularly if a CAS RN is known. Specific data can be selected for retrieval. References for the data are provided, but Beilstein Handbook citations are not included.

SANDRA Computer Search. SANDRA is a computer program that provides information on the general location of where a substance might be found in the Beilstein Handbook, and therefore, enables rapid searching of the handbook. If a CAS On-Line search of a substance does not list the chemical as being available from Beilstein On-Line, one can use SANDRA to draw the structure of the substance, of an analog, or of a fragment of either, and then one can search to locate the range of the structure (system number, home register page(s), and supplement volumes) within the Beilstein Handbook.

Beilstein Handbook. The Beilstein Handbook (see the discussion of Beilstein On-Line and SANDRA, above) can be searched manually using the molecular formula indexes. EPA typically uses SANDRA, as described above, to expedite the search. Physicochemical properties most commonly found in Beilstein are melting point, boiling point, density, and refractive index. Other data such as vapor pressure or water solubility are less commonly reported.

Other Handbooks/Catalogs. EPA also may search various handbooks and commercial chemical catalogs for data on PMN chemicals, although these sources are most useful if the substance in question is relatively simple or if a close structural analog is commercially marketed. Handbooks and catalogs EPA uses include the Aldrich Chemical Company Catalog Handbook of Fine Chemicals, the Merck Index, Hüls Silicon Compounds Register and Review, and the Farm Chemicals Handbook (includes data on pesticide intermediates).

Confidential PMN Database. EPA has an in-house confidential PMN database that contains chemical structures and data from chemistry reports from over 8,000 PMNs submitted since January 1993. Most entries provide physicochemical properties that were either measured by the submitter or estimated by EPA chemists. All information in this database is regarded and treated as confidential business information (CBI), and only EPA personnel with TSCA CBI clearance have access to it.

Water Solubility Database. EPA has developed a water solubility database file that can be searched by structure. At present, this database contains over 6,000 substances with measured water solubility values (expressed as grams per liter at measured temperatures) and contains other measured physical properties for some of these substances as well. It currently contains data from the Arizona database (also known as the AQUASOL DATABASE, see Yalkowsky and Banerjee 1992), the PHYSPROP® database (available from Syracuse Research Corporation, Syracuse, NY), the Merck Index, Beilstein, and other pertinent literature and journal articles. All information is referenced within this database.

Patents. EPA periodically searches for patents that may have useful physicochemical property data, manufacturing information, and use information. The IFIPAT (IFI Patent Database) file in the STN computer network system contains records for granted U.S. chemical and chemically-related patents from 1950 to the present. Patents on some other subjects are also included. Hard copies of U.S. patents can be obtained from the Public Search Room at the U.S. Patent Office in Arlington, Virginia. The location of a patent within the Public Search Room can be found from the classification number (determined from the U.S. Patent number, which can be obtained from a CAS On-Line search).

Scientific Literature. EPA often uses articles published in scientific journals to obtain information on synthetic methods as well as physicochemical and spectral properties.

2.4.3 Methods For Estimating Physicochemical Properties From Structural Analogs

When measured physicochemical property data are unavailable for a specific PMN chemical, EPA attempts to obtain the needed data by extrapolating from measured data available for close structural analogs. EPA searches the same information sources for analogs as for specific chemicals, but the search strategy differs in that compounds that are structurally and functionally similar to the substance under consideration must either be "designed" or found using handbooks and databases.

Confidential PMN Database. EPA's confidential PMN database is searched using a skeletal drawing of the PMN substance, if the structure is not too novel or complex. More often, a fragment that contains the important structural features of the PMN substance is used in the search. The PMN database has evolved to contain numerous classes of chemicals that are structurally very similar, and all entries found that possess the same basic structural and functional features as the PMN substance can be identified and reviewed for useful information.

Designing Structural Analogs. One effective method that EPA uses for searching the enormous expanse of chemicals in the literature is to design appropriate structural analogs that may have been previously reported. By changing functional groups, alkyl chain lengths, ring sizes, or other features in a step-wise fashion, close structural analogs can be created and prioritized for searching. The molecular formula, as well as a chemical name are then determined for each analog. EPA searches CAS On-Line for these analogs, as described below, to determine whether they actually exist and, if they do, whether physicochemical property data are available.

CAS On-Line Search. Searching CAS On-Line for an analog designed for a PMN substance can be accomplished most readily by simply entering the analog's molecular formula. If a relatively small number of entries are obtained from the search, then all are retrieved and reviewed. If a large number of entries are obtained, then the search can be narrowed by using selected name segments. From this narrowed search, any entries that are suitable analogs are retrieved to obtain CAS RNs and to determine if Beilstein data are available. EPA has found that expanding on the molecular formula of pre-designed analogs is successful for finding very close structural analogs.

The Merck Index. EPA periodically uses this comprehensive, interdisciplinary encyclopedia of organic chemicals, pharmaceuticals, and biological substances to scan for new analogs or to search for designed analogs. The Merck Index is an excellent source for obtaining measured physicochemical properties for over 50,000 chemical substances.

2.4.4   Methods For Estimating Physicochemical Properties Using Computer Estimation Programs

If measured property values are unavailable or cannot be found for the PMN substance or for compounds that are structurally analogous to the PMN substance, then EPA tries to estimate the properties using appropriate estimation methods. EPA uses several computerized chemical property estimation programs, including PC-NOMOGRAPH, PC-Graphical Exposure Modeling System (PC-GEMS), Oligo 56, and Estimation Programs Interface (EPI). Values obtained from these estimation programs are scrutinized at CRSS meetings (see chapter 1) by EPA chemists, who exercise professional judgment to determine whether the values are reasonable. Some of the computer estimation programs used by EPA are discussed briefly below.

PC-NOMOGRAPH. This computer program calculates a normal boiling point (boiling point at one atmosphere pressure, 760 torr) from either a measured or estimated boiling point obtained at reduced pressure. The vapor pressure at 25°C also can be calculated from a normal or reduced boiling point. Actual boiling point-pressure nomographs (pressure-temperature alignment charts) can also be used in boiling point estimations by helping to verify the computer calculations. These charts allow the conversion of a reduced pressure boiling point to a boiling point at one atmosphere. Separate vapor pressure nomographs are available for low-boiling and high-boiling compounds.

PC-GEMS. The estimation routines in PC-GEMS represent a computerized version of well-known methods from the Handbook of Chemical Property Estimation Methods (Lyman et al. 1982). Estimation routines are available for the octanol/water partition coefficient, water solubility, soil adsorption coefficient, boiling point, vapor pressure, melting point, and Henry's Law constant.

EPI. EPI, developed by Syracuse Research Corporation, Syracuse, New York, integrates several computer programs. Programs are included for estimating: octanol/water partition coefficient; Henry's Law constant; soil adsorption coefficient; rate of hydrolysis (for substances with a hydrolyzable group); atmospheric oxidation (including half-lives for reaction with hydroxyl radicals and ozone); probability of biodegradation (based on several different models); and removal during wastewater treatment.

OLIGO 56. Oligo 56, developed by the Mitre Corporation, McLean, Virginia, is used to estimate molecular weight and functional group equivalent weight of polymers.

REFERENCES FOR CHAPTER 2

Abramowitz R, Yalkowsky SH. 1990. Estimation of aqueous solubility and melting point of PCB congeners. Chemosphere 21: 1221–1229.

Anders MW, ed. 1985. Bioactivation of Foreign Compounds. New York: Academic Press.

Bahnick DA, Doucette WJ. 1988. Use of molecular connectivity indices to estimate soil sorption coefficients for organic chemicals. Chemosphere 17: 1703–1715.

Banerjee S, Yalkowsky SH, Valvani SC. 1980. Water solubility and octanol/water partition coefficients of organics. Limitations of the solubility-partition coefficient correlation. Environ Sci Technol 14: 1227–1229.

Blum DJ, Speece RE. 1990. Determining chemical toxicity to aquatic species. Environ Sci Technol 24: 284–293.

Bowman BT, Sans WW. 1983. Determination of octanol/water partitioning coefficients (K_{oc}) of 61 organophosphorus and carbamate insecticides and their relationship to respective water solubility (S) values. J Environ Sci Health B18: 667–683.

Brent DA, Sabatka JJ, Minick DJ, Henry DW. 1983. A simplified high-pressure liquid chromatography method for determining lipophilicity for structure-activity relationships. J Med Chem 26: 1014–1020.

Brooke DN, Dobbs AJ, Williams N. 1986. Octanol/water partition coefficients (P): Measurement, estimation, and interpretation, particularly for chemicals with $P > 10^5$. Ecotoxicol Environ Safety 11: 251–260.

Brooke DN, Dobbs AJ, Williams N. 1987. Correlation of octanol/water partition coefficients and total molecular surface area for highly hydrophobic aromatic compounds. Environ Sci Technol 21: 821–824.

Bruggemann R, Munzer B. 1988. Physico-chemical estimation for environmental chemicals. In: Jochum C, Hicks MG, Sunkel J (eds.). Physical Property Prediction in Organic Chemistry, Proceedings of the Beilstein Workshop. Schloss Korb, Italy, May 16–20, 1988. Sponsor: BMFT.

Chiou CT, Porter, PE, Schmedding, DW. 1983. Partition equilibria of nonionic organic compounds between soil organic matter and water. Environ Sci Technol 17: 227–231.

CLOGP computer program; available through Pomona College Medicinal Chemistry Project, Claremont, CA 91711.

Da YZ, Ito K, Fujiwara H. 1992. Energy aspects of oil/water partition leading to the novel hydrophobic parameters for the analysis of quantitative structure-activity relationships. J Med Chem 35: 3382–3387.

de Bruijn J, Hermens J. 1990. Relationships between octanol/water partition coefficients and total molecular surface area and total molecular volume of hydrophobic organic chemicals. Quant Struct-Act Relat 9: 11–21.

de Bruijn J, Busser F, Seinen W, Hermens J. 1989. Determination of octanol/water partition coefficients for hydrophobic organic chemicals with the "slow-stirring" method. Environ Toxicol Chem 8: 499–512.

Dearden JC. 1991. The QSAR prediction of melting point, a property of environmental relevance. In: Hermens JLM, Opperhuizen A (eds.). QSAR in Environmental Toxicology-IV. New York: Elsevier.

De Matteis F, Lock EA, eds. 1987. Selectivity and Molecular Mechanisms of Toxicity. New York: Macmillan.

Doucette WJ, Andren AW. 1987. Correlation of octanol/water partition coefficients and total molecular surface area for highly hydrophobic aromatic compounds. Environ Sci Technol 21: 821–824.

Doucette WJ, Andren AW. 1988. Estimation of octanol/water partition coefficients: Evaluation of six methods for highly hydrophobic aromatic hydrocarbons. Chemosphere 17 (2): 345–359.

Dunn WJ, Block JH, Pearlman RS, eds. 1986. Partition Coefficient, Determination and Estimation. New York: Pergamon Press.

Dunn WJ, Nagy PI, Collantes ER. 1991. A computer-assisted method for estimation of the partition coefficient. Monte Carlo simulations of the chloroform/water log P for methylamine, methanol, and acetonitrile. J Am Chem Soc 113: 7898–7902.

Finlayson-Pitts BJ, Pitts JN. 1986. Atmospheric Chemistry: Fundamentals and Experimental Techniques. New York: John Wiley & Sons.

Garst JE. 1984. Accurate, wide-range, automated, high performance liquid chromatographic method for the estimation of octanol/water partition coefficients II: Equilibrium in partition coefficient measurements, additivity of substituent constants, and correlation of biological data. J Pharm Sci 73: 1623–1629.

Garst JE, Wilson WC. 1984. Accurate, wide-range, automated, high-performance liquid chromatographic method for the estimation of octanol/water partition coefficients I: Effect of chromatographic conditions and procedure variables on accuracy and reproducibility of the method. J Pharm Sci 73: 1616–1622.

Gregus Z, Klaassen CD. 1996. Mechanisms of toxicity. In: Casarett & Doull's Toxicology, The Basic Science of Poisons. Klaassen C (ed.). New York: McGraw Hill, Inc. pp. 35–74.

Grogan J, DeVito SC, Pearlman RS, Korzekwa, KR. 1992. Modeling cyanide release from nitriles: prediction of cytochrome P450 mediated acute nitrile toxicity. Chem Res Toxicol 5: 548–552.

Hansch C, Clayton JM. 1973. Lipophilic character and biological activity of drugs II: The parabolic case. J Pharm Sci 62: 1–20.

Hansch C, Dunn WJ III. 1972. Linear relationships between lipophilic character and biological activity of drugs. J Pharm Sci 61: 1–19.

Hansch C, Leo A. 1979. Substituent Constants for Correlation Analysis in Chemistry and Biology. New York: John Wiley and Sons.

Hermens JLM, Opperhuizen A, eds. 1991. QSAR in Environmental Toxicology-IV. New York: Elsevier. This volume is reprinted from Science of the Total Environment Journal, volume 109/110, 1991.

Hine J, Mookerjee PK. 1975. The intrinsic hydrophilic character of organic compounds. Correlations in terms of structural contributions. J Org Chem 40: 292–298.

Isnard P, Lambert S. 1989. Aqueous solubility and *n*-octanol/water partition coefficients correlations. Chemosphere 18: 1837–1853.

Kamlet MJ, Doherty RM, Carr PW, Mackay D, Abraham MH, Taft RW. 1988. Linear solvation energy relationships. 44. Parameter estimation rules that allow accurate prediction of octanol/water partition coefficients and other solubility and toxicity properties of polychlorinated biphenyls and polycyclic aromatic hydrocarbons. Environ Sci Technol 22: 503–509.

Karcher W, Devillers J, eds. 1990. Practical Applications of Quantitative Structure-Activity Relationships (QSAR) in Environmental Chemistry and Toxicology. Boston: Kluwer, pp. 281–293.

Karickhoff SW, Brown DS, Scott TA. 1979. Sorption of hydrophobic pollutants on natural sediments. Water Res 13: 241–247.

Kenaga EE, Goring CAI. 1980. Relationship between water solubility, soil sorption, octanol-water partitioning, and concentrations of chemicals in biota. In: Aquatic Toxicology, Proceedings of the Third Annual Symposium on Aquatic Toxicology. American Society of Testing Materials (ASTM) STP 707.

Klopman G, Wang S, Balthasar DM. 1992. Estimation of aqueous solubility of organic molecules by the group contribution approach. Application to the study of biodegradation. J Chem Inf Comput Sci 32: 474–482.

Leifer A. 1988. The Kinetics of Environmental Aquatic Photochemistry. Washington, DC: American Chemical Society.

Leo AJ. 1990. Methods of calculating partition coefficients. In: Comprehensive Medicinal Chemistry: The Rationale Design, Mechanistic Study & Therapeutic Application of Chemical Compounds, Volume 4 (Ramsden CA, ed.). New York: Pergamon Press. Chapter 18.7, pp. 295–319.

Leo AJ. 1993. Calculating log *P* from structures. Chem Rev 93: 1289–1306.

Lins CLK, Block JH, Doerge RF, Barnes GJ. 1982. Determination of octanol-water equivalent partition coefficients of indolizine and substituted 2-phenylindolizines by reversed-phase high pressure liquid chromatography and fragmentation values. J Pharm Sci 71: 614–617.

Lu PY, Metcalf RL. 1975. Environmental fate and biodegradability of benzene derivatives as studied in a model aquatic ecosystem. Environ Health Perspect 10: 269–284.

Lyman WJ, Reehl WF, Rosenblatt DH. 1982. Handbook of Chemical Property Estimation Methods. New York: McGraw-Hill.

Lynch DG, Tirado NF, Boethling RS, Huse GR, Thom GC. 1991. Performance of on-line chemical property estimation methods with TSCA premanufacture notice chemicals. Sci Total Environ 109/110: 643–648.

Mackay D, Shiu WY. 1981. A critical review of Henry's Law constants for chemicals of environmental interest. J Phys Chem Ref Data 10: 1175–1199.

Mabey W, Mill T. 1978. Critical review of hydrolysis of organic compounds in water under environmental conditions. J Phys Chem Ref Data 7: 383–415.

Martin YC. 1978. Quantitative Drug Design. New York: Marcel Dekker.

McCall PJ, Laskowski DA, Swann RL, Dishburger HJ. 1983. Estimation of environmental partitioning of organic chemicals in model ecosystems. Residue Rev 85: 231–243.

Means JC, Wood SG, Hassett JJ, Banwart WL. 1982. Sorption of amino- and carboxy-substituted polynuclear aromatic hydrocarbons by sediments and soils. Environ Sci Technol 16: 93–98.

Meylan WM, Howard PH. 1991. Bond contribution method for estimating Henry's law constants. Environ Toxicol Chem 10: 1283–1293.

Meylan WM, Howard PH. 1995. Atom/fragment contribution method for estimating octanol-water partition coefficient. J Pharm Sci 84: 83–92.

Meylan W, Howard PH, Boethling RS. 1992. Molecular topology/fragment contribution method for predicting soil sorption coefficients. Environ Sci Technol 26: 1560–1567.

Moriguchi I, Hirono S, Liu Q, Nakagome I, Matsushita Y. 1992. Simple method of calculating octanol/water partition coefficient. Chem Pharm Bull 40: 127–130.

Minick DJ, Frenz JH, Patrick MA, Brent DA. 1988. A comprehensive method for determining hydrophobicity constants by reversed-phase high performance liquid chromatography. J Med Chem 31: 1923–1933.

Noegrohati S, Hammers WE. 1992. Regression models for octanol-water partition coefficients, and for bioconcentration in fish. Toxicol Environ Chem 34: 155–173.

Pavia DL, Lampman GM, Kriz GS Jr. 1979. Introduction to Spectroscopy: A Guide for Students of Organic Chemistry. Philadelphia: Saunders College.

Sabljic A. 1984. Predictions of the nature and strength of soil sorption of organic pollutants by molecular topology. J Agric Food Chem 32: 243–246.

Sabljic A. 1987. On the prediction of soil sorption coefficients of organic pollutants from molecular structure: Application of molecular topology model. Environ Sci Technol 21: 358–366.

Sasaki Y, Kubodera H, Matuszaki T, Umeyama H. 1991. Prediction of octanol/water partition coefficients using parameters derived from molecular structures. J Pharmacobio-Dyn 14: 207–214.

Shah PV. 1990. Environmental Exposure to Chemicals through Dermal Contact. In: Saxena J (ed.). Hazard Assessment of Chemicals, Volume 7. New York: Hemisphere Publishing Corp., pp. 111–156.

Shiu WY, Doucette W, Gobas FAPC, Andren A, Mackay D. 1988. Physical-chemical properties of chlorinated dibenzo-*p*-dioxins. Environ Sci Technol 22: 651–658.

Shriner RL, Fuson RC, Curtin DY, Morrill TC. 1980. The Systematic Identification of Organic Compounds. New York: John Wiley and Sons, pp. 13, 37–45.

Silverstein RM, Bassler GC, Morrill TC. 1981. Spectrometric Identification of Organic Compounds, 4th ed. New York: John Wiley & Sons.

Swann RL, Laskowski DA, McCall PJ, Vander Kuy K, and Dishburger HJ. 1983. A rapid method for the estimation of the environmental parameters octanol/water partition coefficient, soil sorption constant, water to air ratio, and water solubility. Residue Rev 85: 18–28.

USEPA. 1979. U.S. Environmental Protection Agency. Office of Toxic Substances. Toxic Substances Control Act Premanufacture Testing of New Chemical Substances. (44 FR 16240).

USEPA. 1985. U.S. Environmental Protection Agency. Office of Toxic Substances. Toxic Substances Control Act Test Guidelines. (50 FR 39252).

USEPA. 1986. U.S. Environmental Protection Agency. Measurement of Hydrolysis

Rate Constants for Evaluation of Hazardous Waste and Land Disposal: Volume I. Washington, DC: U.S. Environmental Protection Agency. EPA/600/3-86/043.

USEPA. 1987. U.S. Environmental Protection Agency. Measurement of Hydrolysis Rate Constants for Evaluation of Hazardous Waste and Land Disposal: Volume II. Washington, DC: U.S. Environmental Protection Agency. EPA/600/3-87/019.

USEPA. 1988a. U.S. Environmental Protection Agency. Measurement of Hydrolysis Rate Constants for Evaluation of Hazardous Waste and Land Disposal: Volume III. Washington, DC: U.S. Environmental Protection Agency. EPA/600/3-88/028.

USEPA. 1988b. U.S. Environmental Protection Agency. Interim Protocol for Measuring Hydrolysis Rate Constants in Aqueous Solutions. Washington, DC: U.S. Environmental Protection Agency. EPA/600/3-88/014.

USEPA. 1992. U.S. Environmental Protection Agency. Office of Research and Development. Dermal Exposure Assessment: Principles and Applications. Interim Report. Washington, DC: U.S. Environmental Protection Agency. EPA/600/8-91/011B.

USEPA. 1995. U.S. Environmental Protection Agency. Office of Pollution Prevention and Toxics. Premanufacture Notification; Revisions of Premanufacture Notification Regulations; Final Rule. (60 FR 16298–16352).

USEPA. 1996 (August 28). U.S. Environmental Protection Agency. Office of Pesticides and Toxic Substances. Test guidelines; Notice of Availabilty. (61 FR 44308–44311).

van de Waterbeemd H, Mannhold R. 1996. In: Lipophilicity in Drug Action and Toxicology. Pliska V, Testa B, van de Waterbeemd H, eds. New York: VCH Publishers, Inc. pp. 401–415.

Verscheuren K. 1983. Handbook of Environmental Data on Organic Chemicals, 2nd ed. New York: Van Nostrand Reinhold.

Wakita K, Yoshimoto M, Miyamoto S, and Watanabe H. 1986. A method for calculation of the aqueous solubility of organic compounds by using new fragment solubility constants. Chem Pharm Bull 11: 4663–4681.

Waller CL. 1994. A three-dimensional technique for the calculation of octanol-water partition coefficient. Quant Struct-Act Relat. 13: 172–176.

Yalkowsky SH. 1988. Estimation of the aqueous solubility of organic compounds. In: Jochum C, Hicks MG, Sunkel J (eds.). Physical Property Prediction in Organic Chemistry, Proceedings of the Beilstein Workshop. Schloss Korb, Italy, May 16–20, 1988. Sponsor: BMFT.

Yalkowsky SH, Banerjee S. 1992. Aqueous Solubility: Methods of Estimation for Organic Compounds. New York: Marcel Dekker.

Yalkowsky SH, Valvani SC. 1976. J Med Chem 19: 727–728.

Yalkowsky SH, Valvani SC. 1979. Solubility and partitioning 2. Relationships between aqueous solubilities, partition coefficients, and molecular surface areas of rigid aromatic hydrocarbons. J Chem Eng Data 24: 127–129.

Yalkowsky SH, Valvani SC. 1980. Solubility and partitioning I: Solubility of nonelectrolytes in water. J Pharm Sci 69: 912–922.

Yalkowsky SH, Orr RJ, Valvani SC. 1979. Solubility and partitioning 3. The solubility of halobenzenes in water. Ind Eng Chem Fundam 18: 351–353.

Yalkowsky SH, Sinkula AA, Valvani SC, eds. 1980. Physical Chemical Properties of Drugs. New York: Marcel Dekker.

Yamagami C, Ogura T, Takao N. 1990. Hydrophobicity parameters determined by reversed-phase liquid chromatography. I. Relationship between capacity factors and octanol-water partition coefficients for monosubstituted pyrazines and the related pyridines. J Chromatog 514: 123–136.

LIST OF SELECTED READINGS FOR CHAPTER 2

Use of physicochemical properties in risk assessment: Estimating biological (toxicological) activity

Albert A. 1987. Xenobiosis. Food, Drugs and Poisons in the Human Body. New York: Chapman and Hall.

Albert A. 1985. Selective Toxicity, The Physico-Chemical Basis of Therapy. Seventh Edition. New York: Chapman and Hall.

Amdur MO, Doull J, Klaassen CD, eds. 1991. *Casarett and Doull's* Toxicology, The Basic Science of Poisons. Fourth Edition. New York: Pergamon Press.

Bradbury SJ, Lipnick RL, eds. 1990. Structural Properties for Determining Mechanisms of Toxic Action (papers from an EPA Workshop, October 18–20, 1988, Duluth, MN). Environ Health Perpect 87: 180–271.

Gerrity TR, Henry CJ, eds. 1990. Principles of Route-to-Route Extrapolation for Risk Assessment. New York: Elsevier.

Hermens JLM, Opperhuizen A, eds. 1991. QSAR in Environmental Toxicology-IV. New York: Elsevier.

Hansch C, Clayton JM. 1973. Lipophilic character and biological activity of drugs II: The parabolic case. J Pharm Sci 62: 1–20.

Hansch C, Dunn WJ III. 1972. Linear relationships between lipophilic character and biological activity of drugs. J Pharm Sci 61: 1–19.

Hansch C, Leo A. 1979. Substituent Constants for Correlation Analysis in Chemistry and Biology. New York: John Wiley and Sons.

Karcher W, Devillers J, eds. 1990. Practical Applications of Quantitative Structure-Activity Relationships (QSAR) in Environmental Chemistry and Toxicology. Dordrecht, The Netherlands: Kluwer Academic Publishers.

Kubinyi H. 1979. Lipophilicity and drug activity. Prog Drug Res 23: 97–198.

Lu PY, Metcalf RL. 1975. Environmental fate and biodegradability of benzene derivatives as studied in a model aquatic ecosystem. Environ Health Perspect 10: 269–284.

Lyman WJ, Reehl WF, Rosenblatt DH, eds. 1991. Handbook of Property Estimation Methods. Washington, DC: American Chemical Society.

Martin YC. 1978. Quantitative Drug Design. New York: Marcel Dekker.

National Academy of Sciences. 1983. Risk Assessment in the Federal Government, Managing the Process. Washington, DC: National Academy Press.

Ramsden CA, ed. 1989. Quantitative Drug Design. Comprehensive Medicinal Chemistry: The Rationale Design, Mechanistic Study & Therapeutic Application of Chemical Compounds, Volume 4. New York: Pergamon Press.

USEPA. 1992. U.S. Environmental Protection Agency. Office of Research and Development. Dermal Exposure Assessment: Principles and Applications. Interim Report. Washington, DC: U.S. Environmental Protection Agency. EPA/600/8-91/011B.

Yalkowsky SH, Sinkula AA, Valvani SC, eds. 1980. Physical Chemical Properties of Drugs. New York: Marcel Dekker.

Witschi HP, ed. 1980. The Scientific Basis of Toxicity Assessment. New York: Elsevier.

Octanol-Water Partition Coefficient (log K_{ow} *): Estimation and measurement of*

Bowman BT, Sans WW. 1983. Determination of octanol-water partitioning coefficients (K_{ow}) of 61 organophosphorus and carbamate insecticides and their relationship to respective water solubility (S) values. J Environ Sci Health B18: 667–683.

Brent DA, Sabatka JJ, Minick DJ, Henry DW. 1983. A simplified high-pressure liquid chromatography method for determining lipophilicity for structure-activity relationships. J Med Chem 26: 1014–1020.

Da YZ, Ito K, Fujiwara H. 1992. Energy aspects of oil/water partition leading to the novel hydrophobic parameters for the analysis of quantitative structure-activity relationships. J Med Chem 35: 3382–3387.

Danielsson L-G, Zhang Y-H. 1996. Methods for determining *n*-octanol-water partition constants. Trends Anal Chem 15: 188–196.

de Bruijn J, Busser F, Seinen W, Hermens J. 1989. Determination of octanol/water partition coefficients for hydrophobic organic chemicals with the "slow-stirring" method. Environ Toxicol Chem 8: 499–512.

de Bruijn J, Hermens J. 1990. Relationships between octanol/water partition coefficients and total molecular surface area and total molecular volume of hydrophobic organic chemicals. Quant Struct-Act Relat 9: 11–21.

Doucette WJ, Andren AW. 1988. Estimation of octanol/water partition coefficients: Evaluation of six methods for highly hydrophobic aromatic hydrocarbons. Chemosphere 17: 345–359.

Doucette WJ, Andren AW. 1987. Correlation of octanol/water partition coefficients and total molecular surface area for highly hydrophobic aromatic compounds. Environ Sci Technol 21: 821–824.

Dunn WJ, Block JH, Pearlman RS, eds. 1986. Partition Coefficient, Determination and Estimation. New York: Pergamon Press.

Dunn WJ, Nagy PI. 1992. Relative log P and solution structure for small organic solutes in the chloroform/water system using monte carlo methods. J Computat Chem 13: 468–477.

Dunn WJ, Nagy PI, Collantes ER. 1991. A computer-assisted method for estimation of the partition coefficient. Monte Carlo simulations of the chloroform/water log P for methylamine, methanol, and acetonitrile. J Am Chem Soc 113: 7898–7902.

Dunn WJ, Nagy PI, Collantes ER, Glen WG, Alagona G, Ghio C. 1991. Log P and solute structure. Pharmacochem Libr 16: 59–65.

Garst JE. 1984. Accurate, wide-range, automated, high performance liquid chromatographic method for the estimation of octanol/water partition coefficients II: Equilibrium in partition coefficient measurements, additivity of substituent constants, and correlation of biological data. J Pharm Sci 73: 1623–1629.

Garst JE, Wilson WC. 1984. Accurate, wide-range, automated, high-performance liquid chromatographic method for the estimation of octanol/water partition coefficients. I: Effect of chromatographic conditions and procedure variable on accuracy and reproducibility of the method. J Pharm Sci 73: 1616–1622.

Guesten H, Horvatic D, Sabljic A. 1991. Modeling n-octanol/water partition coefficients by molecular topology: Polycyclic aromatic hydrocarbons and their alkyl derivatives. Chemosphere 23: 199–213.

Hansch C, Leo A. 1979. Substituent Constants for Correlation Analysis in Chemistry and Biology. New York: John Wiley and Sons.

Kamlet MJ, Doherty RM, Carr PW, Mackay D, Abraham MH, Taft RW. 1988. Linear solvation energy relationships. 44. Parameter estimation rules that allow accurate prediction of octanol/water partition coefficients and other solubility and toxicity properties of polychlorinated biphenyls and polycyclic aromatic hydrocarbons. Environ Sci Technol 22: 503–509.

Kishi H, Hashimoto Y. 1989. Evaluation of the procedures for the measurement of water solubility and n-octanol/water partition coefficient of chemicals. Results of a ring test in Japan. Chemosphere 18: 1749–1759.

Leo AJ. 1990. Methods of calculating partition coefficients. In: Comprehensive Medicinal Chemistry: The Rationale Design, Mechanistic Study & Therapeutic Application of Chemical Compounds, Volume 4 (Ramsden CA, ed.). New York: Pergamon Press. Chapter 18.7, pp. 295–319.

Lins CLK, Block JH, Doerge RF, Barnes GJ. 1982. Determination of octanol-water equivalent partition coefficients of indolizine and substituted 2-phenylindolizines by reversed-phase high pressure liquid chromatography and fragmentation values. J Pharm Sci 71: 614–617.

Minick DJ, Frenz JH, Patrick MA, Brent DA. 1988. A comprehensive method for determining hydrophobicity constants by reversed-phase high performance liquid chromatography. J Med Chem 31: 1923–1933.

Moriguchi I, Hirono S, Liu Q, Nakagome I, Matsushita Y. 1992. Simple method of calculating octanol/water partition coefficient. Chem Pharm Bull 40: 127–130.

Niemi GJ, Basak SC, Veith GD, Grunwald G. 1992. Prediction of octanol/water partition coefficient (K_{ow}) with algorithmically derived variables. Environ Toxicol Chem 11: 893–900.

Noegrohati S, Hammers WE. 1992. Regression models for octanol-water partition coefficients, and for bioconcentration in fish. Toxicol Environ Chem 34: 155–173.

Rekker RF. 1977. The Hydrophobic Fragment Constant. Amsterdam: Elsevier.

Rekker RF, Mannhold R. 1992. Calculation of Drug Lipophilicity. Weinheim, Germany: VCH Publishers, Inc.

Sasaki Y, Kubodera H, Matuszaki T, Umeyama H. 1991. Prediction of octanol/water partition coefficients using parameters derived from molecular structures. J Pharmacobio-Dyn 14: 207–214.

Shiu WY, Mackay D. 1986. A critical review of aqueous solubilities, vapor pressures, Henry's law constants, and octanol-water partition coefficients of the polychlorinated biphenys. J Phys Chem Ref Data 15: 911–929.

Shiu WY, Doucette W, Gobas F, Andren A, Mackay D. 1988. Physical-chemical properties of chlorinated dibenzo-p-dioxins. Environ Sci Technol 22: 651–658.

Yalkowsky SH, Valvani SC. 1976. Partition coefficients and surface areas of some alkylbenzenes. J Med Chem 19: 727–728.

Yamagami C, Ogura T, Takao N. 1990. Hydrophobicity parameters determined by reversed-phase liquid chromatography. I. Relationship between capacity factors and octanol-water partition coefficients for monosubstituted pyrazines and the related pyridines. J Chromatog 514: 123–136.

Yamagami C, Takao N, Fujita T. 1991. Hydrophobicity parameter of diazines. II. Analysis and prediction of partition coefficients of disubstituted pyrazines. J Pharm Sci 80: 772–777.

Yamagami C, Takao N, Fujita T. 1990. Hydrophobicity parameter of diazines (1). Analysis and prediction of partition coefficients of monosubstituted diazines. Quant Struct-Act Relat 9: 313–320.

Information on Ecological Risk Assessment

Zeeman M, Gilford J. 1993. Ecological Hazard Evaluation and Risk Assessment under EPA's Toxic Substances Control Act (TSCA): An Introduction. In: Environmental Toxicology and Risk Assessment, Volume 1. Landis W, Hughes J, Lewis M, eds. ASTM STP 1179. Philadelphia, PA: American Society for Testing & Materials. pp. 7–21.

Zeeman M. 1995. EPA's Framework for Ecological Effects Assessment. In: Screening and Testing Chemicals in Commerce. U.S. Congress, Office of Technology Assessment. OTA-BP-ENV-166. Washington, DC: Office of Technology Assessment. pp. 69–78.

Index